Lakshminarayanaiah,
N.

Membrane electrodes

DATE DUE

MEMBRANE ELECTRODES

MEMBRANE ELECTRODES

N. LAKSHMINARAYANAIAH

Department of Pharmacology
University of Pennsylvania School of Medicine
Philadelphia, Pennsylvania

ACADEMIC PRESS New York San Francisco London 1976

A Subsidiary of Harcourt Brace Jovanovich, Publishers

ACADEMIC PRESS, INC.
111 Fifth Avenue, New York, New York 10003

United Kingdom Edition published by
ACADEMIC PRESS, INC. (LONDON) LTD.
24/28 Oval Road, London NW1

Library of Congress Cataloging in Publication Data

Lakshminarayanaiah, N (date)
Membrane electrodes.

Includes bibliographical references and index.
1. Electrodes, Ion selective. I. Title.
[DNLM: 1. Electrodes. 2. Membranes. QD571 L192m]
QD571.L27 541'.3724 75-30469
ISBN 0-12-434240-X

PRINTED IN THE UNITED STATES OF AMERICA

CONTENTS

Part II Solid and Liquid Membrane Electrodes

Part III Glass and Other Related Membrane Electrodes

PREFACE

Considerable research activity has taken place in the last decade resulting in the publication of hundreds of papers dealing with membrane electrodes. This testifies in general to the attention membrane electrodes have attracted because of their usefulness in the estimation of different ionic and other species present in various fluids. A variety of electrodes in different shapes and forms has been developed for analytical applications in various fields. Concern for maintaining the quality of life by the preservation of a clean environment and the desire for better health care through the conquest and control of diseases have given the biggest boost in the form of financial help for the development of selective membrane electrodes and other sensing devices. With the help of these devices, the pollutants of the atmosphere and the constituents of body fluids can be easily monitored. Also, the need to monitor automatically the ionic and nonionic constituents of plasma, blood, sweat, urine, etc., in order to diagnose various deficiencies which cause diseases has driven investigators to explore various possibilities for the construction of elegant sensing equipment. The developments that have taken place in the area of these sensing probes are presented in this book.

The material is divided into three parts. Part I is an introduction to the variety of ion-selective membrane electrodes that have been constructed and with which experiments have been conducted. In Chapter 2, the thermodynamic principles and other concepts underlying the description of the behavior of electrolyte solutions are outlined. The various theories of membrane potential applicable to a variety of solid and liquid membrane electrodes are reviewed in Chapter 3. It is believed that the basic theoretical background covered in Chapters 2 and 3 will help the reader, whether he or she be a student, technician, or researcher with limited or no background in electrochemistry, in the intelligent use of electrodes in his or her special-interest area.

In Part II, the preparation, properties, and uses of the various solid and liquid membrane electrodes are described. A critical approach to the evaluation of published data has been taken wherever possible. It should be realized however, that this critical evaluation is sometimes made difficult by the fact that successful experimentation with a number of electrodes depends on the experience and the skill of the experimenter. This is very true of liquid membrane electrodes whose components are supplied by the manufacturer but must be assembled by the experimenter. This assembling requires great skill. Further, in a number of studies, the investigators have used their own make of electrodes. In view of this, data realized by different investigators for a particular electrode system (in many cases, a system still in its initial stages of development) have been given as such without further comment.

In Part III, the recent work on glass membrane electrodes is presented as a prelude to the description of other membrane systems in which glass electrodes are invariably used as the primary sensing device. Such membrane systems, which have assumed great importance in the sensing of gases in the atmosphere and of body fluids and in the detection of enzymes and the study of their kinetics, form the subject matter of the last two chapters.

It is believed that this attempt at assembling the material that has appeared in the last decade or so pertaining to membrane electrodes will prove useful to students, technologists, and researchers in various fields of science and technology such as analytical chemistry, biochemistry, clinical medicine, biophysics, physiology, biomedical engineering, chemical engineering, water engineering, the pharmaceutical industry, environmental science, and other technologies in which quick analysis of gases or other dissolved species is required.

My thanks are due to Dr. C. Paul Bianchi for his interest in this project and to my wife for her patience.

PART I

Introduction and Theory

Chapter 1

INTRODUCTION

The title "Membrane Electrodes" has been chosen to cover the subject matter presented in this book. It is even appropriate to consider such names as "Ion-Selective Membrane Electrodes," "Ion-Selective Electrodes," or "Reference Electrodes." But the title used has been chosen in the belief that the word membrane is comprehensive enough to include both the conventional and nonconventional electrodes fabricated in recent years. In its broadest sense, the word membrane is used to denote a thin section of conducting material that regulates the movement of charged species across it, thereby creating conditions for the generation of an electric potential. Although this is true of an ion-permeable membrane, the generation of a potential across the membrane may involve other mechanisms. Irrespective of the mechanism involved, the word membrane is used in a phenomenological sense to indicate all types of electrodes that act reversibly as membrane electrodes. In this broad sense, the word membrane includes the various phase boundaries.[1]

Most of the membranes used as electrodes possess some capacity for undergoing ion exchange. Consequently, they are ion selective in that a cation exchanger is selective to cations and an anion exchanger is selective to anions without exhibiting any particular preference to any one cation or anion, i.e., without being ion specific. In recent years some membrane electrodes have been developed which exhibit some specificity to a particular ion over other ions. But no one electrode that is completely specific to a particular ion to the exclusion of all other ions seems to have been discovered. Although the valinomycin-based membrane electrode is highly specific to K ions in the presence of Na ions, it is not so in the presence of Rb ions. Furthermore, the extent of specificity is concentration dependent, and so the use of the word ion specific to describe the performance of a membrane electrode is discounted.[2]

The membrane electrode was first discovered by Cremer in 1906.[3] This was the glass electrode, whose historical development is reviewed briefly by Eisenman.[4] Hills[5] reviewed the subject of membrane electrodes for the first time in 1961. Sollner[1] has given an interesting historical review of the development of membrane electrodes including the work on model systems that were constructed to mimic the electrochemical properties of the living cell. In recent years, a number of books[2, 6–9] and review articles[10–30] have appeared, which emphasize various aspects of ion-selective membrane electrodes. Some of the highlights in the development of nonglass membrane electrodes are presented in Table 1 from which it is interesting to note that Kolthoff and Sanders[33] in 1937 were the first to try coating AgCl onto platinum wires. These coated wire electrodes gave erratic results, but in recent years, wire, platinum, or silver, coated with polyvinyl chloride (PVC) containing various electroactive materials, has been successfully used in a number of studies.[72–75]

The silicone rubber[57–59] and other polymer[44, 65, 76, 77]-based electrodes that have been satisfactorily tested and used to sense particular ions in the last decade owe their existence to the earlier work that emerged from the laboratories of Tendeloo,[31, 43, 51, 52] Marshall, [34–36] and Sollner[37–39] and to the availability of better polymer materials. A list of the electrodes, both experimental and commercial, that respond to particular cations is given in Table 2 together with one or two principal references pertaining to each case. In Table 3 is given a similar list pertaining to anions. The contents of both tables indicate the variety of electrodes available in different forms and show the interest and skill exhibited by the different investigators in constructing them.

TABLE 1

Some Highlights in the Development of Nonglass Membrane Electrodes[a]

Investigator	Active material	Matrix	Selective to	Nernst slope (mV/decade conc.)	Comments	Ref.
Tendeloo (1936)	Fluorite, CaF_2	—	Ca^{2+}	Linear relation between emf and pCa	Anderson[32] found that it did not work as Ca selective	31
Kolthoff and Sanders (1937)	Silver halide disks	—	Cl^- Br^- I^-	57.9 57.4 52	AgI not affected by Cl^-, Br^-; $KMnO_4$ had no effect	33
					Coated Pt wire electrodes reported; could not reproduce Tendeloo's work	
Marshall (1939)	Natural zeolites (chabazite, apophyllite)	—	M^+ M^{2+}	Chabazite M^+ 40–50 M^{2+} 20–30 Apophyllite M^+ 49–58 M^{2+} 21–30	—	34
Marshall and co-workers (1941–1942)	Clay (montmorillinite, bentonite)	—	K^+ NH_4^+		Interference from Na but not from Ca, Mg	35, 36
Sollner and co-workers (1943–1954)	Collodion (oxidized in 1 *M* NaOH and dried)	—	Li^+, Na^+, K^+, NH_4^+, Mg^{2+}		Serious interferences, useful in titrations	37–39
	Collodion (treated with protamine sulfate and dried)		F^-, Cl^-, Ac^-, IO_3^-, NO_3^-, ClO_3^-, ClO_4^-		Serious interferences. First anion-responsive electrode	

TABLE 1 (continued)

Investigator	Active material	Matrix	Selective to	Nernst slope (mV/decade conc.)	Comments	Ref.
Wyllie and Patnode (1950)	Commercial cation exchange resins	Polymethyl methacrylate; polystyrene	Na^+	51–56		40
Sinha (1953–1955)	Commercial cation and anion exchange resins	Polystyrene	Cations, anions	—	Estimation of activities of cations and anions; acid-base titrations	41
Woermann *et al*. (1956)	Ion exchange resin containing dipicrylamine groups	Resin pressed to form a membrane with binder	K^+	—	Poor selectivity to K^+ over Na^+	42
Tendeloo and Krips (1957)	Calcium oxalate and other Ca salts	Paraffin + non-ionic detergent on gauze	Ca^{2+}	—	Acceptable but criticized by Shatkay[44]	43
Gregor and co-workers (1957–1964)	Alkaline earth and iron(III) stearates and/or palmitates	Multilayers formed between edges of cracked glass plate	Ca^{2+}, Ba^{2+}, Mg^{2+}, Sr^{2+}, Fe(III)	—	Fairly selective to alkaline earth ions. Used with Na and K solutions to estimate the activities of alkaline earth ions. Very difficult to fabricate the electrodes	45–48
Parsons (1958)	Commercial cation and anion exchangers	Polystyrene	Na^+	58 in the range pNa 1–3	K^+ interferes; used in titrations	49
Fischer and Babcock (1958)	$BaSO_4$, $BaCrO_4$	Paraffin without gauze	Ba^{2+}, SO_4^{2-}	—	Not selective to cations or anions	50

Tendeloo and Krips (1959)	Potassium tetra-phenyl borate	Polystyrene + gauze	K^+	—	Not selective	51
Tendeloo and Van der Voort (1960)	Calcium stearate	Paraffin as above	Ca^{2+}	—	No response to K; stronger response than CaC_2O_4 electrode	52
Clcos and Fripiat (1960)	CaC_2O_4	Paraffin and detergent	Nonspecific	—	Shows memory effect	53
Pungor and Hollos-Rokosinyi (1961)	AgI	Paraffin	I^-	—	KCl does not interfere	54
Ilani (1963)	Organic liquids	Millipore filter	K^+ over Na^+ only for the liquid toluene + butanol	52 for log K^+ between 0.7 and 1.8	Membrane resistance high	55
Sollner and Shean (1964)	Amberlite LA-2 (lauryl trialkyl-methyl amine salts)	Benzene, xylene, nitrobenzene	Cl^-, CNS^-	58 for KCl	—	56
Pungor *et al.* (1964)	Ion exchange resins	Silicone rubber	SO_4^{2-}, Cl^-, OH^-, H^+, K^+, Zn^{2+}, Ni^{2+}	—	Not specific but selective to valence type	57
Pungor *et al.* (1964–1965)	$BaSO_4$, AgI	Silicone rubber	SO_4^{2-}, I^-	K_2SO_4 = 24–30 KI = 56–60	Phosphate interferes but not 0.1 *M* KCl	57, 58
Pungor *et al.* (1965)	Silver halides, manganese(III) phosphate aluminum oxine nickel dimethyl-glyoxime	Silicone rubber	Ag^+, X^-, PO_4^{3-}, Al^{3+}, Ni^{2+}	—	—	59

TABLE 1 (continued)

Investigator	Active material	Matrix	Selective to	Nernst slope (mV/decade conc.)	Comments	Ref.
Morazzani-Pelletier and Baffier (1965)	Cobalt phosphate nickel dimethylglyoxime	Collodion paraffin	Co^{2+}	20–23 in cobalt solutions	Poor response to SO_4^{2-} porous membranes, highly permeable to KCl	60
	MnC_2O_4					
	NiC_2O_4					
Geyer and Syring (1966)	TiO_2, Fe_2O_3, SnO_2	Polyethylene	H^+, OH^-	—	Studied titrations	61
	ZrO_2	Polypropylene	Na^+		—	
	Al_2O_3	Paraffin				
	K_2SiF_6	Agar or paper	SiF_6^{2-}, K^+			
	$Ag_4Fe(CN)_6$	Agar	Both ions			
	$PbWO_4$	Paraffin				
Bonner and Lunney (1966)	Dinonylnaphthalenesulfonate salts	Nitrobenzene + *o*-dichlorobenzene	Na^+, NH_4^+, Ca^{2+}	Less than theoretical	Useful range 3×10^{-1}–3×10^{-3} *M*	62
	Aliquat 336 (tricaprylmethyl ammonium chloride)	*o*-Dichlorobenzene	Cl^-			
Frant and Ross (1966)	LaF_3	—	F^-	58	OH^- interferes	63
Ross (1967)	Calcium didecyl phosphate in di-*n*-octylphenyl phosphonate	Filter	Ca^{2+}	Theoretical	H^+, Zn^{2+} interfere	64

Shatkay and co-workers (1967)	CaC_2O_4	Paraffin + non-ionic detergent + gauze	Ca^{2+}	15–20	Not specific; not completely permselective.	44, 65
	Theonyltrifluoro-acetone	Polyvinyl chloride + tributyl phosphate	Ca^{2+}	27–28	Very selective; compared well with the commercial electrode	
Coetzee and Freiser (1968)	Aliquat 336 in 1-decanol	Millipore filter	Anions	50–58	Depending on the ionic form, electrodes selective to particular anions	66
Pioda *et al.* (1969)	Valinomycin in diphenyl ether	Filter	K^+	58.3	Highly selective for K^+ over Na^+	67
Ruzicka and co-workers (1970–1972)	Suitable salt in organic liquid	Porous graphite	Particular cation or anion concerned	—	This work resulted in the construction of Selectrodes[69, 70] selective to specific cations and anions	68
Cattrall and Freiser (1971)	Calcium di-decyl phosphate in di-*n*-octylphenyl phosphonate	Platinum wire coated with PVC containing the electroactive material	Ca^{2+}	Theoretical	Found very promising	71

[a] After Covington.[30]

TABLE 2

Principal Membrane Electrode Systems Responding to Particular Cations

Ion	Membrane type	Matrix	Active material	Ref.
H^+, OH^-	Glass bulb			4
	Solid state (heterogeneous)	Silicone rubber	Ion exchanger in OH^- form	78
		Selectrode	Thymoquinhydrone	79
		Coated Pt wire	Stearic acid + methyl tri-*n*-octyl ammonium stearate	73
	Liquid state		Quinhydrone	68
Li^+	Glass bulb			4, 80
Na^+	Glass bulb			4
	Solid state (heterogeneous)	PVC	Special Na ligand	81
	Liquid state		Special Na ligand	81
K^+	Glass bulb			4, 82
	Solid state (heterogeneous)		Biological material	83
		Silicone rubber	Valinomycin	84
		PVC membrane Selectrode	Valinomycin	85
		PVC-coated Ag wire	Valinomycin	74
		PVC-coated Pt wire	Valinomycin	75
		Polyether in PVC	Dimethyl-dibenzo-30-crown-10	86
		Silicone rubber	Potassium zinc ferrocyanide	87
	Liquid state		Valinomycin	88
	Polyether		Dicyclohexyl-18-crown-6	89
Rb^+	Glass bulb			4
Cs^+	Glass bulb			4
	Solid state (heterogeneous)	Silicone rubber	Cesium 12-molybdo-phosphate	90
	Liquid state		Cesium tetraphenyl borate	68

Ag^+	Glass bulb			4, 80
	Solid state (homogeneous)	Disk	Ag_2S	91
			Ag-tcnq$^+$ [a]	92
	Solid state (heterogeneous)	Silicone rubber	Ag_2S	93
		Liquid state	Silver dithizonate	68
NH_4^+	Glass bulb			4, 82
	Solid state (heterogeneous)		Organic material (name not revealed)	94
	Liquid state		Monactin + nonactin	95
Tl^+	Glass bulb			4, 80
	Solid state (heterogeneous)	Epoxy	Tl salts of molybdophosphoric acid and tungstophosphoric acid	96
	Liquid state		Tetrachlorothallium(III) salt of Sevron Red L	97
Acetylcholine	Solid state (heterogeneous)	PVC	Tetra-*p*-chlorophenyl borate	98
	Liquid state		Acetylcholine-tetra-*p*-chlorophenyl borate	99
Mg^{2+}	Glass			4
	Liquid state		$(RO_2)_2PO_2^-$	100, 101
Be^{2+}	Liquid state		$(RO_2)_2PO_2^-$ in Be form	102
Ca^{2+}	Glass			4
	Solid state (heterogeneous)		PVC + tributyl phosphate + theonyltrifluoro acetone	44
		PVC	Didecylphosphoric acid	103
		PVC-coated Pt wire	Calcium didecyl phosphate	71
		PVC membrane Selectrode	Di(*n*-octylphenyl)phosphoric acid	104

TABLE 2 (continued)

Ion	Membrane type	Matrix	Active material	Ref.
(Ca^{2+})	Liquid state		Didecylphosphoric acid	64, 101
			Special Ca ligand	29
Sr^{2+}	Glass			4
Ba^{2+}	Glass			4
	Solid state (heterogeneous)	Parchment	$BaSO_4$	105
	Liquid state		Nonylphenoxy(polyethylene) ethane + $BaCl_2$ + excess sodium tetraphenyl borate	106
Cu^+	Solid state (homogeneous)	Disk	Cu_2S	107
	Solid state (heterogeneous)	Silicone rubber	Cu_2S	108
Cu^{2+}	Solid state (homogeneous)		Ag_2S + CuS	109, 110
			Ag_2S + CuS deposited on Ag wire	111
		Selectrode	Ag_2S + CuS on graphite rod	112
			$Cu(tcnq)_2$[a]	92
	Solid state (heterogeneous)	Polyethylene	CuS	113
	Liquid state		R–S–CH_2COO^- ion exchange resin	109, 114
			Copper diethyldithiocarbonate	68
			Copper dithizonate (primary)	68
			Copper dithizonate (secondary)	68
Cd^{2+}	Solid state (homogeneous)		CdS + Ag_2S	109, 115
			CdS + Ag_2S deposited on Ag wire	111
		Selectrode	CdS	116

	Solid state (heterogeneous)	Polyethylene	$CdS + Ag_2S$	117
	Liquid state		Cadminum dithizonate	68
			Bis(*O*, *O*′-diisobutyldithiophos-phato) cadmium (II)	118
Pb^{2+}	Solid state (homogeneous)		$PbS + Ag_2S$	109, 119
			$PbS + Ag_2S$ deposited on Ag wire	111
		Selectrode	$PbS + Ag_2S$	120
	Solid state (heterogeneous)	Silicone rubber	PbS	121
		Polyethylene	$PbS + Ag_2S$	122
	Liquid state		$R–S–CH_2COO^-$ ion exchange resin	109
			Lead dithizonate	68
			Bis(*O*, *O*′-diisobutyldithio-phosphato)lead(II)	118
Mn^{2+}	Solid state (heterogeneous)	Silicone rubber	$MnPO_4$	58
Ni^{2+}	Solid state (homogeneous)		$NiS + Ag_2S$ deposited on Ag wire	111
	Solid state (heterogeneous)	Silicone rubber	Nickel dimethylglyoxime	78
	Liquid state		Bis(*O*, *O*′-diisobutyldithio-phosphato)nickel(II)	118
Co^{2+} Hg^{2+}	Solid state (homogeneous)		$MS + Ag_2S$ deposited on Ag wire	111
Zn^{2+}	Solid state (homogeneous)		$ZnS + Ag_2S$ deposited on Ag wire	111
	Liquid state		Zinc dithizonate	68
Zinc	Solid state (heterogeneous)	Silicone rubber	Brilliant Green–tetrathiocyanato–zincate(II)	123

TABLE 2 (continued)

Ion	Membrane type	Matrix	Active material	Ref.
(Zinc)	Liquid state		Brilliant Green–tetrathio-cyanatozincate(II)	123
Fe^{3+}	Solid state (heterogeneous)		Chalcogenide glass (60% Se, 28% Ce, 12% Sb) doped with Fe, Co, or Ni	124
		PVC-coated Pt wire	Tetrachloroferrate(III) salt of quaternary ammonium compound (Aliquat 336S)	125
	Liquid state		Hemin	126
Uranyl	Solid state (heterogeneous)	PVC	Uranyl organophosphorus complexes	127
Molyb-denum	Liquid state		Bistetraethyl ammonium-pentathiocyanato-oxomolybdate	128
Selenium	Liquid state		3, 3′-Diaminobenzidine	129
Antimony	Liquid state		Hexachloroantimonate salt of Sevron Red L	97

[a]tcnq = 7, 7, 8, 8,-tetracyanoquinodimethane.

TABLE 3

Principal Membrane Electrode Systems Responding to Particular Anions

Ion	Membrane type	Matrix	Active material	Ref.
F^-	Solid state (homogeneous)		LaF_3 crystal	63
	Solid state (heterogeneous)	Silicone rubber	ThF_4	130
Cl^-	Solid state (homogeneous)		Fused AgCl	33
		Pressed disk	$HgS + Hg_2Cl_2$	130a
		Selectrode	AgCl	79
	Solid state (heterogeneous)	Silicone rubber	AgCl	10
		Polythene	AgCl	77
		PVC-coated Pt wire	Quaternary ammonium chloride	131
	Liquid state		Dimethyldistearyl ammonium chloride	109, 132
Br^-	Solid state (homogeneous)		Fused AgBr	33
		Selectrode	AgBr	69
	Solid state (heterogeneous)	Silicone rubber	AgBr	10
		Polythene	AgBr	77
I^-	Solid state (homogeneous)		Fused AgI	33
		Selectrode	AgI	69
	Solid state (heterogeneous)	Silicone rubber	AgI	10
		Polythene	AgI	77
	Liquid state		Tetraoctylphosphonium	133
CN^-	Solid state (homogeneous)	Disk	$AgI + Ag_2S$	134
	Solid state (heterogeneous)	Silicone rubber	AgI	15
		Polythene	AgI	77
CNS^-	Solid state (heterogeneous)	Phenol + formaldehyde + NH_3	$Ni(NO_3)_2$	135
		Polythene	AgCNS	136
		PVC-coated Pt wire	Quaternary ammonium thiocyanate	131
	Liquid state		Quaternary ammonium salt ion exchanger	137

TABLE 3 (continued)

Ion	Membrane type	Matrix	Active material	Ref.
ClO_4^-	Solid state (homogeneous)		Perchlorates of *p*-diamines	138
			Salt, derived from *N*-ethylbenzothiazole-2, 2′-azaviolene	139
	Solid state (heterogeneous)	Phenol + formaldehyde + NH_3	$Ni(NO_3)_2$	135
		PVC membrane or PVC-coated Pt wire	Iron phenanthroline	140
	Liquid state		Iron phenanthroline	141, 142
			Salts derived from *N*-ethylbenzothiazole-2, 2′-azaviolene	143
			Brilliant Green perchlorate	144
NO_3^-	Solid state (heterogeneous)	Phenol + formaldehyde + NH_3	$Ni(NO_3)_2$	135
		Carbon paste	Nickel phenanthroline	145
		Polymethyl methacrylate-coated Pt wire	Aliquat 336S in NO_3 form	146
	Liquid state		Nickel phenanthroline	147
BF_4^-	Liquid state		Nickel phenanthroline in BF_4^- form	148
			Brilliant Green–tetrafluoroborate	149
S^{2-}	Solid state (homogeneous)		Ag_2S	150
	Solid state (heterogeneous)	Silicone rubber	Ag_2S	10
		Polythene	Ag_2S	151
SO_4^{2-}	Solid state (homogeneous)	Pressed	$Ag_2S + PbS + PbSO_4 + Cu_2S$	152
	Solid state (heterogeneous)		$BaSO_4$	57
SiF_6^{2-}	Solid state (heterogeneous)	Agar	K_2SiF_6	61
IO_4^-	Liquid state		Iron phenanthroline	142
MnO_4^-	Liquid state		Iron phenanthroline	142

$Cr_2O_7^{2-}$	Liquid state		Iron phenanthroline	142
ReO_4^-	Liquid state		Iron phenanthroline	137
$H_2PO_4^-$	Solid state (heterogeneous)	Silicone rubber	$BiPO_4$	78
HPO_4^{2-}	Solid state (heterogeneous)	Silicone rubber	$MnPO_4$	78
PO_4^{3-}	Solid state (heterogeneous)	Silicone rubber	KH_2PO_4, $FePO_4$, $CrPO_4$, $AlPO_4$	153
	Liquid state		Phosphomolybdic and phosphotungstic acids	154
Various anions	Solid state (heterogeneous)	Polymethyl methacrylate-coated Pt wire	Aliquat 336S (tricaprylylmethyl ammonium ion) in various anionic forms	72
	Electrodes selective to Cl^-, Br^-, I^-, NO_3^-, CNS^-, ClO_4^-, SO_4^{2-}, $C_2O_4^{2-}$, CH_3COO^-, formate, propionate, benzoate, toluate, toluene sulfonate, salicylate			
	Liquid state		Aliquat 336S in various anionic forms as above	66, 155
CO_3^{2-}	Liquid state		Aliquat 336S in the solvent tetrafluoroacetyl-*p*-butyl benzene	156
Amino acids	Liquid state		Aliquat 336S converted to amino acid salt of interest	157
Sur-factant	Solid state (heterogeneous)	PVC-coated Pt wire	Detergent anion + quaternary ammonium chloride	158
		Silicone rubber	Surfactant concerned	159
		PVC	Surfactant concerned	160
	Liquid state		Na salt of dodecyl sulfate, etc.,	161 161

REFERENCES

1. K. Sollner, *Ann. N. Y. Acad. Sci.* **148**, 154 (1968).
2. R. A. Durst (ed.), *in* "Ion Selective Electrodes," p.v. Nat. Bur. Std. Spec. Publ. 314, Washington, D. C., 1969.
3. M. Cremer, *Z. Biol.* **47**, 562 (1906).
4. G. Eisenman, *in* "Glass Electrodes for Hydrogen and Other Cations, Principles and Practice" (G. Eisenman, ed.), p. 1. Dekker, New York, 1967.
5. G. J. Hills, *in* "Reference Electrodes" (D. J. G. Ives and G. J. Janz, eds.), p. 411. Academic Press, New York, 1961.
6. M. Lavallee, O. F. Schanne, and N. C. Hebert (eds.), "Glass Microelectrodes." Wiley, New York, 1969.
7. G. J. Moody and J. R. D. Thomas, "Selective Ion Sensitive Electrodes." Merrow Publ., Watford, Hertfordshire, England, 1971.
8. K. Cammann, "Das Arbeiten mit Ionen Selecktion Electroden." Springer-Verlag, Berlin, 1973.
9. E. Pungor (ed.), "Ion Selective Electrodes." Akademiai Kiado, Budapest, 1973.
10. E. Pungor, *Anal. Chem.* **39**, 28A (1967).
11. G. A. Rechnitz, *Chem. Eng. News* **45**(25), 146 (1967); **53**(4), 29 (1975); *Anal. Chem.* **41**, 109A (1969); *Accounts Chem. Res.* **3**, 69 (1970).
12. A. K. Covington, *Chem. Britain* **5**, 388 (1969).
13. G. J. Moody, R. B. Oke, and J. D. R. Thomas, *Lab Practice* **18**, 941, 1056 (1969).
14. E. J. Zinser, *Chem. Canada* **21**, 31 (1969).
15. E. Pungor and K. Toth, *Analyst* **95**, 625 (1970); *Pure Appl. Chem.* **34**, 105 (1973); **36**, 441 (1973).
16. C. Liteanu, E. Hopirtean, and M. Mioscu, *Studii Cercetari. Chim.* **18**, 241 (1970).
17. B. Fleet, *Proc. Soc. Anal. Chem.* **7**, 84 (1970).
18. P. D. Novikov, *in* "Ionoobmen Membrany Elektrodialize" (K. M. Saldadze, ed.), p. 108. Izdatel'stvo 'Khimiya', Leningradskoe Otdelenie, Leningrad, U.S.S.R., 1970.
19. E. C. Toren and R. P. Buck, *Anal. Chem.* **42**, 286R (1970).
20. T. M. Florence, *Proc. Roy. Aust. Chem. Inst.* **37**, 261 (1970).
21. R. P. Buck, *in* "Physical Methods of Chemistry" (A. Weissberger and B. Rossiter, eds.), Vol. 1, Part IIA, Chapter 2. Wiley, New York, 1971; *Anal. Chem.* **44**, 270R (1972); **46**, 28R (1974).
22. J. Bagg, *Proc. Roy. Aust. Chem. Inst.* **38**, 91 (1971).
23. W. Simon, *Pure Appl. Chem.* **25**, 811 (1971).
24. J. Koryta, *Anal. Chim. Acta* **61**, 329 (1972).
25. J. Sandblom and F. Orme, *in* "Membranes" (G. Eisenman, ed.), Vol. 1, p. 125. Dekker, New York, 1972.
26. J. T. Clerc, G. Kahr, E. Pretsch, R. P. Scholer, and H. R. Wuhrmann, *Chimia* **26**, 287 (1972).
27. G. Fiori and L. Formaro, *Chim. Ind.* **54**, 883 (1972).
28. N. Lakshminarayanaiah, *in* "Electrochemistry" (Specialist Periodical Reports, The Chemical Society, London.) (G. J. Hills, ed.), Vol. 2, Chap. 5 (1972); (H. R. Thirsk, ed.), Vol. 4, Chap. 6 (1974); (H. R. Thirsk, ed.), Vol. 5, Chapter 3 (1975).
29. W. E. Morf, D. Ammann, E. Pretsch, and W. Simon, *Pure Appl. Chem.* **36**, 421 (1973).
30. A. K. Covington, *Crit. Rev. Anal. Chem.* **3**, 355 (1974).
31. H. J. C. Tendeloo, *J. Biol. Chem.* **113**, 333 (1936); *Rec. Trav. Chim.* **55**, 227 (1936).
32. R. S. Anderson, *J. Biol. Chem.* **115**, 323 (1936).
33. I. M. Kolthoff and H. L. Sanders, *J. Amer. Chem. Soc.* **59**, 416 (1937).

34. C. E. Marshall, *J. Phys. Chem.* **43**, 1155 (1939).
35. C. E. Marshall and W. E. Bergman, *J. Amer. Chem. Soc.* **63**, 1911 (1941); *J. Phys. Chem.* **46**, 52, 325 (1942).
36. C. E. Marshall and C. A. Krinbill, *J. Amer. Chem. Soc.* **64**, 1814 (1942).
37. K. Sollner, *J. Amer. Chem. Soc.* **65**, 2260 (1943).
38. C. W. Carr and K. Sollner, *J. Gen. Physiol.* **28**, 119 (1944).
39. H. P. Gregor and K. Sollner, *J. Phys. Chem.* **50**, 88 (1946); **58**, 409 (1954).
40. M. R. J. Wyllie and H. W. Patnode, *J. Phys. Chem.* **54**, 204 (1950).
41. S. K. Sinha, *J. Indian Chem. Soc.* **30**, 529 (1953); **31**, 572, 577 (1954); **32**, 35 (1955).
42. D. Woermann, K. F. Bonhoeffer, and F. Helfferich, *Z. Phys. Chem. (Frankfurt)* **8**, 265 (1956).
43. H. J. C. Tendeloo and A. Krips, *Rec. Trav. Chim.* **76**, 703, 946 (1957).
44. A. Shatkay, *Anal. Chem.* **39**, 1056 (1967).
45. H. P. Gregor and H. Schonhorn, *J. Amer. Chem. Soc.* **79**, 1507 (1957); **81**, 3911 (1959).
46. H. Schonhorn and H. P. Gregor, *J. Amer. Chem. Soc.* **83**, 3576 (1961).
47. H. P. Gregor, A. C. Glatz, and H. Schonhorn, *J. Amer. Chem. Soc.* **85**, 3926 (1963).
48. J. Bagg and H. P. Gregor, *J. Amer. Chem. Soc.* **86**, 3626 (1964).
49. J. S. Parsons, *Anal. Chem.* **30**, 1262 (1958).
50. R. B. Fischer and R. F. Babcock, *Anal. Chem.* **30**, 1732 (1958).
51. H. J. C. Tendeloo and A. Krips, *Rec. Trav. Chim.* **78**, 177 (1959).
52. H. J. C. Tendeloo and F. H. Van der Voort, *Rec. Trav. Chim.* **79**, 639 (1960).
53. P. Cloos and J. J. Fripiat, *Bull. Soc. Chim. France* 423 (1960).
54. E. Pungor and E. Hollos-Rokosinyi, *Acta Chim. Acad. Sci. Hung.* **27**, 63 (1961).
55. A. Ilani, *J. Gen. Physiol.* **46**, 839 (1963).
56. K. Sollner and G. M. Shean, *J. Amer. Chem. Soc.* **86**, 1901 (1964).
57. E. Pungor, J. Havas, and K. Toth, *Acta Chim. Acad. Sci. Hung.* **41**, 239 (1964).
58. E. Pungor, K. Toth, and J. Havas, *Hung. Sci. Instrum.* **3**, 2 (1965).
59. E. Pungor, J. Havas, and K. Toth, *Z. Chem.* **5**, 9 (1965).
60. S. Morazzani-Pelletier and M. A. Baffier, *J. Chim. Phys.* **62**, 429 (1965).
61. R. Geyer and W. Syring, *Z. Chem.* **6**, 92 (1966).
62. O. D. Bonner and D. C. Lunney, *J. Phys. Chem.* **70**, 1140 (1966).
63. M. S. Frant and J. W. Ross, Jr., *Science* **154**, 1553 (1966).
64. J. W. Ross, Jr., *Science* **156**, 1378 (1967).
65. R. Bloch, A. Shatkay, and H. A. Saroff, *Biophys. J.* **7**, 865 (1967).
66. C. J. Coetzee and H. Freiser, *Anal. Chem.* **40**, 2071 (1968).
67. L. A. R. Pioda, V. Stankova, and W. Simon, *Anal. Lett.* **2**, 665 (1969).
68. J. Ruzicka and J. C. Tjell, *Anal. Chim. Acta* **49**, 346 (1970).
69. J. Ruzicka and C. J. Lamm, *Anal. Chim. Acta* **54**, 1 (1971).
70. J. Ruzicka, C. J. Lamm, and J. C. Tjell, *Anal. Chim. Acta* **62**, 15 (1972).
71. R. W. Cattrall and H. Freiser, *Anal. Chem.* **43**, 1905 (1971).
72. H. J. James, G. P. Carmack, and H. Freiser, *Anal. Chem.* **44**, 856 (1972).
73. J. H. Wang and E. Copeland, *Proc. Nat. Acad. Sci. U.S.* **70**, 1909 (1973).
74. M. D. Smith, M. A. Genshaw, and J. Geyson, *Anal. Chem.* **45**, 1782 (1973).
75. R. W. Cattrall, S. Tribuzio, and H. Freiser, *Anal. Chem.* **46**, 2223 (1974).
76. E. B. Buchannan and J. L. Seago, *Anal. Chem.* **40**, 517 (1968).
77. M. Mascini and A. Liberti, *Anal. Chim. Acta* **47**, 339 (1969).
78. E. Pungor, K. Toth, and J. Havas, *Acta Chim. Acad. Sci. Hung.* **48**, 17 (1966); *Microchim. Acta* 689 (1966).
79. C. G. Lamm, E. H., Hansen, and J. Ruzicka, *Anal. Lett.* **5**, 451 (1970).
80. G. Mattock and R. Uncles, *Analyst* **87**, 977 (1962).

81. D. Ammann, E. Pretsch, and W. Simon, *Anal. Lett.* **7**, 23 (1974).
82. G. Mattock and R. Uncles, *Analyst* **89**, 350 (1964).
83. I. H. Krull, C. A. Mask, and R. E. Cosgrove, *Anal. Lett.* **3**, 43 (1970).
84. J. Pick, K. Toth, M. Vasak, E. Pungor, and W. Simon, *in* "Ion Selective Electrodes" (E. Pungor, ed.), p. 245. Akademiai Kiado, Budapest, 1973.
85. U. Fiedler and J. Ruzicka, *Anal. Chim. Acta* **67**, 179 (1973).
86. J. Petranek and O. Ryba, *Anal. Chim. Acta* **72**, 375 (1974).
87. A. G. Fogg, A. S. Pathan, and D. T. Burns, *Anal. Lett.* **7**, 539 (1974).
88. L. A. R. Pioda and W. Simon, *Chimia* **23**, 72 (1969).
89. G. A. Rechnitz and E. Eyal, *Anal. Chem.* **44**, 370 (1972).
90. C. J. Coetzee and A. J. Basson, *Anal. Chim. Acta* **57**, 478 (1971).
91. T. M. Hseu and G. A. Rechnitz, *Anal. Chem.* **40**, 1054, 1661 (1968).
92. M. Sharp and G. Johansson, *Anal. Chim. Acta* **54**, 13 (1971).
93. E. Schmidt and E. Pungor, *Magyar. Kem. Foly.* **77**, 397 (1971).
94. R. E. Cosgrove, C. A. Mask, and I. H. Krull, *Anal. Lett.* **3**, 457 (1970).
95. R. P. Scholer and W. Simon, *Chimia* **24**, 372 (1970).
96. C. J. Coetzee and A. J. Basson, *Anal. Chim. Acta* **64**, 300 (1973).
97. A. G. Fogg, A. A. Al-Sibaai, and C. Burgess, *Anal. Lett.* **8**, 129 (1975).
98. G. Baum, M. Lynn, and F. B. Ward, *Anal. Chim. Acta* **65**, 385 (1973).
99. G. Baum, *J. Phys. Chem.* **76**, 1872 (1972).
100. M. E. Thompson, *Science* **153**, 866 (1966).
101. M. E. Thompson and J. W. Ross, Jr., *Science* **154**, 1643 (1966).
102. B. Fleet and G. A. Rechnitz, *Anal. Chem.* **42**, 690 (1970).
103. G. J. Moody, R. B. Oke, and J. D. R. Thomas, *Analyst* **95**, 910 (1970).
104. J. Ruzickz, E. H. Hansen, and J. C. Tjell, *Anal. Chim. Acta* **67**, 155 (1973).
105. C. Liteanu and M. Mioscu, *Rev. Roum. Chim.* **11**, 863 (1966).
106. R. J. Levins, *Anal. Chem.* **43**, 1045 (1971).
107. L. F. Heerman and G. A. Rechnitz, *Anal. Chem.* **44**, 1655 (1972).
108. H. Hirata and K. Date, *Talanta* **17**, 883 (1970).
109. J. W. Ross, Jr., *in* "Ion Selective Electrodes" (R. A. Durst, ed.), Chapter 2. Nat. Bur. Std. Spec. Publ. 314, Washington, D.C., 1969.
110. G. A. Rechnitz and N. C. Kenney, *Anal. Lett.* **2**, 395 (1969).
111. T. Anfalt and D. Jagner, *Anal. Chim. Acta* **56**, 477 (1971).
112. E. H. Hansen, C. G. Lamm, and J. Ruzicka, *Anal. Chim. Acta* **59**, 403 (1972).
113. M. Mascini and A. Liberti, *Anal. Chim. Acta* **53**, 202 (1971).
114. G. A. Rechnitz and Z. F. Lin, *Anal. Lett.* **1**, 23 (1967).
115. M. J. D. Brand, J. J. Militello, and G. A. Rechnitz, *Anal. Lett.* **2**, 523 (1969).
116. J. Ruzicka and E. H. Hansen, *Anal. Chim. Acta* **63**, 115 (1973).
117. M. Mascini and A. Liberti, *Anal. Chim. Acta* **64**, 63 (1973).
118. E. A. Materova, V. V. Muchovikov, and M. G. Grigorjeva, *Anal. Lett.* **8**, 167 (1975).
119. J. W. Ross, Jr. and M. S. Frant, *Anal. Chem.* **41**, 967 (1969).
120. E. H. Hansen and J. Ruzicka, *Anal. Chim. Acta* **72**, 365 (1974).
121. H. Hirata and K. Date, *Anal. Chem.* **43**, 279 (1971).
122. M. Mascini and A. Liberti, *Anal. Chim. Acta* **60**, 405 (1972).
123. A. G. Foff, M. Duzinkewycz, and A. S. Pathan, *Anal. Lett.* **6**, 1101 (1973).
124. C. T. Baker and I. Trachtenberg, *J. Electrochem. Soc.* **118**, 571 (1971).
125. R. W. Cattrall and P. Chin-Poh, *Anal. Chem.* **47**, 93 (1975).
126. P. Gabor-Klatsmanyi, K. Toth, and E. Pungor, *in* "Ion Selective Electrodes" (E. Pungor, ed.), p. 183. Akademiai Kiado, Budapest, 1973.
127. D. L. Manning, J. R. Stokely, and R. W. Magouyrk, *Anal. Chem.* **46**, 1116 (1974).

128. A. G. Fogg, J. L. Kumar, and D. T. Burns, *Anal. Lett.* **7**, 629 (1974).
129. T. L. Malone and G. D. Christian, *Anal. Lett.* **7**, 33 (1974).
130. A. M. G. Macdonald and K. Toth, *Anal. Chim. Acta* **41**, 99 (1968).
130a. J. F. Lechner and I. Sekerka, *J. Electroanal. Chem. Interfacial Electrochem.* **57**, 317 (1974).
131. T. Stworzewicz, J. Czapkiewicz, and M. Leszko, *in* "Ion Selective Electrodes" (E. Pungor, ed.), p. 259. Akademiai Kiado, Budapest, 1973.
132. K. Srinivasan and G. A. Rechnitz, *Anal. Chem.* **41**, 1203 (1969).
133. A. Rouchouse and J. M. M. Porthault, *Anal. Chim. Acta* **74**, 155 (1975).
134. B. Fleet and H. Von Storp, *Anal. Chem.* **43**, 1575 (1971).
135. T. N. Dobbelstein and H. Diehl, *Talanta* **16**, 1341 (1969).
136. M. Mascini, *Anal. Chim. Acta* **62**, 29 (1972).
137. R. F. Hirsch and J. D. Portock, *Anal. Lett.* **2**, 295 (1969).
138. M. Sharp, *Anal. Chim. Acta* **61**, 99 (1972).
139. M. Sharp, *Anal. Chim. Acta* **62**, 385 (1972).
140. T. J. Rohm and G. G. Guilbault, *Anal. Chem.* **46**, 590 (1974).
141. T. M. Hseu and G. A. Rechnitz, *Anal. Lett.* **1**, 629 (1968).
142. R. J. Baczuk and R. J. Dubois, *Anal. Chem.* **40**, 685 (1968).
143. M. Sharp, *Anal. Chim. Acta* **65**, 405 (1973).
144. A. G. Fogg, A. S. Pathan, and D. T. Burns, *Anal. Chim. Acta* **73**, 220 (1974).
145. G. A. Qureshi and J. Lindquist, *Anal. Chim. Acta* **67**, 243 (1973).
146. B. M. Kneebone and H. Freiser, *Anal. Chem.* **45**, 449 (1973).
147. S. S. Potterton and W. D. Shults, *Anal. Lett.* **1**, 11 (1967).
148. R. M. Carlson and J. L. Paul, *Anal. Chem.* **40**, 1292 (1968).
149. A. G. Fogg, A. S. Pathan, and D. T. Burns, *Anal. Lett.* **7**, 545 (1974).
150. T. S. Light and J. L. Swartz, *Anal. Lett.* **1**, 825 (1968).
151. M. Mascini and A. Liberti, *Anal. Chim. Acta* **51**, 231 (1970).
152. M. S. Mohan and G. A. Rechnitz, *Anal. Chem.* **45**, 1323 (1973).
153. G. G. Guilbault and P. J. Brignac, Jr., *Anal. Chem.* **41**, 1136 (1969).
154. G. G. Guilbault and P. J. Brignac, Jr., *Anal. Chim. Acta* **56**, 139 (1971).
155. C. J. Coetzee and H. Freiser, *Anal. Chem.* **41**, 1128 (1969).
156. H. B. Herman and G. A. Rechnitz, *Science* **184**, 1074 (1974).
157. M. Matsui and H. Freiser, *Anal. Lett.* **3**, 161 (1970).
158. T. Fujinaga, S. Okazaki, and H. Freiser, *Anal. Chem.* **46**, 1842 (1974).
159. A. S. Pathan and D. T. Burns, *Anal. Chim. Acta* **69**, 238 (1974).
160. T. Tanaka, K. Hiiro, and K. Kawahara, *Anal. Lett.* **7**, 173 (1974).
161. P. Gavach and P. Seta, *Anal. Chim. Acta* **50**, 407 (1970).

Chapter 2

ELECTROCHEMISTRY OF AQUEOUS ELECTROLYTE SOLUTIONS

The theoretical principles describing those properties of aqueous solutions that can be measured using membrane electrodes are presented in this chapter. One such property is the activity of the constituents, the solvent and the solute, forming the solution. The thermodynamic foundation of this concept of activity will now be reviewed.

A. FREE ENERGY

The free energy or the energy available for doing useful work, represented by the symbol G, is defined by[1]

$$G = H - TS = E + PV - TS \tag{1}$$

where H is the heat content or enthalpy and E is the energy content of the system whose temperature, pressure, volume, and entropy are represented by T, P, V, and S, respectively.

The free energy change upon going from an initial to a final state is given by

$$G_2 - G_1 = (E_2 - E_1) + (P_2V_2 - P_1V_1) - (T_2S_2 - T_1S_1)$$
$$\Delta G = \Delta E + \Delta(PV) - \Delta(TS) \tag{2}$$

At constant temperature and pressure, Eq. (2) becomes

$$\Delta G = \Delta E + P\,\Delta V - T\,\Delta S \tag{3}$$

For a reversible isothermal process, ΔE may be replaced according to the first law of thermodynamics by

$$\Delta E = q_{\text{rev}} - W_{\text{max}} = T\,\Delta S - W_{\text{max}} \tag{4}$$

where q is the heat supplied and W_{max} is the maximum work done since the process is carried out reversibly.

Substitution of Eq. (4) into Eq. (3) leads to

$$-\Delta G = W_{\text{max}} - P\,\Delta V \tag{5}$$

The quantity $P\,\Delta V$ is the work done against the external pressure and so $W_{\text{max}} - P\,\Delta V$ is the net work. Consequently, the increase of free energy is a measure of the net work done on the system.

Differentiation of Eq. (1) gives

$$dG = dE + P\,dV + V\,dP - T\,dS - S\,dT \tag{6}$$

For a reversible process in which work is restricted to the work of expansion ($W = P\,dV$), the quantity of heat absorbed is given by

$$q_{\text{rev}} = dE + P\,dV \tag{7}$$

But $q_{\text{rev}}/T = dS$ and so Eq. (7) becomes

$$dE = T\,dS - P\,dV \tag{8}$$

Substitution of Eq. (8) into Eq. (6) yields

$$dG = V\,dP - S\,dT \tag{9}$$

For a reversible change, i.e., change brought about at an infinitesimally slow rate so that the system as a whole is virtually in temperature and pressure equilibrium with its surroundings, dE, dV, dP, and dT are zero. Thus Eqs. (8) and (9) lead to the criteria

$$(\partial S)_{E,V} = 0 \tag{10}$$

and

$$(\partial G)_{T,P} = 0 \tag{11}$$

for a system at equilibrium.

B. CHEMICAL POTENTIAL

The free energy of an open system in which both composition (energy) and mass may vary (a closed system is one in which only composition varies) can be written as a function of the different variables. Thus

$$G - f(T, P, n_1, n_2, \ldots, n_i)$$

where $n_1, n_2, \ldots, n_i$ are the numbers of moles of the respective constituents 1, 2, ..., i of the system. When all variables change by infinitesimally small amounts, then the change dG in G is given by

$$dG = \left(\frac{\partial G}{\partial T}\right)_{P, n_1, n_2, \ldots, n_i} dT + \left(\frac{\partial G}{\partial P}\right)_{T, n_1, n_2, \ldots, n_i} dP + \left(\frac{\partial G}{\partial n_1}\right)_{T, P, n_2, \ldots, n_i} dn_1 + \cdots + \left(\frac{\partial G}{\partial n_i}\right)_{T, P, n_1, n_2, \ldots} dn_i \quad (12)$$

In general, the derivative $\partial X/\partial n$, where X is an extensive property of the system, is called the partial molar quantity. The partial molar free energy, otherwise called the chemical potential μ of the species i, is given by

$$\mu_i = \left(\frac{\partial G}{\partial n_i}\right)_{T, P, n_j} \quad (13)$$

Although partial molar free energy is the most useful definition of chemical potential, it can be defined in other ways,[2] such as

$$\mu_i = \left(\frac{\partial E}{\partial n_i}\right)_{S, V, n_j} = \left(\frac{\partial H}{\partial n_i}\right)_{S, P, n_j} = \cdots \quad (14)$$

Equation (12) may be written as

$$dG = \left(\frac{\partial G}{\partial T}\right)_{P, n_1, \ldots, n_i} dT + \left(\frac{\partial G}{\partial P}\right)_{T, n_1, \ldots, n_i} dP + \mu_1\, dn_1 + \mu_2\, dn_2 + \cdots + \mu_i\, dn_i \quad (15)$$

If the system is a closed one, i.e., constant composition N, Eq. (15) becomes

$$dG = \left(\frac{\partial G}{\partial T}\right)_{P, N} dT + \left(\frac{\partial G}{\partial P}\right)_{T, N} dP \quad (16)$$

Comparing this equation with Eq. (9) gives

$$\left(\frac{\partial G}{\partial T}\right)_P = -S \quad (17)$$

and

$$\left(\frac{\partial G}{\partial P}\right)_T = V \quad (18)$$

Substitution of Eqs. (17) and (18) into Eq. (15) yields

$$dG = -S\, dT + V\, dP + \sum \mu_i\, dn_i \quad (19)$$

This equation, unlike Eq. (9), which does not take concentration of the species comprising the system into account, is applicable to any open system.

For a change brought about under isothermal and isobaric conditions, Eq. (19) becomes

$$(dG)_{T,P} = \sum \mu_i \, dn_i \tag{20}$$

which on integration, for a system of given chemical composition, becomes

$$G_{T,P,N} = \sum \mu_i n_i \tag{21}$$

Differentiation of Eq. (21) gives

$$(dG)_{T,P} = \sum (\mu_i \, dn_i + n_i \, d\mu_i) \tag{22}$$

but substitution of Eq. (20) into Eq. (22) leads to the relation

$$\sum n_i \, d\mu_i = 0 \tag{23}$$

Equation (23) is called the Gibbs–Duhem equation and has an important place in the study of thermodynamics of solutions as discussed later in this chapter.

Free energy G is an exact differential; it is a smooth function of the variables T, P, and N. Hence, according to the principles of calculus, the second differential of G with respect to any pair of variables is independent of the order of differentiation.[2, 3] The equations thus would be of the general form

$$\frac{\partial^2 G}{\partial n_i \, \partial n_j} = \frac{\partial^2 G}{\partial n_j \, \partial n_i} \tag{24}$$

Accordingly, the chemical potentials at constant temperature and pressure [see Eq. (13)] are related by equations of the type

$$\left(\frac{\partial \mu_1}{\partial n_2}\right)_{n_1, n_3, \ldots, n_i} = \left(\frac{\partial \mu_2}{\partial n_1}\right)_{n_2, n_3, \ldots, n_i} = \cdots \tag{25}$$

These equations are known as the reciprocity relations or cross-differentiation identities.[2]

Upon differentiation of Eq. (18) with respect to n_i, the number of moles of i in the system, it is seen that

$$\frac{\partial^2 G}{\partial P \, \partial n_i} = \overline{V}_i \tag{26}$$

where $\overline{V}_i$ is the partial molar volume of the constituent i of the system.

Differentiation of Eq. (13) with respect to pressure gives the effect of pressure on the chemical potential. Thus

$$\left(\frac{\partial \mu_i}{\partial P}\right)_{T,N} = \frac{\partial^2 G}{\partial n_i \partial P} \tag{27}$$

Equations (26) and (27) satisfy Eq. (24) and therefore lead to the relation

$$\left(\frac{\partial \mu_i}{\partial P}\right)_{T,N} = \overline{V}_i \tag{28}$$

For a system of ideal gases,

$$V_i = (RT/P_i)n_i$$

and so

$$\overline{V}_i = RT/P_i \tag{29}$$

Therefore Eq. (28) becomes

$$\left(\frac{\partial \mu_i}{\partial P}\right)_{T,N} = \frac{RT}{P_i} \tag{30}$$

The total pressure of a gaseous mixture is given by the sum of the partial pressures of the components of the gaseous system:

$$P = P_1 + P_2 + \cdots + P_i = \sum P_i \tag{31}$$

where the P_i's are the partial pressures. If the partial pressures of all other components except i remain constant, then

$$\partial P = \partial P_i \tag{32}$$

So Eq. (30) can be written as

$$d\mu_i = RT\, d \ln P_i \tag{33}$$

C. CONDITION FOR EQUILIBRIUM BETWEEN TWO PHASES

In the case of a pure liquid or a liquid mixture in equilibrium with its vapors, the chemical potential of any constituent in the liquid phase must be equal to that in the vapor phase. If the liquid is not in equilibrium with its vapors, then there will be a flow of dn_i molecules from the liquid phase (′) to the vapor phase (″) to establish equilibrium. The free energy change accompanying this transfer at constant temperature and pressure is given by

$$(dG)_{T,P} = \mu_i'\, dn_i' + \mu_i''\, dn_i''$$

but $dn_i' = -dn_i''$, since loss is equal to the gain, and at equilibrium

$$(dG)_{T,P} = 0$$

[see Eq. (11)]. Therefore,

$$dn_i''(\mu_i'' - \mu_i') = 0$$

But $dn_i'' \neq 0$, and so

$$\mu_i'' = \mu_i' \tag{34}$$

It follows that the condition for stable equilibrium between the two phases so that no phase vanishes is

$$d\mu_i'' = d\mu_i' \tag{35}$$

D. ACTIVITY AND ACTIVITY COEFFICIENT

In the case of an ideal solution, the vapors of component i at partial pressure P_i will be in equilibrium with the solution. But the partial pressure, according to Raoult's law (i.e., $P_i = N_i P_i^\circ$ where N_i is the mole fraction of i and P_i° is the vapor pressure in the pure state), is proportional to its mole fraction N_i in solution. Therefore, Eq. (33) for an ideal solution may be written as

$$\mu_i = \mu_N^\circ + RT \ln N_i \tag{36}$$

where μ_N° is a constant for the particular constituent and is independent of composition but depends on temperature and pressure. If the solution is not ideal, Eq. (36) is not applicable. So it is arbitrarily modified as

$$\mu_i = \mu_N^\circ + RT \ln N_i f_i \tag{37}$$

where f_i is a correction factor and is known as the activity coefficient. The product $N_i f_i$ is called the activity a_i. Thus μ_N° is the chemical potential when the activity of i (a_i) is unity. Activity is not an absolute quantity. Its full significance is realized in reference to a standard state, which is an arbitrarily chosen state of unit activity. Any convenient state irrespective of experimental difficulties may be chosen.

Two standard states are used in practice. For solid, liquid, or gas, and mixtures in which the solute is completely miscible with the solvent, the standard state chosen is usually the pure substance itself. In the case of dilute solutions, the standard state is expressed in three different ways, depending on the system employed for expressing the concentration of the solute. The concentration of a species i may be expressed on the molal (m_i, moles per 1000 grams of solvent), molar (C_i, moles per liter of solution), or mole fraction (N_i, the ratio of the number of moles of i to the total number of moles in solution) scale. The standard state for each concentration scale is so chosen that the activity coefficient is unity at infinite dilution. This applies to every temperature and pressure. For any particular solution, the three activity coefficients will be different, but they will all approach unity as the solution becomes more dilute. Equation (37) may therefore be written as

$$\begin{aligned}
\mu_i &= \mu_N^\circ + RT \ln N_i(f_i)_N = \mu_N^\circ + RT \ln(a_i)_N \\
\mu_i &= \mu_C^\circ + RT \ln C_i(f_i)_C = \mu_C^\circ + RT \ln(a_i)_C \\
\mu_i &= \mu_m^\circ + RT \ln m_i(f_i)_m = \mu_m^\circ + RT \ln(a_i)_m
\end{aligned} \tag{38}$$

$(f_i)_N$, $(f_i)_C$, and $(f_i)_m$ are called the rational, molar, and molal activity coefficients. $(f_i)_C$ and $(f_i)_m$ are the practical activity coefficients. $(f_i)_m$ is generally represented by γ_i, which is sometimes called the stoichiometric activity coefficient when the total molality of the electrolyte is considered without correction for its incomplete dissociation.

The interrelationships between N, C, and m and between f_N, f_C, and f_m are

$$N = \frac{mM_1}{1000 + mM_1} = \frac{CM_1}{1000\rho - C(M_2 - M_1)} \tag{39}$$

$$f_N = f_C \frac{1000\rho - C(M_2 - M_1)}{1000\rho_0} = f_m \frac{1000 + mM_1}{1000} \tag{40}$$

where M_1 and M_2 are the molecular weights of the solvent and the solute, respectively, and ρ and ρ_0 the densities of the solution and the solvent, respectively.

E. MEAN ACTIVITY OF THE ELECTROLYTE

In dealing with electrolyte solutions, it would be convenient to use the activities of the different ionic species present in the solution. But there are serious difficulties in such a procedure. The requirement of electroneutrality in the solution prevents adding either excess negative or excess positive ions and so there is no way of separating the effects caused by positive or negative ions. Nevertheless, it is convenient to have an expression for the activity of an electrolyte in terms of the ions into which it dissociates. The standard state of each ionic species is chosen so that the ratio of its activity to its concentration becomes unity at infinite dilution.

An electrolyte $M_{\nu+}A_{\nu-}$ which dissociates in solution to yield ν_+ positive ions (M^{z+}) of valence z_+ and ν_- negative ions (A^{z-}) of valence z_- may be considered. That is, it dissociates to

$$\mathrm{M}_{\nu_+}\mathrm{A}_{\nu_-} \rightleftarrows \nu_+ \mathrm{M}^{z+} + \nu_- \mathrm{A}^{z-} \tag{41}$$

The chemical potential of each of these ions is given by

$$\mu_+ = \mu^\circ_+ + RT \ln a_+ \tag{42}$$

$$\mu_- = \mu^\circ_- + RT \ln a_- \tag{43}$$

The chemical potential of the electrolyte as a whole is given by

$$\mu_2 = \mu^\circ_2 + RT \ln a_2 \tag{44}$$

For the electrolyte that dissociates completely, μ_2 may be written in terms of its constituent ions as

$$\mu_2 = \nu_+ \mu_+ + \nu_- \mu_- \tag{45}$$

Similarly, the chemical potential in the standard state μ_2° may be written as

$$\mu_2^\circ = \nu_+ \mu_+^\circ + \nu_- \mu_-^\circ \tag{46}$$

Thus Eq. (45) becomes

$$\mu_2^\circ + RT \ln a_2 = \nu_+ (\mu_+^\circ + RT \ln a_+) + \nu_- (\mu_-^\circ + RT \ln a_-) \tag{47}$$

Using Eq. (46) and simplifying, Eq. (47) becomes

$$a_+^{\nu+} a_-^{\nu-} = a_2 \tag{48}$$

The mean activity $a_\pm$ of the electrolyte is defined by

$$a_\pm^\nu = a_+^{\nu_+} a_-^{\nu_-} \tag{49}$$

where $\nu = \nu_+ + \nu_-$, the total number of ions produced by the electrolyte. Thus Eqs. (48) and (49) give the relation

$$a_2 = a_\pm^\nu \tag{50}$$

The activity of each ion may be written as

$$a_+ = m_+ \gamma_+ \quad \text{and} \quad a_- = m_- \gamma_- \tag{51}$$

The mean activity coefficient $\gamma_\pm$, in accordance with Eq. (49) is defined by

$$\gamma_\pm^\nu = \gamma_+^{\nu_+} \gamma_-^{\nu_-} \tag{52}$$

Thus

$$\gamma_\pm^\nu = \frac{a_+^{\nu_+}}{m_+^{\nu_+}} \frac{a_-^{\nu_-}}{m_-^{\nu_-}} = \frac{a_\pm^\nu}{m_\pm^\nu}$$

or

$$\gamma_\pm = a_\pm / m_\pm \tag{53}$$

The mean ionic molality is defined by

$$m_\pm^\nu = m_+^{\nu_+} m_-^{\nu_-} \tag{54}$$

If the molality of the solution is m, then

$$m_+ = \nu_+ m \quad \text{and} \quad m_- = \nu_- m$$

and thus Eq. (54) becomes

$$m_\pm^\nu = (\nu_+ m)^{\nu_+} (\nu_- m)^{\nu_-} = m^\nu (\nu_+^{\nu_+} \nu_-^{\nu_-})$$

or

$$m_{\pm} = m(\nu_{+}^{\nu_{+}} \nu_{-}^{\nu_{-}})^{1/\nu} \tag{55}$$

The interrelationships between the mean ionic activity coefficients are as follows [see also Eq. (40)]:

$$\begin{aligned} f_N &= f_C[(\rho + 0.001\nu C M_1 - 0.001 C M_2)/\rho_0] \\ &= \gamma(1 + 0.001\nu m M_1) \end{aligned} \tag{56}$$

where the ± subscripts are dropped from the mean activity coefficients.

In the special case of mixtures of two or more electrolytes, it is convenient to express the concentrations of electrolytes 1, 2, 3, . . . , i in the mixture (note the earlier notation where 1 and 2 referred to solvent and solute) on the molal scale since the number of molecules of the solvent remains constant. Thus $m_1, m_2, \ldots, m_i$, the molalities of the electrolytes, become independent variables. In this case, Eq. (25) may be written with the help of Eq. (38) in the form

$$\nu_1\left(\frac{\partial \log \gamma_1}{\partial m_2}\right)_{m_1, m_3, \ldots, m_i} = \nu_2\left(\frac{\partial \log \gamma_2}{\partial m_1}\right)_{m_2, m_3, \ldots, m_i}, \quad \ldots \tag{57}$$

F. OSMOTIC COEFFICIENT

Sometimes it is convenient to express nonideal behavior of electrolyte solutions in terms of the properties of the solvent. One such property is the osmotic coefficient which is the ratio of the real ($\Delta\Pi_{obs}$) to the ideal ($\Delta\Pi_{id}$) osmotic pressure.[4] The rational osmotic coefficient g may therefore be written as

$$g = \Delta\Pi_{obs}/\Delta\Pi_{id} \tag{58}$$

In the case of a semipermeable membrane separating the solvent phase (′) from the solution phase (″), only the solvent molecules move through the membrane. At equilibrium, in accordance with Eq. (34), one can write

$$\mu'_{1(P^\circ)} = \mu''_{1(P^\circ+\Delta\Pi)} = [\mu_1^{\circ\prime\prime} + RT \ln a_1]_{P^\circ+\Delta\Pi} \tag{59}$$

where P° is the pressure at which the solvent represented by 1 exists. Equation (35) applied to Eq. (59) gives

$$d\mu'_1 = d\mu''_1 = d\mu_1^{\circ\prime\prime} + RT\, d \ln a_1 \tag{60}$$

But $d\mu'_1 = 0$, since the pressure P° is constant, and therefore Eq. (60) becomes

$$d\mu_1^{\circ\prime\prime} = -RT\, d \ln a_1 \tag{61}$$

But $d\mu_1^{\circ\prime\prime} = (\partial\mu_1^{\circ\prime\prime}/\partial P)dP$ and $\partial\mu_1^{\circ\prime\prime}/\partial P$ according to Eq. (28) is equal to $\bar{V}_1$. Thus

$$\bar{V}_1\, dP = -RT\, d \ln a_1 \tag{62}$$

Integration of Eq. (62) yields

$$\left[\bar{V}_1 P\right]_{P=P^\circ}^{P=P^\circ+\Delta\Pi} = -\left[RT \ln a_1\right]_1^{a_1}$$

Thus the osmotic pressure $\Delta\Pi$ is given by

$$\Delta\Pi = -\frac{RT}{\bar{V}_1} \ln a_1 = -\frac{RT}{\bar{V}_1} \ln N_1 f_1 \tag{63}$$

For the ideal case when $f_1 = 1$, Eq. (63) becomes

$$\Delta\Pi_{\mathrm{id}} = -\frac{RT}{\bar{V}_1} \ln N_1 \tag{64}$$

Thus Eq. (58) becomes

$$g = \frac{\ln a_1}{\ln N_1} \tag{65}$$

But

$$N_1 = \frac{1000/M_1}{(1000/M_1) + \nu m} = \frac{1}{1 + (\nu m M_1/1000)}$$

Equation (65) therefore becomes

$$\begin{aligned} \ln a_1 &= -g \ln\left(1 + \frac{\nu m M_1}{1000}\right) \\ &= -g\left[\frac{\nu m M_1}{1000} - \frac{1}{2}\left(\frac{\nu m M_1}{1000}\right)^2 + \cdots\right] \end{aligned} \tag{66}$$

The practical molal osmotic coefficient ϕ is defined by the relation[5]

$$\ln a_1 = -\frac{\nu m M_1}{1000}\phi \tag{67}$$

Equations (66) and (67) give an approximate relation between the two osmotic coefficients. Thus

$$\phi \approx g\left[1 - \frac{1}{2}\frac{\nu m M_1}{1000}\right] \tag{68}$$

The relation between osmotic coefficient and osmotic pressure is given by equating Eqs. (63) and (67). Thus

$$\Delta\Pi = \frac{\nu RTM_1}{1000\bar{V}_1}\phi m \tag{69}$$

whereas for a dilute solution, the relation between $\Delta\Pi$ and g is given approximately by

$$\Delta\Pi = \frac{gRT\nu mM_1}{1000\bar{V}_1} \approx \nu gRTC \tag{70}$$

The osmotic coefficient is thus related to the Van't Hoff factor i by $\nu g \approx i$.

The Gibbs–Duhem equation (23) may be used to establish the relation between the molal osmotic coefficient and the activity coefficient. Equation (23) may be written in the form

$$\frac{1000}{M_1} d \ln a_1 + \nu m\, d \ln(m\gamma_\pm) = 0 \tag{71}$$

Differentiation of Eq. (67) and substitution for $d \ln a_1$ in Eq. (71) gives, on simplification, the relation

$$d[m(1-\phi)] + m\, d \ln \gamma_\pm = 0 \tag{72}$$

In integrated form, Eq. (72) is expressed as

$$\phi = 1 + \frac{1}{m}\int_0^m m\, d \ln \gamma_\pm \tag{73}$$

Equation (72) can also be written as

$$(\phi - 1)\frac{dm}{m} + d\phi = d \ln \gamma_\pm \tag{74}$$

which on integration gives

$$\ln \gamma_\pm = [\phi]_{m=0,\,\phi=1}^{m,\,\phi} + \int_0^m [(\phi - 1) d \ln m] \tag{75}$$

Thus substituting the limits of ϕ yields

$$\ln \gamma_\pm = (\phi - 1) + \int_0^m (\phi - 1) d \ln m \tag{76}$$

The activities of solutions can be determined by several methods. The most important among them are measurements of colligative properties of solutions such as depression of freezing point and osmotic pressure. The solubilities of sparingly soluble salts and methods based on the measurement of the emf of electrochemical cells have also been used. For details of these methods, standard books[1, 3, 5–7] may be consulted.

The mean activity coefficients of several salts have been derived from a number of measurements and these have been listed in various standard texts[4, 5] dealing with electrolyte solutions. All these results indicate that typically the activity coefficients decline markedly with an increase in the concentration in dilute solutions, then pass through minima, and rise again in more concentrated solutions. The interpretation of this behavior constitutes one of the important problems in the theory of strong electrolytes.

G. THE DEBYE–HÜCKEL THEORY

The activity or the activity coefficient of an electrolyte has been considered hitherto as a purely thermodynamic quantity which may be evaluated from measurable properties of the solution. The treatment involved no theory. Generally theories involving the behavior of matter lie outside the precincts of pure thermodynamics. However, a brief description is given of the Debye-Hückel theory, which has made it possible to calculate the mean activity coefficients of dilute electrolyte solutions without recourse to experiments. Although the treatment has a number of limitations, the theory represents the most significant advance in the electrochemistry of dilute electrolyte solutions. The important feature of the theory is the calculation of the electrical potential V at a point in the solution in terms of the charges of the ions and their concentrations and of the properties of the solvent. This is accomplished by combining the Poisson equation of electrostatic theory with a statistical distribution formula. The form of the Poisson equation for a spherically symmetrical situation is given by

$$\frac{1}{r^2}\frac{d}{dr}\left(r^2\frac{dV}{dr}\right) = -\frac{4\pi\rho}{\epsilon} \tag{77}$$

where ρ is the charge density and ϵ is the dielectric constant of the solvent. This applies to time-average values of potential and charge density at a distance r from the central ion.

Consider one particular ion, say a k ion, as the center of the coordinate system. The condition of electroneutrality

$$\sum_{i=1}^{s} n_i z_i \tag{78}$$

(n_i is the number of ions per unit volume) tells us that the net charge in the solution surrounding the kth ion is $-z_k e$ (e is the electronic charge). If a positive ion is chosen as the center of coordinates in a spherical shell, at a distance r from it, there will be on the average more negative ions than positive ions. The shell will therefore carry a negative charge. The total charge of the whole solution outside the central positive ion will be negative and equal to the positive charge of the cation. This is expressed as

$$\int_a^\infty 4\pi r^2 \rho_k \, dr = -z_k e \tag{79}$$

where a is the limit within which no other ion can approach the central ion. It is called the distance of closest approach.

The distribution of ions is governed by the Boltzmann law. The electrical potential energy of an ion i is given by $z_i e V_k$. The average local concentration n_i' of i ions at the point under consideration is given by

$$n_i' = n_i \exp(-z_i e V_k / KT) \tag{80}$$

where K is the Boltzmann constant. Each ion carries a charge $z_i e$. The net charge density at the point considered is (summing for all ionic species)

$$\rho_k = \sum_i z_i e n_i \exp(-z_i e V_k / KT) \tag{81}$$

Expanding e^{-x} $(e^{-x} = 1 - x + (x^2/2!) - (x^3/3!) + \cdots)$

$$\rho_k = \sum n_i z_i e - \sum n_i z_i e \left(\frac{z_i e V_k}{KT} \right) + \sum \frac{n_i z_i e}{2!} \left[\frac{z_i e V_k}{KT} \right]^2 - \cdots \tag{82}$$

According to Eq. (78) the first term in Eq. (82) is zero, and if $z_i e V_k \ll KT$ ($z_i e V_k$ is the potential energy term and KT is the kinetic energy term), only the linear term in V is appreciable. This linearization gives the result

$$\rho_k = -\sum_{i=1}^{s} \frac{n_i z_i^2 e^2}{KT} V_k \tag{83}$$

Substituting this value into Eq. (77) gives

$$\frac{1}{r^2} \frac{d}{dr} \left(r^2 \frac{dV_k}{dr} \right) = \frac{4\pi e^2}{\epsilon KT} \sum n_i z_i^2 V_k = \kappa^2 V_k \tag{84}$$

where

$$\kappa^2 = \frac{4\pi e^2}{\epsilon KT} \sum_i n_i z_i^2 \tag{85}$$

κ has the dimensions of reciprocal length and $1/\kappa$ is therefore called the radius or thickness of the ionic atmosphere. Substituting $n_i = C_i \mathfrak{N}/1000$ and $I = \frac{1}{2} \sum_i C_i z_i^2$ (C_i is in moles per liter, $\mathfrak{N}$ is Avogadro's number, and I

is called ionic strength), Eq. (85) becomes

$$\kappa = \sqrt{\frac{8\pi e^2 \mathcal{N}}{1000\epsilon KT} I} \tag{86}$$

Introducing numerical values for the universal constants, Eq. (86) becomes

$$\frac{1}{\kappa} = 1.988 \times 10^{-10} \sqrt{\frac{\epsilon T}{I}} \quad \text{cm} \tag{87}$$

The radius of the ionic atmosphere thus becomes bigger with decreasing I. For aqueous solutions at 25°C ($\epsilon = 78.30$), Eq. (87) becomes

$$\frac{1}{\kappa} = 3.04 \times 10^{-8} \sqrt{\frac{1}{I}} \quad \text{cm} \tag{88}$$

The $1/\kappa$ values calculated for different salt types and concentrations are given in Table 1, from which it is seen that the radius of the ionic atmosphere is much greater than the radius of the central ion itself.

TABLE 1

Thicknesses of the Ionic Atmosphere[a] as Functions of Concentration and Salt Types

Concentration (mole/liter)	Salt type			
	1 : 1	1 : 2	2 : 2	1 : 3
10^{-1}	9.6	5.5	4.8	3.9
10^{-2}	30.4	17.6	15.2	12.4
10^{-3}	96	55.5	48.1	39.3
10^{-4}	304	176	152	124

[a] At 25°C (in angstroms).

Equation (84) is a linear second-order differential equation between V and r and the solution has the general form

$$V_k = A\frac{e^{-\kappa r}}{r} + B\frac{e^{\kappa r}}{r} \tag{89}$$

The constants A and B have to be evaluated from the physical conditions of the problem. At high values of r, V_k must have a finite value, and for this to be realized B must equal zero.

Therefore

$$V_k = A\frac{e^{-\kappa r}}{r} \tag{90}$$

Substitution of this value into Eq. (83) gives

$$\rho_k = -A \frac{e^{-\kappa r}}{r} \sum_i \frac{n_i z_i^2 e^2}{KT} = -A \frac{\kappa^2 \epsilon}{4\pi} \frac{e^{-\kappa r}}{r} \tag{91}$$

Introducing Eq. (91) into Eq. (79) gives

$$A\kappa^2 \epsilon \int_a^\infty e^{-\kappa r} r \, dr = z_k e \tag{92}$$

and integration by parts yields

$$A = \frac{z_k e}{\epsilon} \frac{e^{\kappa a}}{1 + \kappa a} \tag{93}$$

Substituting this value of A into Eq. (90) gives the value for the potential:

$$V_k = \frac{z_k e}{\epsilon} \frac{e^{\kappa a}}{1 + \kappa a} \frac{e^{-\kappa r}}{r} \tag{94}$$

An isolated ion of valence z_k in a medium of dielectric constant ϵ gives rise to a field V_k'' which is

$$V_k'' = \frac{z_k e}{\epsilon r} \tag{95}$$

The total potential V_k at r can be treated as the sum of V_k'' due to the central ion and V_k' due to all the remaining ions. Thus

$$V_k = V_k' + V_k'' \tag{96}$$

Therefore

$$V_k' = \frac{z_k e}{\epsilon} \frac{e^{\kappa a}}{1 + \kappa a} \frac{e^{-\kappa r}}{r} - \frac{z_k e}{\epsilon r}$$

Thus

$$V_k' = \frac{z_k e}{\epsilon r} \left[\frac{e^{\kappa a}}{1 + \kappa a} e^{-\kappa r} - 1 \right] \tag{97}$$

This equation holds for all ions up to $r = a$. If $r < a$, no other ion can penetrate. Therefore for $r = a$, Eq. (97) becomes

$$V_k' = -\frac{z_k e}{\epsilon} \frac{\kappa}{1 + \kappa a} \tag{98}$$

This is the potential due to the ionic atmosphere. Because of this, the electrical energy of the central ion will be reduced by the product of its charge $z_k e$ and this potential because of its interaction with its neighbors. This argument can be applied to all ions in solution. In this case, each ion gets counted twice, once as the central ion and a second time as part of the ionic atmosphere. Therefore, the change in electrical energy of the kth ion due to its interaction with the surrounding ions is given by

$$\Delta G = -\frac{z_k^2 e^2}{2\epsilon} \frac{\kappa}{1 + \kappa a} \tag{99}$$

For 1 mole of k ions, Eq. (99) becomes

$$\Delta G_{\mathrm{el}} = -\frac{z_k^2 e^2 \mathfrak{N}}{2\epsilon} \frac{\kappa}{1 + \kappa a} \tag{100}$$

where ΔG_{el} is the change in electrical energy. The partial free energy of a mole of k ions can be written as

$$\mu_k = \mu_k(\text{ideal}) + \mu_k(\text{electrical})$$

but

$$\mu_k = \mu_k^\circ + RT \ln N_k + RT \ln(f_k)_N$$

$$\mu_k(\text{ideal}) = \mu_k^\circ + RT \ln N_k, \qquad \mu_k(\text{electrical}) = \Delta G_{\mathrm{el}}$$

Therefore

$$\ln(f_k)_N = \frac{\Delta G_{\mathrm{el}}}{RT} = -\frac{z_k^2 e^2}{2\epsilon KT} \frac{\kappa}{1 + \kappa a} \tag{101}$$

This is the expression for the individual ionic activity coefficient. The mean rational activity coefficient $f_\pm$ of an electrolyte dissociating into ν_1 cations of valence z_1 and ν_2 anions of valence z_2, according to Eq. (52), is given by

$$\ln f_\pm = (1/\nu)(\nu_1 \ln f_+ + \nu_2 \ln f_-) \tag{102}$$

Substituting the value of $(f_k)_N$ for each ion from Eq. (101) and employing the electroneutrality condition

$$\nu_1 z_1 = -\nu_2 z_2$$

leads to

$$\ln f_\pm = -\frac{|z_1 z_2| e^2}{2\epsilon KT} \frac{\kappa}{1 + \kappa a} \tag{103}$$

Substituting the value for κ from Eq. (86),

$$\log f_\pm = -\frac{A|z_1 z_2|\sqrt{I}}{1 + Ba\sqrt{I}} \tag{104}$$

where

$$A = \sqrt{\frac{2\pi \mathfrak{N}}{1000}} \frac{e^3}{2.303 K^{3/2}} \frac{1}{(\epsilon T)^{3/2}}$$

$$= \frac{1.8246 \times 10^6}{(\epsilon T)^{3/2}} \tag{105}$$

$$B = \left(\frac{8\pi \mathfrak{N} e^2}{1000 K}\right)^{1/2} \frac{1}{(\epsilon T)^{1/2}} = \frac{50.29 \times 10^8}{(\epsilon T)^{1/2}} \tag{106}$$

At very low values of I, i.e., dilute solutions, $Ba\sqrt{I}$ becomes small compared to unity and therefore Eq. (104) can be written as

$$\log f_{\pm} = -A|z_1 z_2|\sqrt{I} \tag{107}$$

where A has a value of 0.512 for water at 25°C. Equation (107) is called the Debye–Hückel limiting law and is a very useful guide to the behavior of activity coefficients at high dilutions. In Eq. (104), the numerator gives the effect of the long-range Coulombic forces and the denominator shows how these are affected by short-range interactions between ions. In any solution there will be short-range ion–solvent interactions to consider. These are assumed to give linear variations of $\log f_{\pm}$ with concentration. In an empirical fashion, these may be included by adding a term linear in concentration. Thus

$$\log f \pm = -\frac{A|z_1 z_2|\sqrt{I}}{1 + Ba\sqrt{I}} + bI \tag{108}$$

The parameters b and a can be adjusted to fit any experimental curve. A number of authors have written these equations in different empirical forms to describe the behavior of electrolytes. These equations are useful in calculating activity coefficients for salts for which no experimental values are available. Other important empirical equations are the following[5]: The first is the Guntelberg equation

$$\log f_{\pm} = -\frac{A|z_1 z_2|\sqrt{I}}{1 + \sqrt{I}} \tag{109}$$

Here a has a value of 3.04 Å for values of I up to 0.1 m. The Guggenheim formula has a linear term in concentration. Thus

$$\log f_{\pm} = -\frac{A|z_1 z_2|\sqrt{I}}{1 + \sqrt{I}} + bI \tag{110}$$

where b is an adjustable parameter. The other useful equation is due to Davies[8]:

$$\log f_{\pm} = -\frac{A|z_1 z_2|\sqrt{I}}{1 + \sqrt{I}} + 0.1|z_1 z_2|I$$

or

$$\log f_{\pm} = -\frac{A|z_1 z_2|\sqrt{I}}{1 + \sqrt{I}} + 0.15|z_1 z_2|I \tag{111}$$

The limiting law, Eq. (107), predicts that the logarithm of the activity coefficient must decrease linearly with the square root of ionic strength.

The slope of the plot of $\log f_\pm$ vs. $I^{1/2}$ can be evaluated and this should be equal to the product $-A|z_1z_2|$. Thus the slope depends on the valence type of the electrolyte. Experimental results have been found to agree remarkably well with these predictions at extremely low electrolyte concentrations. But at higher concentrations, the limiting law fails in that the plot of $\log f_\pm$ vs. $I^{1/2}$ is not a straight line but a curve. This deviation from a straight line up to a certain concentration can be accounted for by Eq. (104), which differs from the limiting law in having the term $(1 + Ba\sqrt{I})$. The parameter a assigns a finite size to the ion. Consequently, the ion is no longer considered to be a point charge. Definite values for the parameter a are unknown. However, a numerical value can be derived on the basis of an experiment. Equation (104) can be approximately represented by

$$\log f_\pm \approx -A|z_1z_2|I^{1/2}(1 - BaI^{1/2}) \qquad (112)$$

Since the values of A and B are known, the experimental determination of a value for $f_\pm$ for any given electrolyte (z_1z_2 are known) at a known concentration ($I^{1/2}$ known) enables calculation of a value for a. The suitability of this derived value can be judged by calculating the value for $f_\pm$ at another concentration and comparing it with the experimental value. This type of check has been carried out by a number of investigators, who have found the values of a to lie around 3–5 Å, which is greater than the sum of the crystallographic radii of positive and negative ions. This tends to show that the values pertain to the hydrated ion. In many cases, Eq. (104) has been found to give a very good fit with experiment by choosing a reasonable value for a, independent of concentration. The fit has been found often for ionic strengths up to 0.1. For example, a value of 4.0 Å for a in Eq. (104) gives good agreement up to 0.02 m with the value of $f_\pm$ determined by experiment in the case of NaCl. Some of the values of a determined this way for some electrolytes are given in Table 2.

TABLE 2

Values of a, the Ion-Size Parameter, for Some Electrolytes

Electrolyte	a (Å)
HCl	4.5
HBr	5.2
LiCl	4.3
NaCl	4.0
KCl	3.6

An unfortunate feature of the ion-size parameter a is that it is concentration dependent. As the electrolyte concentration is changed, the value of a

has to be modified (from about 6 Å at $m = 0.1$ to about 14 Å at $m = 1.0$ in the case of NaCl). In addition a has to assume impossible values for some electrolytes to fit the theory to experimental data[9] (see Table 3). Obviously the ion-size parameter a has been forced to take care of a number of other short-range interactions. Ions in solution exist in various states of interaction with the solvent molecules, some of which form part and parcel of the ions themselves, i.e., no longer solvent molecules, thereby decreasing the amount of effective solvent available for dissolving other ions. This has the effect of increasing the concentration of the solution. This aspect of ion hydration has not been taken care of by the parameter a although it took care of the increase in size of the point charge due to association of some water molecules with the ion (considered point charge). Attempts have been made to incorporate this aspect into the theory. It is based on the simple consideration that the total free energy of a fixed quantity of solution is fixed regardless of how much solvent exists as solvated with the ion. If, in a certain quantity of solution, 1 mole of the unhydrated solute 2 dissociates into ν_+ moles of cations and ν_- moles of anions and dissolves in s moles of solvent 1, the total free energy can be calculated for the system in two ways: first when the solute exists unsolvated, and second when the solute is solvated with h moles of the solvent. Thus the total free energy is given by

$$\begin{aligned} G &= s\mu_1 + \nu_+\mu_+ + \nu_-\mu_- \qquad \text{(unsolv)} \\ &= (s-h)\mu_1 + \nu_+\mu_+^{\mathrm{s}} + \nu_-\mu_-^{\mathrm{s}} \qquad \text{(solv)} \end{aligned} \tag{113}$$

Substituting from Eq. (38), Eq. (113) becomes, on rearrangement,

$$\frac{h\mu_1^\circ}{RT} + \frac{\nu_+(\mu_+^\circ - \mu_+^{\mathrm{s}\circ})}{RT} + \frac{\nu_-(\mu_-^\circ - \mu_-^{\mathrm{s}\circ})}{RT}$$
$$= -h\ln a_1 + \nu_+\ln a_+^{\mathrm{s}} - \nu_+\ln a_+ + \nu_-\ln a_-^{\mathrm{s}} - \nu_-\ln a_- \tag{114}$$

TABLE 3

Values of a, the Ion-Size Parameter, at Higher Concentrations[a]

	Molality				
Electrolyte	1.0	1.8	2.0	2.5	3.0
HCl	13.8	85.0	− 411.2	− 27.9	− 14.8
LiCl			41.3	− 141.9	− 26.4

[a]To fit experimental data with Eq. (104); values in angstroms.

At infinite dilution (i.e., $s \to \infty$), all the a_i's become unity and thus all logarithmic terms become zero. This makes the left-hand side of Eq. (114) zero. Substituting the mole fractions for cation and anion and using Eq.

(102) in Eq. (114) gives

$$\ln f_{\pm}^{s} = \ln f_{\pm} + \frac{h}{\nu} \ln a_1 + \ln \frac{s - h + \nu}{s + \nu} \tag{115}$$

Substituting for $s[= 1000/mM_1]$ and expressing $f_{\pm}$ in terms of the molal activity coefficient $\gamma_{\pm}$ [see Eq. (56)] and a_1 in terms of the osmotic coefficient [Eq. (67)], Eq. (115) becomes

$$\ln f_{\pm}^{s} = \ln \gamma_{\pm} + \frac{h}{\nu} \ln a_1 + \ln[1 + 0.001 mM_1(\nu - h)] \tag{116}$$

$$\ln f_{\pm}^{s} = \ln \gamma_{\pm} - 0.001 hmM_1\phi + \ln[1 + 0.001 mM_1(\nu - h)] \tag{117}$$

In these equations, it is assumed that h is independent of concentration. Assuming that Eq. (104) is referred to the solvated ions and ignoring other effects, such as molecular size and heat of mixing, Eq. (116) can be written as

$$\log \gamma_{\pm} = -\frac{A|z_1 z_2|\sqrt{I}}{1 + Ba\sqrt{I}} - \frac{h}{\nu} \log a_1 - \log[1 + 0.001 mM_1(\nu - h)] \tag{118}$$

This contains only two parameters, a and h, whose exact values are difficult to determine by experimentation. However, Eq. (118) has been successfully tested for a number of aqueous electrolyte solutions. In Table 4 are given the values of a and h for a number of electrolytes[5] whose activity coefficients were calculated with the help of Eq. (118), in excellent agreement with those derived by experiments. For example, in the case of NaCl, the activity coefficient calculated from Eq. (118) (a = 3.97 Å, h = 3.5) agreed with experimental values for solutions as concentrated as 5 m. In summary it may be said that the incorporation of a term to describe

TABLE 4

Values of a and h of a Molar Solution

Electrolyte	a (Å)	h
HCl	4.47	8.0
LiCl	4.32	7.1
NaCl	3.97	3.5
KCl	3.63	1.9
NH_4Cl	3.75	1.6
RbCl	3.49	1.2
$MgCl_2$	5.02	13.7
$CaCl_2$	4.73	12.0
$SrCl_2$	4.61	10.7
$BaCl_2$	4.45	7.7

the influence of an ion–solvent interaction on an ion–ion interaction has extended the range of electrolyte concentration over which the theory becomes applicable, although the theory has two parameters that defy exact experimental solution.

H. MIXTURE OF ELECTROLYTE AND NONELECTROLYTE

So far only single electrolyte solutions have been considered. But the solutions or rather the fluids of biology (i.e., the internal and external environments of a single biological cell) are more complex in that the contents of the external world of the cell are made up of a mixture of electrolytes and the internal fluid is composed of a mixture of electrolytes and nonelectrolyte-like substances. In addition, the concentrations are high enough to generate osmotic pressures of the order of about 7 atm in the case of humans and nearly three times that for marine forms. The quantitative description of the ion–ion and ion–solvent interactions in these mixtures is a formidable task. However, for very dilute solutions, a brief description of what happens when an electrolyte is added to a nonelectrolyte and one electrolyte is mixed with another electrolyte has been given.

Addition of a salt to an aqueous solution containing a neutral molecule generally decreases the nonelectrolyte solubility (salting out) and thereby increases its activity coefficient. Examples of reverse behavior (salting in) are also known. Debye and McAulay[10] calculated the electrical work done in reversibly discharging an ion of radius a_i and charge e_i in a solvent of dielectric constant ϵ (ionic atmosphere effect considered zero). This is given by

$$W_1 = -\int_0^{e_i} V''\, de_i = \frac{e_i^2}{2\epsilon a_i} \tag{119}$$

where $V'' = e_i/\epsilon a_i$ according to Eq. (95). The work involved in changing ϵ to ϵ_s by adding a nonelectrolyte and then recharging the ion reversibly is

$$W_2 = \frac{e_i^2}{2\epsilon_s a_i} \tag{120}$$

The difference in the two quantities of work corresponds to the change in free energy of the system brought about by the change in the dielectric constant. If the number of ions per milliliter is n_i, the total change in free energy amounts to

$$\Delta G = \sum \frac{n_i e_i^2}{2\epsilon_s a_i} - \sum \frac{n_i e_i^2}{2\epsilon a_i} \tag{121}$$

If the dielectric constant of a mixture is written as

$$\epsilon_s = \epsilon(1 - \beta n - \beta' n') \tag{122}$$

where β and β' are constants, and n and n' are the number of moles of electrolyte and nonelectrolyte, then Eqs. (121) and (122) give

$$\Delta G = \sum \frac{n_i e_i^2 (1 - \beta n - \beta' n')^{-1}}{2a_i \epsilon} - \sum \frac{n_i e_i^2}{2\epsilon a_i} \tag{123}$$

As a first approximation, Eq. (123) becomes

$$\Delta G = \frac{\beta n}{2\epsilon} \sum \frac{n_i e_i^2}{a_i} + \frac{\beta' n'}{2\epsilon} \sum \frac{n_i e_i^2}{a_i} \tag{124}$$

Contribution to the chemical potential of a nonelectrolyte ($\Delta\mu_s$ per mole) because of the addition of an electrolyte, or vice versa, is given by the differentiation of Eq. (124) with respect to n'. Thus

$$\Delta\mu_s = \frac{\partial(\Delta G)}{\partial n'} = \frac{\beta'}{2\epsilon} \sum \frac{n_i e_i^2}{a_i} = KT \ln f_s$$

or

$$\ln f_s = \frac{\beta'}{2KT\epsilon} \sum \frac{n_i e_i^2}{a_i} \tag{125}$$

When β' is positive, the dielectric constant is decreased, leading to an increase in f_s and thus the occurrence of salting out. When β' is negative, salting in takes place. Equation (125) is a limiting equation when the approximations on which it is based are considered.

The "salt effect" of an electrolyte upon itself can be obtained from Eq. (124) by differentiating it with respect to n. Since $n_i = \nu_i n$ and $e_i = z_i e$ (where e is the electronic charge), Eq. (124), upon differentiation, becomes

$$\nu KT \ln f_{\pm(s)} = \frac{\partial(\Delta G)}{\partial n} = \frac{\partial}{\partial n}\left[\frac{\beta n^2 e^2}{2\epsilon} \sum \frac{\nu_i z_i^2}{a_i}\right]$$

$$= \frac{\beta n e^2}{\epsilon} \sum \frac{\nu_i z_i^2}{a_i} \tag{126}$$

But $\beta n = (\epsilon - \epsilon_s)/\epsilon$, according to Eq. (122). Therefore Eq. (126) becomes

$$\ln f_{\pm(s)} = \frac{(\epsilon - \epsilon_s) e^2}{\epsilon^2 \nu KT} \sum \frac{\nu_i z_i^2}{a_i} \tag{127}$$

Equation (127) describes the salt effect of one electrolyte upon another at high dilution. In this case, the summation must be taken over all ions of both electrolytes.[11]

The effect of a nonelectrolyte upon the activity coefficient of an electrolyte in very dilute solution can be described by Eq. (121). The free energy change involved in transferring charges of n molecules of electrolyte from water of dielectric constant ϵ to a water–nonelectrolyte solution of dielectric constant ϵ_s is given by

$$\Delta G = \frac{ne^2}{2\epsilon_s} \sum \frac{\nu_i z_i^2}{a_i} - \frac{ne^2}{2\epsilon} \sum \frac{\nu_i z_i^2}{a_i} \tag{128}$$

Differentiating Eq. (128) with respect to n and rearranging gives

$$\ln f_{\pm} = \frac{(\epsilon - \epsilon_s)e^2}{2\nu KT\epsilon\epsilon_s} \sum \frac{\nu_i z_i^2}{a_i} \tag{129}$$

This equation was derived long ago by Born.[12]

All of these equations describing the effects of the addition of salt to a solution of nonelectrolyte or vice versa are valid only for extremely dilute solutions. As a result they cannot be used to evaluate theoretically the ionic activity coefficients of K^+ and/or Na^+ ions of some solutions (mixture of electrolyte and nonelectrolyte nearly 1000 milliosmoles/liter) used in electrophysiological work pertaining to the control of internal and external environments of some cells such as squid axon or barnacle muscle fiber. The ion-selective electrodes discussed in this book should prove helpful in these determinations.

I. MIXTURE OF TWO ELECTROLYTES

Although no theory developed from first principles exists to describe the interionic effects in mixtures of two or more electrolytes at higher concentrations, extensive experimental work on mixtures of electrolytes at constant total molality has been carried out. Two experimental methods proved important in this area.[11] One is based on the determination of the solubility of salts in salt solutions and the other on the determination of the activity coefficient of one electrolyte in the presence of a second electrolyte by emf measurements. Harned's rule has been formulated from the data so derived.[5, 11] According to this rule, the logarithm of the activity coefficient of one electrolyte in a mixture of constant total ionic strength is directly proportional to the ionic strength of the other electrolyte. That is,

$$\log \gamma_{1(\text{mix})} = \log \gamma_{1(\text{pure})} - \alpha_{12} I_2 \tag{130}$$

$$\log \gamma_{2(\text{mix})} = \log \gamma_{2(\text{pure})} - \alpha_{21} I_1 \tag{131}$$

where $\gamma_{1(\text{mix})}$, $\gamma_{2(\text{mix})}$ are the mean activity coefficients of electrolytes 1 and 2 in the mixture and $\gamma_{1(\text{pure})}$, $\gamma_{2(\text{pure})}$ refer to the activity coefficients of the electrolytes in pure solutions of ionic strength equal to that of the mixture.

The calculation of the mean activity coefficients of electrolytes 1 and 2 in their mixtures requires, according to Eqs. (130) and (131), determination of the constants α_{12} and α_{21}. If either of these can be determined experimentally, the other can be derived by one of two methods. The first method, due to McKay[13] and Harned and Owen,[11] is based on the cross differentiation equation [see Eq. (57)]. Equations (130) and (131) may be written as

$$\log \gamma_{1(\text{mix})} = \log \gamma_{1(\text{pure})} - \alpha_{12}(I - I_1) \tag{132}$$

$$\log \gamma_{2(\text{mix})} = \log \gamma_{2(\text{pure})} - \alpha_{21}(I - I_2) \tag{133}$$

where $I = I_1 + I_2 = \text{const}$. Thus $I_1 = I - I_2 = pm_1$ and $I_2 = I - I_1 = qm_2$, where p and q are constants characteristic of the valence type of the electrolyte. Consequently, Eq. (57) can be written as

$$\nu_1 q\left(\frac{\partial \log \gamma_1}{\partial I}\right)_{I_1} = \nu_2 p\left(\frac{\partial \log \gamma_2}{\partial I}\right)_{I_2} \tag{134}$$

Hence differentiation of Eqs. (132) and (133) and use of Eq. (134) gives

$$\nu_1 q\left[\left(\frac{\partial \log \gamma_{1(\text{pure})}}{\partial I}\right)_{I_1} - \alpha_{12} - I_2\left(\frac{\partial \alpha_{12}}{\partial I}\right)_{I_1}\right]$$

$$= \nu_2 p\left[\left(\frac{\partial \log \gamma_{2(\text{pure})}}{\partial I}\right)_{I_2} - \left\{\frac{\partial\,(\alpha_{21} I_1)}{\partial I}\right\}_{I_2}\right] \tag{135}$$

Since $\gamma_{1(\text{pure})}$, $\gamma_{2(\text{pure})}$, and α_{12} are functions of I only, the partial differentials can be replaced by total differentials except the one containing α_{21}. Thus

$$\nu_1 q\left[\frac{d \log \gamma_{1(\text{pure})}}{dI} - \alpha_{12} - I_2 \frac{d\alpha_{12}}{dI}\right]$$

$$= \nu_2 p\left[\frac{d \log \gamma_{2(\text{pure})}}{dI} - \left\{\frac{\partial\,(\alpha_{21} I_1)}{\partial I}\right\}_{I_2}\right] \tag{136}$$

Integration of Eq. (136) between the limits $I = I_2$ and $I = I$ (at constant I_2) gives, on rearrangement,

$$\nu_2 p \alpha_{21} I_1 = \left[\nu_2 p \log \gamma_{2(\text{pure})}\right]_{I_2}^{I} - \left[\nu_1 q \log \gamma_{1(\text{pure})}\right]_{I_2}^{I}$$

$$+ \nu_1 q \int_{I_2}^{I} \alpha_{12}\, dI + \left[\nu_1 q I_2 \alpha_{12}\right]_{I_2}^{I} \tag{137}$$

If α_{12} at a given total ionic strength I is known, α_{21} can be computed for each value of I_1 provided the values of $\gamma_{1(\text{pure})}$ and $\gamma_{2(\text{pure})}$ are known. This method of evaluation of α_{21} is general and is not based on the linearity or validity of Eq. (133).

The second method is based on the osmotic coefficient (ϕ) data for pure solutions of electrolytes 1 and 2. The Gibbs–Duhem equation (23) for two electrolytes of any valence types can be written as

$$\nu_1 \frac{I_1}{p} d \ln \gamma_1 + \nu_2 \frac{I_2}{q} d \ln \gamma_2 + \frac{\nu_1}{p} dI_1 + \frac{\nu_2}{q} dI_2 = 55.51\, d \ln a_w \quad (138)$$

where a_w is the activity of water.

If x is the fraction of electrolyte 1 present in the mixture ($0 \leqslant x \leqslant 1$), then

$$I_1 = xI \quad \text{and} \quad I_2 = (1 - x)I \quad (139)$$

Differentiation of Eq. (139) gives

$$dI_1 = I\, dx \quad \text{and} \quad dI_2 = -I\, dx \quad (140)$$

Thus

$$dI_1 = -dI_2$$

Differentiation of Eqs. (130) and (131) and using Eq. (140) gives

$$d \log \gamma_{1(\text{mix})} = -\alpha_{12}\, dI_2 = \alpha_{12} I\, dx \quad (141)$$

$$d \log \gamma_{2(\text{mix})} = -\alpha_{21}\, dI_1 = -\alpha_{21} I\, dx \quad (142)$$

Substitution of Eqs. (140)–(142) into Eq. (138) yields

$$\left[\frac{\nu_1}{p}\alpha_{12} + \frac{\nu_2}{q}\alpha_{21}\right] x\, dx + \left[\frac{\nu_1}{2.303 Ip} - \frac{\nu_2}{2.303 Iq} - \frac{\nu_2}{q}\alpha_{21}\right] dx = \frac{55.51}{2.303 I^2} d \ln a_w \quad (143)$$

Integration of this equation between the limits $x = 0$ and x gives

$$\left[\frac{\nu_1}{p}\alpha_{12} + \frac{\nu_2}{q}\alpha_{21}\right] \frac{x^2}{2} + \left[\frac{\nu_1}{2.303 Ip} - \frac{\nu_2}{2.303 Iq} - \frac{\nu_2}{q}\alpha_{21}\right] x = -\frac{55.51}{2.303 I^2} \ln \frac{a_{w(x)}}{a_{w(0)}} \quad (144)$$

For the two electrolytes, the following relations are valid:

$$I_1 = xI = (m_1/2)\left(\nu_{+(1)} z_{+(1)}^2 + \nu_{-(1)} z_{-(1)}^2\right)$$

$$I_2 = (1 - x)I = (m_2/2)\left(\nu_{+(2)} z_{+(2)}^2 + \nu_{-(2)} z_{-(2)}^2\right)$$

$$\nu_{+(1)} z_{+(1)} = \nu_{-(1)} z_{-(1)}; \qquad \nu_{+(2)} z_{+(2)} = \nu_{-(2)} z_{-(2)}$$

$$\nu_1 = \nu_{+(1)} + \nu_{-(1)}, \qquad \nu_2 = \nu_{+(2)} + \nu_{-(2)}$$

$$m_1 = \frac{2xI}{\nu_1 z_{+(1)} z_{-(1)}}; \qquad m_2 = \frac{2(1 - x)I}{\nu_2 z_{+(2)} z_{-(2)}}$$

$$\sum m_i = \nu_1 m_1 + \nu_2 m_2 = 2I\left[\frac{x}{z_{+(1)} z_{-(1)}} + \frac{1 - x}{z_{+(2)} z_{-(2)}}\right]$$

Expressing water activity in terms of an osmotic coefficient, Eq. (67) can be written as

$$-\log a_{\mathrm{w}} = \frac{\phi}{2.303 \times 55.51} \sum m_i \tag{145}$$

Substituting the value of Σm_i, Eq. (145) becomes

$$-\log a_{\mathrm{w}(x)} = \frac{2\phi_x I}{2.303 \times 55.51}\left[\frac{x}{z_{+(1)}z_{-(1)}} + \frac{1-x}{z_{+(2)}z_{-(2)}}\right] \tag{146}$$

Substitution of Eq. (146) into Eq. (144) yields

$$\left(\frac{\nu_1}{p}\alpha_{12} + \frac{\nu_2}{q}\alpha_{21}\right)\frac{x^2}{2} + \left(\frac{\nu_1}{2.303pI} - \frac{\nu_2}{2.303qI} - \frac{\nu_2}{q}\alpha_{21}\right)x$$
$$= \frac{2\phi_2}{2.303I}\,\frac{1}{z_{+(2)}z_{-(2)}} + \frac{2\phi_x}{2.303I}\left[\frac{x}{z_{+(1)}z_{-(1)}} + \frac{1-x}{z_{+(2)}z_{-(2)}}\right] \tag{147}$$

This equation assumes the validity of Eqs. (132) and (133).

In order to determine α_{21} from α_{12} (evaluated experimentally as the slope of the plot of $\log \gamma_{1(\mathrm{mix})}$ vs. I_2) and the osmotic coefficients ϕ_1 and ϕ_2 of electrolytes 1 and 2 in water, Eq. (147) is integrated from $x = 0$ (solution of electrolyte 2 alone) to $x = 1$ (solution of electrolyte 1). The final result on substitution of the values for p and q (i.e., I_1/m_1 and I_2/m_2) is

$$\frac{\alpha_{21}}{z_{+(2)}z_{-(2)}} = \frac{\alpha_{12}}{z_{+(1)}z_{-(1)}} - \frac{2}{2.303I}\left[\left(\frac{\phi_1}{z_{+(1)}z_{-(1)}} - \frac{\phi_2}{z_{+(2)}z_{-(2)}}\right) - \left(\frac{1}{z_{+(1)}z_{-(1)}} - \frac{1}{z_{+(2)}z_{-(2)}}\right)\right] \tag{148}$$

If the mixture is made up of 1 : 1 and 2 : 1 electrolytes, Eq. (148) becomes

$$\frac{\alpha_{21}}{2} = \alpha_{12} - \frac{2}{2.303I}\left[\left(\phi_1 - \frac{\phi_2}{2}\right) - \frac{1}{2}\right] \tag{149}$$

whereas if the mixture is composed of 1 : 1 electrolytes, Eq. (148) becomes

$$\alpha_{21} = \alpha_{12} + \frac{2}{2.303m}(\phi_2 - \phi_1) \tag{150}$$

where $m = m_1 + m_2$.

In terms of activity coefficients [see Eq. (73)], Eq. (150) can be written as

$$\alpha_{21} = \alpha_{12} + \frac{2}{m^2}\left[\int_0^m m\, d\log\frac{\gamma_{2(\mathrm{pure})}}{\gamma_{1(\mathrm{pure})}}\right] \tag{151}$$

Harned's rule [Eqs. (130) and (131)] has been confirmed for a number of electrolyte mixtures such as HCl + KCl; NaCl + HCl; LiCl + HCl; HCl + $NaClO_4$; HCl + $HClO_4$; HCl + $(Na_2S_2)_6$; HCl + $BaCl_2$; and HCl + $AlCl_3$.[5, 11] However, NaCl + NaOH, KCl + KOH, and $CaCl_2$ + $ZnCl_2$ mixtures do not conform to the rule. Detailed discussion of the variation of activity coefficients in mixtures of two electrolytes can be found in standard texts.[5, 11] Most of the data are derived at total ionic strengths considerably higher than those encountered in biological fluids, which contain mostly K^+ and Na^+ cations with small quantities of Ca^{2+} and Mg^{2+} ions. Calculation or determination of activity coefficients of the components of biological fluids is a very difficult problem. Moore and Ross[14] studied NaCl + $CaCl_2$ aqueous solutions at total ionic strengths ranging from 0.05 to $0.5m$ by potentiometric measurements using two electrodes: a Na glass electrode reversible and selective to Na^+ ions and a Ag–AgCl electrode reversible to Cl^- ions. The results were consistent with Harned's rule. The slope of a plot of log $\gamma_{NaCl(mix)}$ (measured) against the ionic strength of $CaCl_2$ in a mixture of total ionic strength 0.5 m was $-0.012(= -\alpha_{12})$. It was -0.122 at $I = 0.05m$. The values of α_{21} calculated from Eq. (149) ranged from 0.4 at $I = 0.05m$ to 0.14 at $I = 0.5m$. Since the value of α_{21} is known at any given value of I, $\gamma_{CaCl_2(mix)}$ can be calculated from Eq. (131). The values for $\gamma_{CaCl_2(mix)}$ were considerably less than those corresponding to solutions of $CaCl_{2(pure)}$. Thus NaCl had a depressing effect on the $CaCl_2$ activity in their mixtures. Although these facts are of great importance in human physiology, it is surprising that few data exist about physiological solutions. The development of ion-selective electrodes and their use in the study of these fluids should correct this situation.

REFERENCES

1. S. Glasstone, "Thermodynamics for Chemists." Van Nostrand-Reinhold, Princeton, New Jersey, 1947.
2. K. Denbigh, "The Principles of Chemical Equilibrium." Cambridge Univ. Press, London and New York, 1961.
3. G. N. Lewis and M. Randall, "Thermodynamics" (rev. by K. S. Pitzer and L. Brewer, 2nd ed.). McGraw-Hill, New York, 1961.
4. G. Kortum and J. O'M. Bockris, "Text Book of Electrochemistry," Vols. I and II. Elsevier, Amsterdam, 1951.
5. R. A. Robinson and R. H. Stokes, "Electrolyte Solutions." Butterworths, London and Washington, D. C., 1959.
6. S. Glasstone, "An Introduction to Electrochemistry." Van Nostrand-Reinhold, Princeton, New Jersey, 1949.
7. S. Glasstone, "Text Book of Physical Chemistry." Van Nostrand-Reinhold, Princeton, New Jersey, 1947.

8. C. W. Davies, "Ion Association." Butterworths, London and Washington, D. C., 1962.
9. J. O'M. Bockris and A. K. N. Reddy, "Modern Electrochemistry," Vol. 1. Plenum Press, New York, 1970.
10. P. Debye and J. McAulay, *Z. Phys.* **26**, 22 (1925).
11. H. S. Harned and B. B. Owen, "The Physical Chemistry of Electrolytic Solutions." Van Nostrand-Reinhold, Princeton, New Jersey, 1958.
12. M. Born, *Z. Phys.* **1**, 45 (1920).
13. H. A. C. McKay, *Nature* (London) **169**, 464 (1952).
14. E. W. Moore and J. W. Ross, *J. Appl. Physiol.* **20**, 1332 (1965).

Chapter 3

THEORIES OF MEMBRANE ELECTRODE POTENTIALS

A voltaic cell is composed of two electrodes arranged in such a way that an electric current will flow in the metallic wire connecting the two electrodes. At each electrode, two phases, one acting as an electronic and the other as an electrolytic (salt solution) conductor, exist. At the surface of separation of the two phases there is a potential difference called the electrode potential. In the absence of any other potential, the emf of the cell is equal to the algebraic sum of the two electrode potentials. The emf of the cell (electrical energy) arises from the chemical reactions occurring at the two electrodes. If the reactions take place spontaneously, even if there is no passage of current, the cell is said to be irreversible in the thermodynamic sense, since thermodynamic reversibility indicates a state of equilibrium at every stage. This implies passing through or drawing from the cell infinitesimally small currents to keep the cell virtually at equilibrium.

A. REVERSIBLE ELECTRODES

A reversible cell is always composed of at least two reversible electrodes. Several kinds of such electrodes are known. The first category consists of electodes of the first kind and are reversible to ions of the electrode material. Some examples are

$$M \rightleftharpoons M^+ + e \qquad \text{(metal)}$$

$$A + e \rightleftharpoons A^- \qquad \text{(nonmetal, halides)}$$

$$\tfrac{1}{2}H_2 \rightleftharpoons H^+ + e \qquad \text{(hydrogen electrode)}$$

$$\tfrac{1}{2}O_2 + H_2O + 2e \rightleftharpoons 2OH^- \qquad \text{(oxygen electrode)}$$

The second category consists of electrodes of the second kind, involving a metal, a sparingly soluble salt of the metal, and a solution of a soluble salt of the same anion. An example of these valuable electrodes is

$$\mathrm{Ag{-}AgX + X\ solution}$$

The chemical reactions are

$$\mathrm{Ag(s) \rightleftarrows Ag^+ + e}$$

$$\mathrm{Ag^+ + X^- \rightleftarrows AgX(s)}$$

and the net reaction is

$$\mathrm{Ag(s) + X^- \rightleftarrows AgX(s) + e}$$

where X could be a halogen. Mercury–mercurous chloride in contact with KCl solution (calomel half-cell) is another example. Electrodes reversible to other anions such as sulfate and oxalate, can be formed in an indirect manner.

The third category, electrodes of the third kind, consists of a metal, one of its insoluble salts, another insoluble salt of the same anion, and a solution of a soluble salt with the same cation as the latter salt. Thus

$$\mathrm{Pb|PbC_2O_4(s)|CaC_2O_4(s)|CaCl_2\ solution}$$

The electrode reactions are

$$\mathrm{Pb(s) \rightleftarrows Pb^{2+} + 2e}$$

$$\mathrm{Pb^{2+} + C_2O_4^{2-} \rightleftarrows PbC_2O_4(s)}$$

The removal of oxalate ions causes calcium oxalate to dissolve and ionize to maintain its solubility product

$$\mathrm{CaC_2O_4 \rightleftarrows Ca^{2+} + C_2O_4^{2-}}$$

Thus the overall reaction is

$$\mathrm{Pb(s) + CaC_2O_4(s) \rightleftarrows PbC_2O_4 + Ca^{2+} + 2e}$$

The system behaves as an electrode reversible to calcium ions.

Another type of reversible electrode consists of an unattackable metal such as platinum or gold, serving as a metallic conductor, immersed in a solution containing species in their oxidized and reduced forms. The general form of the electrode process may be written as

$$\text{Reduced state} \rightleftarrows \text{oxidized state} + ne$$

where n is the number of electrons involved in the change of state from reduced to oxidized forms.

B. RELATION BETWEEN FREE ENERGY AND POTENTIAL OF A GALVANIC CELL

For an electrochemical cell whose emf is E volts and which is operated reversibly at constant temperature and pressure, the electrical work done on the passage of an infinitesimally small quantity of electricity (δq coulombs) is $E\delta q$ (volt-coulombs or joules). If $-\Delta G$ is the decrease in free energy of the reacting species in the cell due to passage of n faradays (F) of electricity, the change in free energy brought about by δq C of electricity is $-\Delta G \delta q / nF$, and this is equivalent to the electrical work $E\delta q$. Thus

$$-\Delta G = nEF \tag{1}$$

In general if the net process occurring in a reversible cell during the passage of n faradays is

$$a\mathrm{A} + b\mathrm{B} + \cdots \rightleftarrows l\mathrm{L} + m\mathrm{M} + \cdots \tag{2}$$

then the change in free energy according to Eq. (38) of Chapter 2 is given by

$$\begin{aligned} \Delta G &= l\mu_{\mathrm{L}} + m\mu_{\mathrm{M}} + \cdots - a\mu_{\mathrm{A}} - b\mu_{\mathrm{B}} - \cdots \\ &= RT \ln \frac{a_{\mathrm{L}}^{l} a_{\mathrm{M}}^{m} \cdots}{a_{\mathrm{A}}^{a} a_{\mathrm{B}}^{b} \cdots} + l\mu_{\mathrm{L}}^{\circ} + b\mu_{\mathrm{M}}^{\circ} + \cdots - a\mu_{\mathrm{A}}^{\circ} - b\mu_{\mathrm{B}}^{\circ} - \cdots \end{aligned} \tag{3}$$

Since the terms $l\mu_{\mathrm{L}}^{\circ} + m\mu_{\mathrm{M}}^{\circ} + \cdots - a\mu_{\mathrm{A}}^{\circ} - b\mu_{\mathrm{B}}^{\circ} - \cdots$ refer to the standard free energy change ΔG° (i.e., the components are in their standard state), Eq. (3) can be written as

$$\Delta G = \Delta G^{\circ} + RT \ln \frac{a_{\mathrm{L}}^{l} a_{\mathrm{M}}^{m} \cdots}{a_{\mathrm{A}}^{a} a_{\mathrm{B}}^{b} \cdots} \tag{4}$$

The condition for equilibrium, namely $\Delta G = 0$ [see Eq. (11) of Chapter 2], can be realized if and only if

$$\left[\frac{a_{\mathrm{L}}^{l} a_{\mathrm{M}}^{m} \cdots}{a_{\mathrm{A}}^{a} a_{\mathrm{B}}^{b} \cdots} \right]_{\mathrm{equil}} = \exp\left(\frac{-\Delta G^{\circ}}{RT} \right) = K \tag{5}$$

Equation (5) is the law of mass action and K is called the thermodynamic equilibrium constant. Equation (4), also called the Van't Hoff isotherm, can therefore be written as

$$-\Delta G = RT \ln K - RT \ln \frac{a_{\mathrm{L}}^{l} a_{\mathrm{M}}^{m} \cdots}{a_{\mathrm{A}}^{a} a_{\mathrm{B}}^{b} \cdots} \tag{6}$$

Substituting from Eq. (1), Eq. (6) becomes

$$E = E^{\circ} - \frac{RT}{nF} \ln \frac{a_{\mathrm{L}}^{l} a_{\mathrm{M}}^{m} \cdots}{a_{\mathrm{A}}^{a} a_{\mathrm{B}}^{b} \cdots} \tag{7}$$

where E° is related to the equilibrium constant K and can be regarded as equal to the difference between two constants E_1° and E_2°. These are characteristic of the separate electrode reactions occurring at the two electrodes composing the cell, so Eq. (7) can be written as

$$E = \left(E_1^\circ - \frac{RT}{nF}\sum \ln a_1^{\nu_1}\right) - \left(E_2^\circ - \frac{RT}{nF}\sum \ln a_2^{\nu_2}\right) \tag{8}$$

where a_1 and a_2 are the terms of activity applicable to the two electrodes and ν_1 and ν_2 are the number of molecules or ions of the corresponding species taking part in the cell reaction. In general, it is possible to write an equation of the form

$$E_i = E_i^\circ - \frac{RT}{nF}\sum \ln a_i^{\nu_i} \tag{9}$$

for the potential of an electrode in terms of its standard potential E_i° and the activities of the molecules and/or ions taking part in the electrode process. It is evident that E_i° is equal to E_i where all the a_i's are at unit activity (i.e., in their standard states).

C. STANDARD POTENTIALS

The application of these equations may be illustrated for the following reversible cell:

$$H_2(1\text{ atm})|HCl(m)|AgCl\text{–}Ag \tag{10}$$

in which the cell reaction for the passage of 1 faraday of current is

$$\tfrac{1}{2}H_2(1\text{ atm}) + AgCl(s) \rightleftarrows H^+ + Cl^- + Ag(s)$$

For this case, Eq. (7) becomes

$$E = E^\circ - \frac{RT}{F}\ln\frac{a_{H^+}a_{Cl^-}a_{Ag}}{a_{H_2}^{1/2}a_{AgCl}} \tag{11}$$

The individual electrode reactions are

$$\tfrac{1}{2}H_2(1\text{ atm}) \rightleftarrows H^+ + e$$

$$AgCl(s) + e \rightleftarrows Ag(s) + Cl^-$$

Equation (11) therefore may be split into

$$E = \left(E_1^\circ - \frac{RT}{F}\ln\frac{a_{H^+}}{a_{H_2}^{1/2}}\right) - \left(E_2^\circ - \frac{RT}{F}\ln\frac{a_{AgCl}}{a_{Ag}a_{Cl^-}}\right)$$

Thus

$$E_1 = E_1^\circ - \frac{RT}{F}\ln\frac{a_{H^+}}{a_{H_2}^{1/2}} \quad \text{and} \quad E_2 = E_2^\circ - \frac{RT}{F}\ln\frac{a_{AgCl}}{a_{Ag}a_{Cl^-}} \tag{12}$$

Since H_2(1 atm), Ag(s), and AgCl(s) are in their standard states (i.e.,

$a_i = 1$), it follows that

$$E_{H_2, H^+} = E^\circ_{H_2, H^+} - \frac{RT}{F} \ln a_{H^+} \tag{13}$$

$$E_{Ag-AgCl, Cl^-} = E^\circ_{Ag-AgCl, Cl^-} + \frac{RT}{F} \ln a_{Cl^-} \tag{14}$$

where the E°'s are the standard potentials of the H_2(1 atm), H^+ and Ag–AgCl, Cl^- electrodes. Thus for an electrode reversible to cations or anions of valence $z_\pm$, the general form of equation for the electrode potential is

$$E_\pm = E^\circ \mp \frac{RT}{z_\pm F} \ln a_i \tag{15}$$

The potential of the reversible cell is given by

$$E = E^\circ - \frac{RT}{F} \ln a_{H^+} a_{Cl^-} \tag{16}$$

In terms of mean activity $a_\pm$ [see Eq. (49) of Chapter 2], Eq. (16) becomes

$$E = E^\circ - \frac{2RT}{F} \ln a_\pm \tag{17}$$

Cells of type (10) may be used to evaluate E°. In addition, Eq. (17) may be used to evaluate the activity coefficient γ for HCl provided E is determined as a function of the molality m of HCl using the galvanic cell (10).[1, 2]

Rewriting Eq. (17) in terms of m and γ ($a = \gamma m$, note that the $\pm$ sign is dropped) and rearranging gives

$$E + \frac{2RT}{F} \ln m - E^\circ = -\frac{2RT}{F} \ln \gamma \tag{18}$$

Expressing γ in terms of molality [see Eq. (108) of Chapter 2], i.e.,

$$\log \gamma = -A\sqrt{m} + Cm \tag{19}$$

Eq. (18) becomes

$$E + \frac{4.606RT}{F} \log m - E^\circ = -\frac{4.606RT}{F} (-A\sqrt{m} + Cm) \tag{20}$$

Substituting the numerical values for R, F, T, and A (8.313 J/deg-mole, 96,500 C, 298°K, and 0.509 for water as solvent), Eq. (20) becomes

$$E + 0.1183 \log m - 0.0602\sqrt{m} = E^\circ - 0.1183Cm \tag{21}$$

According to Eq. (21), a plot of $E + 0.1183m - 0.0602\sqrt{m}$ should be a linear function of m and extrapolation of the plot to $m = 0$ should give E°. Although the plot is not exactly linear,[2] extrapolation nevertheless is possible. In this manner a value of 0.22234 V for E° has been derived.[3] Thus, from Eq. (18) values for the activity coefficient γ can be determined directly using the measured values of E of cell (10).

D. CONCENTRATION CELL WITHOUT TRANSFERENCE

Two cells of type (10) containing HCl at concentrations m_1 and m_2 can be connected in opposition to form the cell

$$H_2(1\ \text{atm})|HCl(m_1)|AgCl\text{–}Ag\text{–}AgCl|HCl(m_2)|H_2(1\ \text{atm}) \tag{22}$$

The emf of this cell is given by $E = E_1 - E_2$, where E_1 and E_2 are the emfs of the two cells forming the concentration cell (22). According to Eq. (17), E therefore is given by

$$E = \frac{2RT}{F} \ln \frac{a_2}{a_1} \tag{23}$$

In general the emf of any concentration cell without transport can be expressed by means of the equation

$$E = \pm \frac{\nu}{\nu_\pm} \frac{RT}{z_\pm F} \ln \frac{a_2}{a_1} \tag{24}$$

where z_+ or z_- is the valence of the ion to which the extreme electrodes are reversible. If the ion is positive, as in cell (22), positive signs apply throughout, but if it is negative, the negative signs apply.

E. CONCENTRATION CELL WITH TRANSFERENCE

Removal of the Ag–AgCl bridge in cell (22) allows one solution to contact the other and direct transfer of HCl from the more concentrated solution (m_2) to the more dilute (m_1) takes place. An example of a concentration cell with transference is

$$H_2(1\ \text{atm})|HCl(m_1)\,\vdots\,HCl(m_2)|H_2(1\ \text{atm}) \tag{25}$$

vertical dashed line represents the liquid junction formed by the two solutions. Passage of a faraday of current results in the following:

(a) 1 g-atom of hydrogen dissolves to give 1 g-ion of hydrogen ions at the left-hand electrode,

(b) at the right-hand electrode, the same 1 g-ion is discharged to form 1 g-atom of hydrogen, and

(c) t_{H^+} g-ions of H^+ ions migrate across the liquid boundary from left to right and t_{Cl-} g-ions of Cl^- ions migrate in the opposite direction.

t_i is the transference number of ion i defined for a general case as $t_i = (z_i C_i u_i)/\Sigma_j(z_j C_j u_j)$ and $\Sigma t_i = 1$. The net result due to passage of a faraday of current is the transfer of $1 - t_{H+}$ or t_{Cl-} g-ions of H^+ ions and t_{Cl-} g-ions of Cl^- ions from right to left, i.e., from m_2 to m_1. The total free energy change is given by

$$\Delta G = t_{Cl^-}[\mu_{H^+(1)} - \mu_{H^+(2)} + \mu_{Cl^-(1)} - \mu_{Cl^-(2)}] \tag{26}$$

Since the t_i's are concentration dependent, it is preferable to consider concentrations that differ by a small amount dm. Under such conditions, Eq. (26) can be written as

$$\begin{aligned} dG &= -t_{Cl^-}(d\mu_{H^+} + d\mu_{Cl^-}) = -t_{Cl^-}(RT\,d\ln a_{H^+} + RT\,d\ln a_{Cl^-}) \\ &= -2RTt_{Cl^-}\,d\ln a = -dE\,F \end{aligned}$$

Therefore

$$dE = 2t_{Cl^-}\frac{RT}{F}\,d\ln a \tag{27}$$

Integrating Eq. (27) between the limits a_1 and a_2 gives

$$E = \frac{2RT}{F}\int_{a_1}^{a_2} t_{Cl^-}\,d\ln a \tag{28}$$

For the general case, Eq. (28) can be written as

$$E = \pm\frac{\nu}{\nu_\pm}\frac{RT}{z_\pm F}\int_{a_1}^{a_2} t_\mp\,d\ln a \tag{29}$$

where ν, $\nu_\pm$, and $z_\pm$ have the same significance as in Eq. (24). The transference number $t_\mp$ refers not to the ion to which the electrodes are reversible but to the other ion.

If the transference number is assumed constant in the concentration range m_1–m_2, Eq. (29) takes the form

$$E = \pm t_\mp\frac{\nu}{\nu_\pm}\frac{RT}{z_\pm F}\ln\frac{a_2}{a_1} \tag{30}$$

Thus in the special case of cell (25), Eq. (30) becomes

$$E = 2t_-\frac{RT}{F}\ln\frac{a_2}{a_1} \tag{31}$$

Dividing Eq. (31) by Eq. (23) gives a value for the transport number of the ion other than the one to which the end electrodes are reversible. If the t_i's are dependent on the concentration, then Eq. (27) and the differential form of Eq. (24) may be considered. Equation (27) may be written for the general case [differential of Eq. (30)] as

$$dE_t = \pm t_{\mp} \frac{\nu}{\nu_{\pm}} \frac{RT}{z_{\pm} F} d \ln a$$

or

$$\frac{dE_t}{d \ln a} = \pm t_{\mp} \frac{\nu}{\nu_{\pm}} \frac{RT}{z_{\pm} F} \tag{32}$$

The corresponding emf of the cell without transport [Eq. (24)] is

$$\frac{dE}{d \ln a} = \pm \frac{\nu}{\nu_{\pm}} \frac{RT}{z_{\pm} F} \tag{33}$$

It follows therefore from Eqs. (32) and (33) that

$$t_{\mp} = \frac{dE_t / d \log a}{dE / d \log a} \tag{34}$$

F. LIQUID JUNCTION OR DIFFUSION POTENTIAL: GENERAL EQUATION

Consider a cell containing a solution in which there are several ions of concentrations C_1, $C_2, \ldots, C_i$ and which forms a liquid junction with another solution containing corresponding ionic concentrations $C_1 + dC_1$, $C_2 + dC_2, \ldots, C_i + dC_i$, the ionic valences and transport numbers being $z_1, z_2, \ldots, z_i$ and $t_1, t_2, \ldots, t_i$. The transport numbers are assumed to be invariant with concentration.

For the passage of a faraday of current through the cell, t_i/z_i g-ions of each species will be transferred across the boundary, the positive ions moving from left to right and the negative ions moving in the opposite direction. The change in free energy due to transfer of the ith ion from $C_i + dC_i$ concentration to C_i concentration is given by

$$dG = \frac{t_i}{z_i} [(\mu_i + d\mu_i) - \mu_i] = \frac{t_i}{z_i} d\mu_i$$

For the transfer of all the ions, the total free energy change is

$$\Delta G = \sum_i \frac{t_i}{z_i} d\mu_i = \sum \frac{t_i}{z_i} RT \, d \ln a_i$$

But $\Delta G = -dE_{\mathrm{L}}F$, so

$$dE_{\mathrm{L}} = -\frac{RT}{F} \sum \frac{t_i}{z_i} d \ln a_i \tag{35}$$

Any liquid junction between solutions 1 and 2 may be considered to be made up of a series of layers with infinitesimal concentration differences. The resultant potential E_L is obtained by the integration of Eq. (35) between the limits 1 and 2 representing the two solutions in the cell; thus

$$E_L = -\frac{RT}{F}\int_{I}^{II}\sum \frac{t_i}{z_i}\, d\ln a_i \tag{36}$$

In order to integrate Eq. (36), approximations have to be made about a_i, a single ion property that cannot be measured, about the concentration dependence of t_i, and about the properties of the boundary itself. The various approximations and/or assumptions are detailed below.

G. DIFFUSION POTENTIAL BETWEEN TWO SOLUTIONS OF THE SAME ELECTROLYTE

The total potential due to the liquid junction formed between HCl solutions of concentration m_1 and m_2, e.g.,

$$HCl(m_1) \,\vdots\, HCl(m_2)$$

is given according to Eq. (36) by

$$E_L = -\frac{RT}{F}\int_{I}^{II}\frac{t_+}{z_+}\, d\ln a_+ + \frac{RT}{F}\int_{I}^{II}\frac{t_-}{z_-}\, d\ln a_-$$

where t_+ and t_- are assumed to be independent of concentration. Then

$$E_L = -\frac{RT}{F}\frac{t_+}{z_+}\ln\frac{(a_+)_2}{(a_+)_1} + \frac{RT}{F}\frac{t_-}{z_-}\ln\frac{(a_-)_2}{(a_-)_1}$$

Since $t_+ + t_- = 1$ and $z_+ = z_- = 1$

$$E_L = -2t_+\frac{RT}{F}\ln\frac{a_2}{a_1} + \frac{RT}{F}\ln\frac{(a_-)_2}{(a_-)_1} \tag{37}$$

where a_2 and a_1 are the mean activities of the two solutions. If it is further assumed that $(a_-)_2/(a_-)_1 = a_2/a_1$, Eq. (37) reduces to

$$E_L = (1 - 2t_+)\frac{RT}{F}\ln\frac{a_2}{a_1} \tag{38}$$

or alternatively,

$$E_L = (t_- - t_+)\frac{RT}{F}\ln\frac{a_2}{a_1} \tag{38a}$$

If the liquid junction potential forms part of the concentration cell as in cell (25), the emf of the complete cell is given by Eq. (31). Dividing Eq. (38) by Eq. (31) gives the relation

$$E_{\mathrm{L}} = \frac{1 - 2t_{+}}{2t_{+}} E \tag{39}$$

In the case of two solutions of the same electrolyte forming a liquid junction, the electrolyte solution is the same at any point in the boundary layer with a definite concentration and therefore the ionic species will have a definite transference number and activity. But if different electrolytes exist to form the boundary, the ionic concentration at any point in it will be determined by its structure. Therefore, the values of t_i and a_i will depend on the nature of the boundary.

Guggenheim[4] has divided the different liquid junctions into four distinct types. They are (i) continuous mixture boundary, (ii) constrained diffusion junction, (iii) flowing junction, and (iv) free diffusion junction. Types (i) and (ii) have been considered by Henderson[5] and Planck,[6] respectively. Types (iii) and (iv), which are systems widely used in practice, are too complex to be treated theoretically.

H. DIFFUSION EQUATIONS

The type of boundary postulated by Henderson[5] consists of a continuous series of mixtures of the two solutions 1 and 2. At any particular point in the diffusion zone, the concentration C_i of the species i is given by

$$C_i = C_{i(1)}(1 - x) + xC_{i(2)}$$

where $1 - x$ is the fraction of solution 1 at the given point in the boundary and x is the fraction of solution 2. $C_{i(1)}$ and $C_{i(2)}$ are the bulk concentrations of i in solutions 1 and 2, respectively. Using the expression just given, Eq. (36), in which activities were replaced by concentrations as an approximation, was integrated by Henderson. The final result, given in the more familiar form, is[7]

$$E_{\mathrm{L}} = \frac{RT}{F} \frac{(U_1 - V_1) - (U_2 - V_2)}{(U_1' + V_1') - (U_2' + V_2')} \ln\left(\frac{U_1' + V_1'}{U_2' + V_2'} \right) \tag{40}$$

where U_1, V_1, are defined by

$$\begin{aligned} U_1 &= \sum_n (C_+ u_+)_1, & V_1 &= \sum_n (C_- u_-)_1 \\ U_1' &= \sum_n (C_+ z_+ u_+)_1, & V_1' &= \sum_n (C_- z_- u_-)_1 \end{aligned} \tag{41}$$

where C_+ and C_- refer to the concentrations of cations and anions in g-ions per liter, u_+ and u_- are the corresponding ionic mobilities, and z_+ and z_- are their valences. The subscript 1 refers to ions in solution 1 and similar relations hold for U_2, V_2, . . . , for ions in solution 2.

In order to integrate the equation for the liquid junction potential, Planck[6] assumed a "constrained diffusion" boundary which can be produced by flowing solutions 1 and 2 over two surfaces of a porous plug or a membrane through which ions diffuse freely. The junction is of a finite thickness across which a constant junction potential arises when the diffusion of ions reaches a steady state.

Planck carried out the integration for the case of univalent ions using only concentration terms and the assumption that the ionic mobilities are constant. His treatment is based on a kinetic approach which gives an equation similar to Eq. (35). The treatment is complicated and based on the manipulation of the Nernst–Planck flux equation which, with the activity coefficient term included, is given by

$$J_i = -\frac{RT}{F} u_i \left[\frac{dC_i}{dx} + C_i \frac{d \ln \gamma_i}{dx} + z_i \frac{F}{RT} C_i \frac{dE}{dx} \right] \tag{42}$$

where J_i is the flux of species i (mole/cm^2 sec). Equation (42) applies to all mobile species which are subject to the condition of electroneutrality, namely,

$$\sum_i z_i C_i = 0 \tag{43}$$

Furthermore in the steady state of diffusion, the continuity equation (Fick's second law)

$$\frac{\partial C}{\partial t} = \frac{\partial}{\partial x}\left(D \frac{\partial C}{\partial x}\right)$$

where the diffusion coefficient D_i is related to u_i by thc Nernst–Einstein relation

$$D_i = \frac{RT}{F} u_i \tag{44}$$

becomes

$$\frac{\partial C_i}{\partial t} = -\operatorname{div} J_i = 0 \tag{45}$$

The total current I carried by the mobile charge species is given by

$$I = F \sum_i z_i J_i \tag{46}$$

For ideal solutions $\gamma_i = 1$; thus Eq. (42) becomes

$$J_i = -\frac{RT}{F} u_i \left[\frac{dC_i}{dx} + z_i \frac{F}{RT} C_i \frac{dE}{dx} \right] \tag{47}$$

which for current I_i (i.e., flux per g-ion) carried by the ith ion is

$$I_i = -z_i u_i \left[RT \frac{dC_i}{dx} + z_i F C_i \frac{dE}{dx} \right] \tag{48}$$

Equation (48) has been integrated by Planck (see MacInnes[8]) to give the transcendental equation

$$\frac{\xi U_2 - U_1}{V_2 - \xi V_1} = \frac{\ln(C_2/C_1) - \ln \xi}{\ln(C_2/C_1) + \ln \xi} \frac{\xi C_2 - C_1}{C_2 - \xi C_1} \tag{49}$$

where U_1, U_2, V_1, and V_2 are defined by Eq. (41) and

$$\xi = \exp(E_L F/RT) \tag{50}$$

Application of Eqs. (40) and (49) to two special cases, first two solutions of the same uni-univalent electrolyte at two different concentrations and second two uni-univalent electrolytes with a common ion at the same concentration, leads to the same result. Both Eqs. (40) and (49) become

$$E_L = \frac{RT}{F} \frac{u_+ - u_-}{u_+ + u_-} \ln \frac{C_2}{C_1} \tag{51}$$

in the first case, and

$$E_L = \frac{RT}{F} \ln \frac{u_{+(1)} + u_-}{u_{+(2)} + u_-} \tag{52}$$

in the second case.

Equation (51) is the same as Eq. (38) except that the ratio of the activities has been replaced by the ratio of the concentrations. Equation (52) can be rewritten as

$$E_L = \frac{RT}{F} \ln \frac{\lambda_{\infty(1)}}{\lambda_{\infty(2)}} \tag{53}$$

where $\lambda_{\infty(1)}$ and $\lambda_{\infty(2)}$ are the limiting equivalent conductances of the two solutions. Lewis and Sargent[9] wrote Eq. (53) as

$$E_L = \frac{RT}{F} \ln \frac{\lambda_1}{\lambda_2} \tag{54}$$

where λ_1 and λ_2 are the equivalent conductances of the two solutions at the same concentration. In the two cases considered, the junction potential E_L is independent of the nature of the boundary formed from the two solutions.

The other two types of liquid junctions, namely flowing junction and free diffusion, although too difficult to be considered theoretically, may be described by either Eq. (40) or Eq. (49) provided the liquid junctions are so constructed as to conform in operation to the details of cases considered by Henderson and Planck. Usually in the practice of emf measurements, these liquid junction potentials are eliminated or at least minimized by using a salt bridge between the two solutions that would normally constitute the junction. Generally saturated solutions of KCl–agar bridges are used. Where KCl cannot be used, as in solutions of $AgNO_3$, NH_4NO_3 may be used. These operate on the theoretical principle that since the ions of the salt in the bridge are concentrated, they carry almost all of the current across the junction, and since the mobilities of K^+ and Cl^- and of NH_4^+ and NO_3^- are nearly equal, the diffusion potential is very small.

Planck's integration of Eq. (48) is applicable to solutions containing only univalent ions. Johnson,[10] on the other hand, has given a solution applicable to mixture of salts in which all cations have one valency z_c and all anions have valency z_a. The equation, transcendental in form, appears as

$$\frac{U_2\xi^{z_c} - U_1}{V_2 - \xi^{z_a}V_1} = \frac{z_a}{z_c}\,\frac{\ln(C_2/C_1) - z_a \ln \xi}{\ln(C_2/C_1) + z_c \ln \xi}\,\frac{C_2\xi^{z_c} - C_1}{C_2 - \xi^{z_a}C_1} \tag{55}$$

A very general formula applicable to ions of several valences has been given by Pleijel.[11] In addition Behn,[12] soon after Planck, also integrated Eq. (48) to determine the flux across the boundary of any one particular species contained in the two solutions forming the junction. The final form of the equation is

$$I_i = -u_i\left[\frac{RT}{d}\,\frac{C_2 - C_1}{C_2\xi - C_1}\,\frac{\ln(C_2\xi/C_1)}{\ln(C_2/C_1)}\right][C_{i(2)}\xi - C_{i(1)}] \tag{56}$$

where C_2 and C_1 are the total concentrations in the two solutions, $C_{i(2)}$ and $C_{i(1)}$ are the concentrations of i in the two solutions, and d is the thickness of the junction.

I. PRINCIPLES UNDERLYING MEMBRANE ELECTRODE BEHAVIOR

Any phase that separates two other phases to prevent mass movement between them but allows passage with various degrees of restriction of one or several species of the external phases may be defined as a membrane[13] which when used as an electrode in an electrochemical cell constitutes a membrane electrode. The behavior of the membrane electrode will be determined by the properties of the membrane, which can be a solid or a liquid containing ionized or ionizable groups. It is possible that these solid and liquid membranes could be uncharged (site free). A completely gaseous membrane has not yet been discussed in the literature, although membrane electrodes indirectly responding to gases have been described.

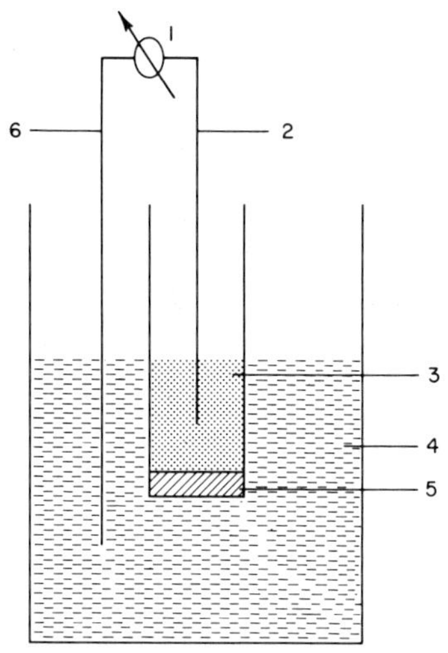

Fig. 1. Schematic representation of membrane electrode cell assembly: 1, membrane potential to be recorded; 2, internal reference electrode; 3, internal filling solution; 4, sample solution to be measured; 5, membrane; and 6, external reference electrode.

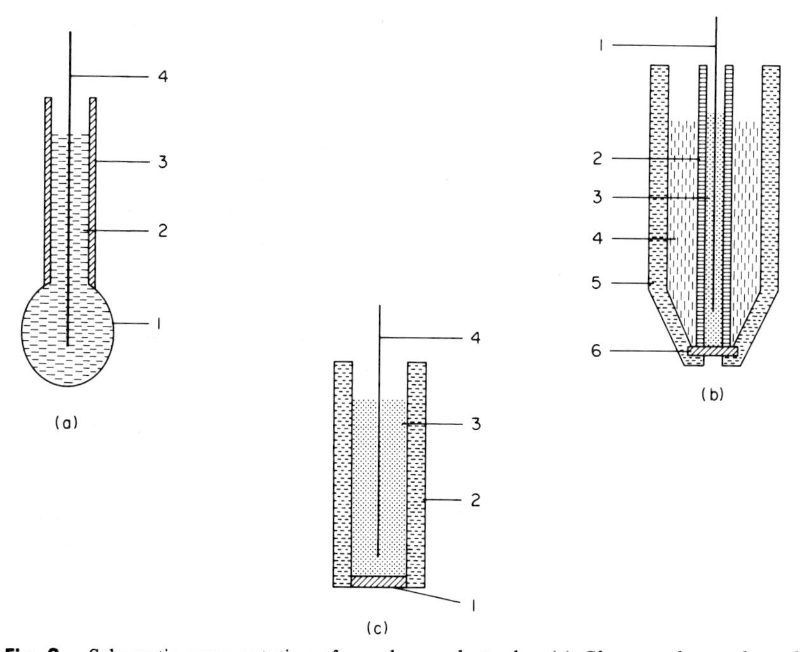

Fig. 2. Schematic representation of membrane electrodes. (a) *Glass membrane electrode*: 1, glass membrane; 2, internal filling solution; 3, glass stem; and 4, internal reference electrode. (b) *Liquid membrane electrode*: 1, internal reference electrode; 2, cylinder to receive internal filling solution; 3, internal filling solution; 4, ion-selective liquid; 5, electrode shaft; and 6, filter paper impregnated with ion-selective liquid. (c) *Solid state homogeneous and heterogeneous membrane electrodes*: 1, homogeneous or heterogeneous solid state membrane; 2, electrode shaft; 3, internal filling solution; and 4, internal reference electrode.

In view of the foregoing discussion, membrane electrodes may be roughly classified into three groups, according to the type of membrane used in their fabrication.[14] These electrodes are (1) glass membrane, (2) solid state membrane, which may be homogeneous or heterogeneous, and (3) liquid membrane, which may contain either electrically charged or neutral ligand groups as components of the membrane. A schematic representation of the cell assembly is shown in Fig. 1. Usually the membrane is held in a compact unit containing the internal filling solution and the reference electrode. Such electrode units are shown schematically in Fig. 2.

1. Membrane Potential

Electrical potentials arising across membranes when they separate two electrolyte solutions may be called membrane potentials. As discussed previously these may arise as a diffusion potential across the membrane due to differences in the mobilities of the ions. There are also other ways in which a potential might arise across the membrane. The simplest way is to have it arise as an ohmic potential drop by passing electric current from an external source of emf through the system: compartment 1–membrane–compartment 2. Another way would be to have it arise as a static potential by adding to one of the compartments some charged species that cannot pass through the membrane. This is the Gibbs–Donnan system.[15, 16] If the membrane has no fixed charges, the membrane potential would be equivalent to a diffusion potential. As a result, the integrations of the Nernst–Planck flux equation (48) carried out by Planck,[6] Behn,[12] Pleijel,[11] Johnson,[10] and Schlogl[17] become applicable. On the other hand, if the membrane carries some fixed charges, a net Donnan potential in addition to a diffusion potential constitutes the membrane potential. These concepts, illustrated by a magnified view of Fig. 1 [see cells of type (57)], were formulated simultaneously by Teorell[18] and Meyer and Sievers[19] (TMS theory).

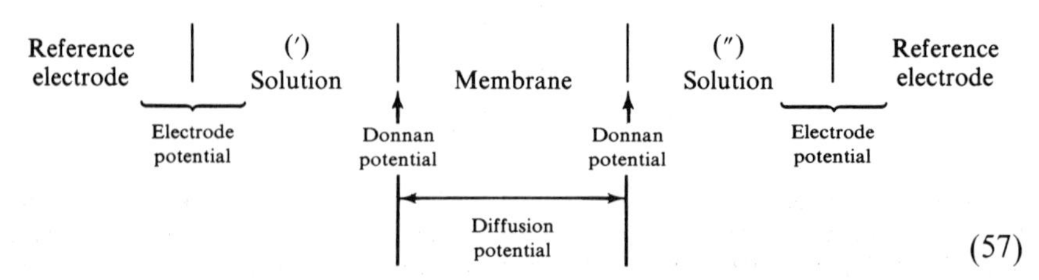

(57)

The reference electrode may be Ag–AgCl in chloride solutions (liquid junctionless cell) or calomel connected to solutions by KCl–agar bridges (liquid junction cell). In the first case, the cell potential measured is made up of the electrode potential and the membrane potential. In the second case, the cell potential is equal to the membrane potential. These measurements of membrane potential are not without ambiguity since they involve some nonthermodynamic assumptions about single ion activity coefficients, liquid junction, etc.,[20] typical of the electrochemical cells with transport already considered. However, these procedures have become routine in the study of membrane phenomena.[13, 21]

The several theoretical approaches made to calculate the emfs of cells of type (57) fall roughly into three categories. This division is based on the nature of the flux equation used in the treatment.[22] In category 1 may be placed those theories which are based on the Nernst–Planck flux equation (48) or its refinements.[23] The theories[24–34] that use the principles of pseudothermostatics[35] and irreversible thermodynamics[36, 37] may be placed in category 2. Category 3 contains those theories[38–45] that are based on the concepts of the theory of absolute rate processes.[46] All of the theories have some common features and so they may be considered to supplement one another, although, depending on the system under consideration, one particular approach may prove more suitable than another.[28, 42, 47–49; 18, 19, 50] The theories under categories 1 and 2 have proved useful in the consideration of the membrane electrodes. The theories of category 3, despite the general and unified view they provide for membrane systems of varying complexities, contain parameters whose physical and/or chemical significance is difficult to understand and whose quantitative evaluation is beyond the usual experimental procedures. Consequently, only those theories that have proved their usefulness by providing insights into the behavior of membrane electrodes classified above will be considered.

2. The TMS Theory

The membrane potential of cell (57) is the algebraic sum of two Donnan potentials and a diffusion potential. The two membrane surfaces are assumed to be in a state of equilibrium. The condition for equilibrium between two phases such as the aqueous and the membrane phases is that the electrochemical potentials of any mobile species i in the two phases are equal. Accordingly, for a univalent species i

$$\mu_i^\circ + RT \ln a_i + PV_i + FE = \mu_1^\circ + RT \ln \bar{a}_i + \bar{P}V_i + F\bar{E} \tag{58}$$

where the terms with overbars refer to the membrane phase. Thus when the membrane is in a 1 : 1 electrolyte solution of activity $a_\pm$, Eq. (58) can be written as

$$F(E - \bar{E}) = RT \ln \frac{\bar{a}_+}{a_+} + V_+ (\bar{P} - P) \tag{59}$$

$$-F(E - \bar{E}) = RT \ln \frac{\bar{a}_-}{a_-} + V_- (\bar{P} - P) \tag{60}$$

Addition of Eqs. (59) and (60) gives the relation

$$\frac{\bar{a}_+\bar{a}_-}{a_+\,a_-} = \exp\frac{\Pi V}{RT} \tag{61}$$

where $\Pi = P - \bar{P}$, the difference between the hydrostatic pressure in the outside solution and the swelling pressure in the membrane, and $V = V_+ + V_-$, the molal volume of the electrolyte. It has been shown that the term $\exp(\Pi V/RT)$ is approximately unity,[51, 52] and so Eq. (61) becomes

$$\bar{m}_+\bar{\gamma}_+\bar{m}_-\bar{\gamma}_- = a_\pm^2 = a^2 \tag{62}$$

If the concentration of ionogenic groups in the membrane is $\bar{X}$, then for a negatively charged membrane, the electroneutrality condition gives

$$\bar{m}_+ = \bar{m}_- + \bar{X} \tag{62a}$$

Substituting Eq. (62a) into Eq. (62) and solving the quadratic gives

$$\bar{m}_+ = \frac{\bar{X}}{2} + \sqrt{\frac{\bar{X}^2}{4} + \frac{a^2}{\bar{\gamma}_\pm^2}} \tag{63}$$

$$\bar{m}_- = -\frac{\bar{X}}{2} + \sqrt{\frac{\bar{X}^2}{4} + \frac{a^2}{\bar{\gamma}_\pm^2}} \tag{64}$$

Teorell[18, 53] and Meyer and Sievers[19] assumed $\bar{\gamma}_+$ and $\bar{\gamma}_-$ to be unity, and so Eqs. (63) and (64) become

$$\bar{m}_+ = \frac{\bar{X}}{2} + \sqrt{\frac{\bar{X}^2}{4} + a^2} \tag{65}$$

$$\bar{m}_- = -\frac{\bar{X}}{2} + \sqrt{\frac{\bar{X}^2}{4} + a^2} \tag{66}$$

The Donnan ratio r given by Eq. (62) becomes

$$r = \frac{\bar{m}_+}{a_+} = \frac{a_-}{\bar{m}_-} = \cdots = \frac{\bar{m}_+}{a} = \frac{a}{\bar{m}_-} \tag{67}$$

When the ionic membrane is bounded by a 1 : 1 electrolyte of activities a' and a'', the two Donnan potentials, E'_{Don} and E''_{Don} at the two interfaces (′) and (″) are given according to Eqs. (59) and (60) by (the PV terms are ignored)

$$E'_{\text{Don}} = E' - \bar{E}' = \frac{RT}{F} \ln \frac{\bar{a}'_+}{a'_+} = \frac{RT}{F} \ln \frac{a'_-}{\bar{a}'_-} \tag{68}$$

$$E''_{\text{Don}} = E'' - \bar{E}'' = \frac{RT}{F} \ln \frac{\bar{a}''_+}{a''_+} = \frac{RT}{F} \ln \frac{a''_-}{\bar{a}''_-} \tag{69}$$

The net Donnan potential therefore is given by

$$E'_{\text{Don}} - E''_{\text{Don}} = \frac{RT}{F} \ln \frac{\bar{a}'_+}{a'_+} - \frac{RT}{F} \ln \frac{\bar{a}''_+}{a''_+} \tag{70}$$

$$= \frac{RT}{F} \ln \frac{a'_-}{\bar{a}'_-} - \frac{RT}{F} \ln \frac{a''_-}{\bar{a}''_-} \tag{71}$$

The diffusion potential $\psi = \bar{E}'' - \bar{E}'$ within the membrane is assumed to be that existing in a constrained liquid junction; Teorell[53] used the expression

$$\psi = \frac{\bar{u}_+ - \bar{u}_-}{\bar{u}_+ + \bar{u}_-} \frac{RT}{F} \ln \frac{\bar{u}_+ \bar{m}'_+ + \bar{u}_- \bar{m}'_-}{\bar{u}_+ \bar{m}''_+ + \bar{u}_- \bar{m}''_-} \tag{72}$$

Substituting for $\bar{m}_+$ and $\bar{m}_-$ from Eqs. (65) and (66), Eq. (72) becomes on rearrangement

$$\psi = \bar{U} \frac{RT}{F} \ln \frac{\bar{U}\bar{X}' + \sqrt{\bar{X}'^2 + 4a'^2_\pm}}{\bar{U}\bar{X}'' + \sqrt{\bar{X}''^2 + 4a''^2_\pm}} \tag{73}$$

where $\bar{U} = (\bar{u}_+ - \bar{u}_-)/(\bar{u}_+ + \bar{u}_-)$. Assuming that $\bar{X}$ is independent of the external electrolyte solution, the total membrane potential is given by the sum of Eq. (70) or (71) and Eq. (73). Thus for a highly idealized membrane system $\bar{\gamma}_+ = \bar{\gamma}_- = 1$ the total membrane potential when agar–KCl salt bridges are used in the measurement is given by

$$E = E'_{\text{Don}} - E''_{\text{Don}} + \psi$$

So

$$E = \frac{RT}{F} \left[\ln \frac{a''}{a'} \frac{(4a'^2 + \bar{X}^2)^{1/2} + \bar{X}}{(4a''^2 + \bar{X}^2)^{1/2} + \bar{X}} + \bar{U} \ln \frac{(4a'^2 + \bar{X}^2)^{1/2} + \overline{UX}}{(4a''^2 + \bar{X}^2)^{1/2} + \overline{UX}} \right] \tag{74}$$

An equation exactly similar to Eq. (74) has been derived by Kobatake and co-workers.[48, 50] However, their equation has a parameter $\phi\bar{X}$ ($0 < \phi$

< 1, the thermodynamically effective charge density) in place of $\bar{X}$ and mobility values corresponding to the bulk aqueous phase in place of $\bar{U}$, i.e., $U = (u_+ - u_-)/(u_+ + u_-)$.

Three special cases of Eq. (74) are of interest:

(i) When $a \ll \bar{X}/2$, Eq. (74) reduces to the Nernst equation (23)

$$E = \frac{RT}{F} \ln \frac{a''}{a'} \tag{75}$$

(ii) When $a \gg \bar{X}/2$, Eq. (74) reduces to Eq. (38) which gives the value for the diffusion potential between two solutions of activities a' and a''. The mobility values would correspond to those prevailing in the aqueous solution although diffusion would be occurring across the membrane. When $a \gg \bar{X}/2$, the sorption of the electrolyte by the membrane is so high that the ionogenic groups (i.e., $\bar{X}$) in the membrane are unable to distinguish between counterions and coions.

(iii) When $a \cong \bar{X}/2$, the ionogenic groups are able to distinguish between counterions and coions to some extent so that the mobility values correspond to the membrane phase. Thus Eq. (74) reduces to

$$E = \frac{RT}{F} \frac{\bar{u}_+ - \bar{u}_-}{\bar{u}_+ + \bar{u}_-} \ln \frac{a'}{a''}$$

or

$$E = \frac{RT}{F} (\bar{t}_+ - \bar{t}_-) \ln \frac{a'}{a''} \tag{76}$$

where $\bar{t}_+$ and $\bar{t}_-$ are the transport numbers of counterion and coion for a negatively charged membrane in the membrane phase. Equation (76) may be rearranged to give

$$E/E_{\max} = 2\bar{t}_+ - 1 \qquad \text{or} \qquad \bar{t}_+ = [E/(2E_{\max})] + 0.5 \tag{77}$$

where $E_{\max} = (RT/F)\ln(a'/a'')$. Equation (77) has often been used to calculate the transport numbers in the membrane phase from measurements of membrane potential.

3. Thermodynamic Theories of Isothermal Membrane Potential

Unlike the TMS theory of membrane potential, which depends on a knowledge of the internal structure and properties of the membrane, the thermodynamic theories do not need this information. Staverman[25] derived the relation

$$-F\,dE = \sum \frac{t_i}{Z_i} d\mu_i \tag{78}$$

using the principles of irreversible thermodynamics. In a similar way,

Scatchard,[24, 35] discarding considerations of membrane properties or structure, expressed membrane potential as

$$E = -\frac{RT}{F}\int_{a_i'}^{a_i''}\sum_i t_i \, d\ln a_i \tag{79}$$

The solutions of activity a_i' and a_i'' extend up to each interface. Equation (79) may be applied to all components moving across the membrane. Those for a 1 : 1 electrolyte are counterion, coion, water, and the fixed charges of the membrane. Regarding the polymer network to which the fixed charges are attached as the reference framework to which the movements of all other species are referred, the summation in Eq. (79) refers only to three species, i.e., counterion (+), coion (−), and water (w). Thus Eq. (79) becomes

$$E = -\frac{RT}{F}\int_{\mathrm{I}}^{\mathrm{II}}\left(\bar{t}_+ d\ln a_+ - \bar{t}_- d\ln a_- + \bar{t}_\mathrm{w}\, d\ln a_\mathrm{w}\right) \tag{80}$$

If anion-reversible electrodes are used in the membrane cell to measure the cell emf E, then the electrode potential E_{ref} between two such electrodes is given by Eq. (15), i.e.,

$$E_{\mathrm{ref}} = \frac{RT}{F}\ln\frac{a_-^{\mathrm{I}}}{a_-^{\mathrm{II}}} \tag{81}$$

Substitution of Eq. (81) together with the use of relations $\bar{t}_+ + \bar{t}_- = 1$ and $d\ln a_\mathrm{w} = -2 \times 10^{-3} m M_1 \, d\ln a_\pm$ [see Eq. (71), Chapter 2] into Eq. (80) gives[32]

$$E = -\frac{2RT}{F}\int_{\mathrm{I}}^{\mathrm{II}}\left(\bar{t}_+ - 10^{-3} m M_1 \bar{t}_\mathrm{w}\right) d\ln a_\pm \tag{82}$$

where M_1 is the molecular weight of the solvent and m is the molality of the solution.

Equation (82) has been derived by Lorimer *et al.*[27] using the principles of irreversible thermodynamics. An experimental test of Eq. (82) carried out by Lakshminarayanaiah[7, 54, 55] has shown that it satisfactorily describes the electrical potentials arising across membranes when they separate solutions of the same 1 : 1 electrolyte but of different concentration.

4. Bi- and Multi-Ionic Potentials

Membrane potential arising across a membrane that separates two electrolyte solutions AX and BX is called a bi-ionic potential (BIP).[56] On the other hand, if the membrane separates two solutions containing mixtures of electrolytes (AX, BX, CX, . . . and PX, QX, RX, . . .), the membrane potential observed is called a multi-ionic potential (MIP).[57, 58]

The total BIP has been considered by Helfferich[59, 60] in accordance with the concepts of the TMS theory,[18, 19] as being the algebraic sum of three potentials—two Donnan potentials and one diffusion potential. A complete mathematical discussion under conditions of (1) membrane diffusion control, (2) film diffusion control, and (3) coupled membrane–film diffusion control has been presented.

For a general case involving complete membrane diffusion control, the total BIP is shown to be given by

$$E = \frac{RT}{F}\left[\frac{\bar{D}_B - \bar{D}_A}{\bar{D}_A z_A - \bar{D}_B z_B} \ln \frac{\bar{D}_B z_B}{\bar{D}_A z_A} + \frac{1}{z_A z_B} \ln K'_{BA} + \frac{z_A - z_B}{z_A z_B} \ln \frac{\bar{C}}{C'_A} + \frac{1}{z_B} \ln \frac{C'_A}{C''_B} + \ln \frac{\gamma'^{1/z_A}_A}{\gamma''^{1/z_B}_B}\right] \tag{83}$$

where the $\bar{D}$'s are the diffusion coefficients, $\bar{C} = z_A \bar{C}_A + z_B \bar{C}_B$ is the total counterion concentration, and C'_A and C''_B are the bulk concentrations of solutions on sides (′) and (″), respectively. It is assumed in the derivation that $\bar{D}_A/\bar{D}_B$, $\ln K'_{BA}$, and $\bar{C}$ are constants, there are no coions and convection in the membrane, and the boundary condition $\bar{C}''_B/C''_B = \bar{C}'_A/C'_A = \bar{C}/C$ holds. K'_{BA} is the corrected molar selectivity coefficient for the exchange reaction

$$\bar{B} + A \rightleftharpoons \bar{A} + B \tag{84}$$

and is given by the ratio of the activity coefficients in the membrane

$$K'_{BA} = \frac{\bar{\gamma}_B^{z_A}}{\bar{\gamma}_A^{z_B}} = \left(\frac{\bar{C}_A}{a_A}\right)^{z_B} \left(\frac{a_B}{\bar{C}_B}\right)^{z_A} \tag{85}$$

When the valences of the counterions are equal ($z_A = z_B$), Eq. (83) reduces to

$$E = \frac{RT}{z_A F} \ln \frac{\bar{D}_A a'_A \bar{\gamma}_B}{\bar{D}_B a''_B \bar{\gamma}_A} \tag{86}$$

Equation (86) has been found to predict the values of BIP reasonably well provided $\bar{\gamma}_B/\bar{\gamma}_A$ remains constant[61, 62] or its variation with the membrane composition is suitably corrected.[63, 64] Other aspects of Eq. (86) have been reviewed by Lakshminarayanaiah.[7, 65]

An expression for the MIP can be derived from Eqs. (43)–(47) applied to the membrane. The interdiffusion potential in the membrane for equal counterion valences, ideal permselectivity (i.e., the absence of coions from the membrane phase), complete membrane diffusion control, constant activity coefficients, and absence of convection has been obtained from Eqs. (43)–(47) by solving for the electrical potential gradient and integrating across the membrane.[60] Thus

$$\psi = \frac{RT}{z_i F} \ln \frac{\sum_i \bar{D}_i \bar{C}_i'}{\sum_j \bar{D}_j \bar{C}_j''} \tag{87}$$

The total MIP is given by

$$E = \psi + E_{\text{Don}}' - E_{\text{Don}}''$$

$$E = \frac{RT}{z_i F} \ln \frac{\sum_i \bar{D}_i \bar{C}_i (a_i'/\bar{a}_i')}{\sum_j \bar{D}_j \bar{C}_j (a_j''/\bar{a}_j'')} = \frac{RT}{z_i F} \ln \frac{\sum_i \bar{D}_i (a_i'/\bar{\gamma}_i)}{\sum_j \bar{D}_j (a_j''/\bar{\gamma}_j)} \tag{88}$$

where i refers to A, B, C, . . . and j to P, Q, R, . . . ions.

Determination of the activity of an ion A in a solution of AY, BY, CY, . . . by means of an ion exchange membrane electrode may be considered on the basis of Eq. (88) which when applied to the multi-ionic system

AY(′)	Membrane	AY, BY, CY, . . . (″)	(89)
Reference solution		Sample mixture	

becomes

$$E = \frac{RT}{z_i F} \ln \frac{\bar{D}_A a_A' / \bar{\gamma}_A}{\sum_i \bar{D}_i a_i'' / \bar{\gamma}_i}$$

$$= \frac{RT}{z_i F} \ln \frac{a_A'}{a_A'' + (\bar{D}_B \bar{\gamma}_A / \bar{D}_A \bar{\gamma}_B) a_B'' + (\bar{D}_C \bar{\gamma}_A / \bar{D}_A \bar{\gamma}_C) a_C'' + \cdots} \tag{90}$$

If the membrane is impermeable to B, C, . . . , Eq. (90) reduces to the Nernst equation. Thus a_A'' can be calculated from the observed membrane potential and the reference activity a_A'. If B, C, . . . become permeable, Eq. (90) reduces to the Nernst equation only if $\bar{D}_i \bar{\gamma}_A / \bar{D}_A \bar{\gamma}_i$ ($i \neq$ A) disappears. However, the values of these quotients may be evaluated by measuring the BIPs of the respective systems BY/AY, CY/AY, . . . with equal activities in both solutions. Thus the specificity of the membrane electrode can be characterized.[66] Other aspects of bi- and multi-ionic potentials are presented elsewhere.[7, 65]

5. Integration of the Nernst–Planck Flux Equation

Planck's integration of Eq. (48) as applied to a "porous" membrane with no fixed charges has been referred to in Section H. In this section reference is made to the theoretical work related to ionic membranes. For these membranes, Eq. (43) becomes

$$\sum z_i \bar{C}_i + \bar{X} = 0 \tag{91}$$

Goldman[67] in 1943 discussed the application of Eq. (48) to both charged and neutral membranes. In the case of charged membranes, he derived an equation which for the case $z_i = \pm 1$ and $\bar{X} = 0$ gave an expression equivalent to the Behn equation (56). But when $\bar{X} \neq 0$, the equation was very complex.

Teorell[18, 53] also integrated Eq. (48) for univalent ions and gave a solution which was equivalent to that of Goldman. The formula derived for free diffusion in the membrane was transcendental in nature and similar to that of Planck. Thus

$$E = (RT/F) \ln \xi \tag{92}$$

$$\bar{k} = \frac{\bar{C}_{+(2)} + 0.5\bar{X}\,(\ln \bar{k}\xi/\ln \xi)}{\bar{C}_{+(1)} + 0.5\bar{X}\,(\ln \bar{k}\xi/\ln \xi)} \tag{93}$$

$$\frac{\bar{U}_2\xi - \bar{U}_1}{\bar{V}_2 - \bar{V}_1\xi} = \frac{\bar{C}_{+(2)}\xi - \bar{C}_{+(1)}}{\bar{C}_{-(2)} - \bar{C}_{-}(1)\xi}\;\frac{\ln \bar{k} - \ln \xi}{\ln \bar{k} + \ln \xi} \tag{94}$$

where $\bar{U}$ and $\bar{V}$ referred to the membrane phase are given by Eq. (41), and $\bar{C}_+$ and $\bar{C}_-$ refer to the total concentrations of univalent cations and anions, respectively.

In the case of biological systems, the cell membrane is bounded by solutions of equal concentrations, i.e., $C_1 = C_2$. In this case, Eq. (94) transforms into a simple expression, namely,

$$\xi = \frac{\bar{U}_1 + \bar{V}_2}{\bar{U}_2 + \bar{V}_1} \tag{95}$$

and the two Donnan potentials, being equal and opposite, cancel out and the potential is given by

$$E = \frac{RT}{F} \ln \frac{\sum_n (\bar{C}_+\bar{u}_+)_1 + \sum_n (\bar{C}_-\bar{u}_-)_2}{\sum_n (\bar{C}_+\bar{u}_+)_2 + \sum_n (\bar{C}_-\bar{u}_-)_1} \tag{96}$$

The membrane concentrations of ions may be related to the concentrations in the outside solutions by the use of the Donnan relation (67). Thus

$$E = \frac{RT}{F} \ln \frac{r\sum_n(C_+\bar{u}_+)_1 + (1/r)\sum_n(C_-\bar{u}_-)_2}{r\sum_n(C_+\bar{u}_+)_2 + (1/r)\sum_n(C_-\bar{u}_-)_1} \tag{97}$$

where

$$r = \frac{\bar{X}}{2a} + \sqrt{\frac{\bar{X}^2}{4a^2} + 1} \quad \text{and} \quad \frac{1}{r} = -\frac{\bar{X}}{2a} + \sqrt{\frac{\bar{X}^2}{4a^2} + 1} \tag{67}$$

and thus expresses the influence of charge in the membrane. On the other hand, if $\bar{X} = 0$ (uncharged membrane and $r = 1$), Eq. (95) reduces to the Planck formula for a liquid junction:

$$\xi = \frac{U_1 + V_2}{U_2 + V_1} \tag{98}$$

Equation (96) has also been derived by Goldman[67] and Hodgkin and Katz[68] who assumed that the electric field acting across the membrane was constant. This simplified the mathematics and enabled direct integration of the Nernst–Planck flux equation (48). Unlike Teorell,[53] who used the Donnan ratio (67) to relate the membrane concentrations of ions to the concentration of ions in the aqueous phases bounding the membrane, Hodgkin and Katz[68] assumed that the concentrations of the ions at the edges of the membrane were directly proportional to those in the aqueous solutions contacting the membrane faces. Thus $\bar{C}_{i(1)} = C_{i(1)}\beta_i$ and $\bar{C}_{i(2)} = C_{i(2)}\beta_i$, where β_i may be called the partition coefficient of ion i between the membrane and the aqueous solution.

In biological systems, the inside of the cell usually has a high concentration of potassium (K) ions and the outside has a high concentration of sodium (Na) ions. There will be on either side of the cell membrane which separates the interior world of the cell from its exterior a number of anions, chief among which are the chloride (Cl) ions. These three principal ions (K, Na, and Cl) control the electrical activity of most living cells. Since the total concentrations of solutions outside (o) and inside (i) the cell membrane are equal, Eq. (96) applied to the three principal ions gives the electrical potential across the cell membrane while at rest. In order to get Eq. (96) into the form in which it appears in the biological literature as the constant field or Goldman–Hodgkin–Katz equation, the following relations are required.

The electrochemical mobility $\bar{u}_i$ is related to the diffusion coefficient $\bar{D}_i$ by the Nernst–Einstein equation (44). The permeability P_i(cm/sec) is often used in the biological literature[69] instead of $\bar{D}_i$(cm^2/sec) to describe the movement of molecules across the cell membranes; the relation between these two parameters, as usually defined, is given by

$$P_i = \bar{D}_i/d = RT\bar{u}_i/Fd \tag{99}$$

where d is the thickness of the membrane.

Hodgkin and Katz[68] introduced β_i to define P_i. Thus Eq. (99) becomes $P_i = \bar{u}_i\beta_i(RT/Fd)$. Substituting this and the relation $\bar{C}_i = C_i\beta_i$ into Eq. (96) gives

$$E = \frac{RT}{F} \ln \frac{\sum (C_+ P_+)_1 + \sum (C_- P_-)_2}{\sum (C_+ P_+)_2 + \sum (C_- P_-)_1} \tag{100}$$

Equation (100), applied to the cell membrane, appears in the biological literature in the familar form

$$E = \frac{RT}{F} \ln \frac{(C_K)_o P_K + (C_{Na})_o P_{Na} + (C_{Cl})_i P_{Cl}}{(C_K)_i P_K + (C_{Na})_i P_{Na} + (C_{Cl})_o P_{Cl}} \tag{101}$$

Equation (101) has been shown by Eisenman and co-workers[70–74] to correspond to a special case of their equation for the total potential arising across a cation exchange membrane that is completely impermeable to anions. They applied Eq. (42) to analyze the electrical potentials in a cation exchange membrane separating two solutions and permeable only to cations A and B. This problem has also been analyzed by Helfferich[59, 60] and Mackay and Meares.[75] In these treatments, empirical equations relating to activity coefficients of counterions within the membrane were included. Eisenman and co-workers[71, 72] also used an empirical equation governing nonideal behavior of counterions to derive expressions for the electrical potentials. The empirical relationship was for the ion exchange equilibrium shown in Eq. (84). Application of the law of mass action gives the equilibrium constant K. Thus

$$K = \frac{\bar{a}_A a_B}{a_A \bar{a}_B} \tag{102}$$

A number of ion exchangers have been found[76–78] to conform to the following empirical relationship, first formulated by Rothmund and Kornfeld[79]:

$$K = \frac{a_B}{a_A}\left(\frac{\bar{x}_A}{\bar{x}_B}\right)^n \tag{103}$$

where the $\bar{x}$'s are the mole fractions and n depends on the membrane properties only.

Equations (102) and (103) give the relations

$$\bar{a}_A = \bar{C}_A\bar{\gamma}_A = p(\bar{x}_A)^n \quad \text{and} \quad \bar{a}_B = \bar{C}_B\bar{\gamma}_B = p(\bar{x}_B)^n \tag{104}$$

where p is a proportionality factor depending on the properties of the membrane and $n = 1$ for ideal behavior.

The electroneutrality condition gives

$$\bar{C}_A + \bar{C}_B = \bar{X} \tag{105}$$

and

$$\bar{x}_A + \bar{x}_B = 1 \tag{106}$$

where

$$\bar{x}_A = \bar{C}_A/\bar{X} \quad \text{and} \quad \bar{x}_B = \bar{C}_B/\bar{X} \tag{107}$$

Equations (104) and (107) give

$$\bar{\gamma}_A = (p/\bar{X})(\bar{x}_A)^{n-1} \quad \text{and} \quad \bar{\gamma}_B = (p/\bar{X})(\bar{x}_B)^{n-1} \tag{108}$$

Equation (42) may be used to calculate the total current carried by ions A and B and for the condition $I = I_A + I_B = 0$, in terms of the $\bar{x}$'s [Eq. (107)], the relation

$$d\bar{E} = -\frac{RT}{F}\frac{\bar{u}_A\,d\bar{x}_A + \bar{u}_B\,d\bar{x}_B + \bar{u}_A\bar{x}_A\,d\ln\bar{\gamma}_A + \bar{u}_B\bar{x}_B\,d\ln\bar{\gamma}_B}{\bar{u}_A\bar{x}_A + \bar{u}_B\bar{x}_B} \tag{109}$$

is obtained. Substitution of the values for $\bar{\gamma}_A$ and $\bar{\gamma}_B$ from Eq. (108) into Eq. (109) gives

$$d\bar{E} = -\frac{nRT}{F}\,d\ln(\bar{u}_A\bar{x}_A + \bar{u}_B\bar{x}_B) \tag{110}$$

Integration of Eq. (110) between $x = 0$ and $x = d$ (the thickness of the membrane) gives the diffusion potential in the membrane. That is,

$$\psi = \bar{E}_{(d)} - \bar{E}_{(0)} = \frac{nRT}{F}\ln\frac{\bar{u}_A\bar{x}_A(0) + \bar{u}_B\bar{x}_B(0)}{\bar{u}_A\bar{x}_A(d) + \bar{u}_B\bar{x}_B(d)} \tag{111}$$

But Eq. (103) together with Eq. (106) gives the relationship between the membrane concentrations and the outside concentrations. Thus

$$\bar{x}_A = \frac{K^{1/n}a_A^{1/n}}{a_B^{1/n} + K^{1/n}a_A^{1/n}} \tag{112}$$

$$\bar{x}_B = \frac{a_B^{1/n}}{a_B^{1/n} + K^{1/n}a_A^{1/n}} \tag{113}$$

Substitution of these values into Eq. (111) gives the diffusion potential.

Thus

$$\psi = \frac{nRT}{F}\ln\left[\frac{(a'_{\mathrm{B}})^{1/n} + (\bar{u}_{\mathrm{A}}/\bar{u}_{\mathrm{B}})K^{1/n}(a'_{\mathrm{A}})^{1/n}}{(a'_{\mathrm{B}})^{1/n} + K^{1/n}(a'_{\mathrm{A}})^{1/n}}\right] - \frac{nRT}{F}\ln\left[\frac{(a''_{\mathrm{B}})^{1/n} + (\bar{u}_{\mathrm{A}}/\bar{u}_{\mathrm{B}})K^{1/n}(a''_{\mathrm{A}})^{1/n}}{(a''_{\mathrm{B}})^{1/n} + K^{1/n}(a''_{\mathrm{A}})^{1/n}}\right] \quad (114)$$

The calculation of the two phase boundary or Donnan potentials could be made again using the nth power relationship of Eq. (104). For the ion exchange reaction (84), the electrochemical potential of A and that of B in the aqueous phase must be equal to the corresponding electrochemical potential in the membrane phase at equilibrium, i.e.,

$$FE' = \mu_{\mathrm{A}} - \bar{\mu}_{\mathrm{A}} = \mu_{\mathrm{B}} - \bar{\mu}_{\mathrm{B}} = \mu_{\mathrm{A}}^{\circ} - \bar{\mu}_{\mathrm{A}}^{\circ} + RT\ln(a_{\mathrm{A}}/\bar{a}_{\mathrm{A}}) = \mu_{\mathrm{B}}^{\circ} - \bar{\mu}_{\mathrm{B}}^{\circ} + RT\ln(a_{\mathrm{B}}/\bar{a}_{\mathrm{B}}) \quad (115)$$

where E' is the difference in potential (boundary potential) between the membrane and the aqueous phase. Using Eqs. (104), (112), and (113) in Eq. (115) gives on simplification

$$E' = \text{const} + \frac{nRT}{F}\ln\left(a_{\mathrm{B}}^{1/n} + K^{1/n}a_{\mathrm{A}}^{1/n}\right) \quad (116)$$

Equation (116) was derived earlier by Dole[80] and Nicolsky[81] when $n = 1$ for the ion exchange reaction at a glass surface. The total potential E is the sum of two phase boundary potentials and a diffusion potential—$E = E' - E'' + \psi$. Thus adding Eqs. (114) and (116) gives for ions i(B) and j(A)

$$E = \frac{nRT}{F}\ln\frac{(a'_i)^{1/n} + (K^{\mathrm{pot}}a'_j)^{1/n}}{(a''_i)^{1/n} + (K^{\mathrm{pot}}a''_j)^{1/n}} \quad (117)$$

where $K^{\mathrm{pot}} = K(\bar{u}_j/\bar{u}_i)^n$ has been called the selectivity constant which includes both the chemical and the mobility factors.

For the case $n = 1$, $P_j/P_i = (\bar{u}_j/\bar{u}_i)K$, and $P_{\mathrm{anion}} = 0$, Eq. (117) becomes the same as Eq. (101). This means that $K = \beta_j/\beta_i$.

The different aspects of Eq. (117) and its applicability to various systems are discussed in detail elsewhere.[70, 77, 78, 82, 83]

If the concentrations on side (″) are held constant, as in a membrane electrode unit, Eq. (117) can be written as

$$E = \text{const} + \frac{nRT}{F}\ln\left[a_i^{1/n} + (K_{ij}^{\mathrm{pot}}a_j)^{1/n}\right] \quad (118)$$

Equation (118) is valid for a monovalent ion in the presence of another monovalent ion. Garrels *et al.*[84] have given equations applicable to divalent ions in the presence of monovalent ions, namely

$$E = \text{const} + \frac{nRT}{2F} \ln\left\{ \left(a_i^{2+}\right)^{1/n} + \left[K_{ij}^{\text{pot}} \left(a_j^{+}\right)^2 \right]^{1/n} \right\} \tag{119}$$

and divalent ions in the presence of other divalent ions, namely

$$E = \text{const} + \frac{nRT}{2F} \ln\left\{ \left(a_i^{2+}\right)^{1/n} + \left(K_{ij}^{\text{pot}} a_j^{2+} \right)^{1/n} \right\} \tag{120}$$

These equations for $n = 1$ may be written in the general form of the extended Nicolsky equation[84–89]:

$$E = \text{const} + \frac{RT}{z_i F} \ln\left[a_i + \sum_{j \neq i} K_{ij}^{\text{pot}} (a_j)^{z_i/z_j} \right] \tag{121}$$

where i is the primary ion of valence z_i to which the membrane electrode is selective and j is the interfering ion of valence z_j.

6. Liquid Membrane with Electrically Charged Ligands

A liquid ion exchange membrane is usually formed by dissolving a liquid ion exchanger in a water-immiscible solvent. Unlike solid ion exchangers, which have their ionogenic groups fixed to the membrane matrix, the ionogenic groups (sites) of liquid ion exchangers are mobile. Depending on the solvent used to form the ion-exchange membrane, the sites would be completely dissociated[90, 91] (dielectric constant of the solvent is high) or highly associated[92–94] (dielectric constant of the solvent is low) into ion pairs. A historical account of the classical concepts reviewed in the preceding sections as applied to liquid membranes has been published by Sollner.[95] In recent years, the behavior of liquid membranes has been described quantitatively by Conti, Eisenman, Sandblom, and Walker[96–101] whose work relating to the problem of electrical potentials arising across liquid membranes under conditions of zero applied potential is outlined below.

The conditions existing in a liquid membrane system are shown in Fig. 3. The species $\overline{\text{A}}^+$ and $\overline{\text{X}}^-$ are in chemical equilibrium with the species $\overline{\text{AX}}$ in the membrane. Thus

$$\overline{\text{A}}^+ + \overline{\text{X}}^- \rightleftarrows \overline{\text{AX}} \tag{122}$$

The chemical potentials of the species everywhere in the membrane are related as

$$\mu_{\text{AX}} = \mu_{\text{A}} + \mu_{\text{X}}$$

Assuming $\bar{\gamma}_i = 1$, the law of mass action applied to Eq. (122) gives

$$K_{\text{AX}} = \frac{\bar{C}_{\text{AX}}}{\bar{C}_{\text{A}} \bar{C}_{\text{X}}} \tag{123}$$

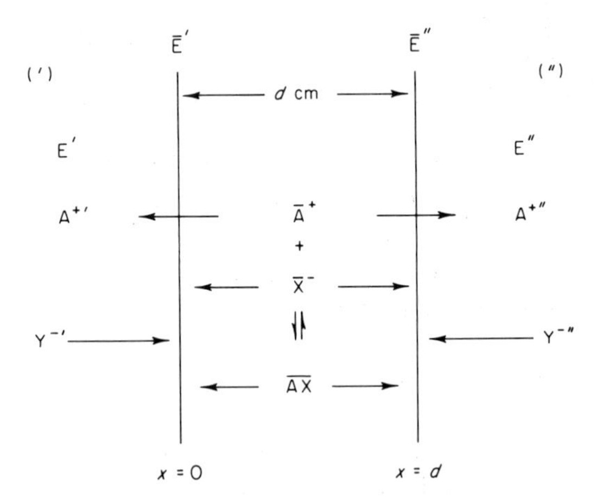

Fig. 3. Schematic diagram of liquid membrane system containing a charged ligand. $\bar{A}^+$, $\bar{X}^-$, and Y^- refer to counterions, sites, and coions, respectively. $\overline{AX}$ is the mobile ion pair. A^+ freely permeates the membrane–solution interfaces, whereas $\bar{X}^-$ is confined to the membrane phase resulting in the exclusion of Y^- from the membrane phase. On the x coordinate, $x = 0$ and $x = d$ indicate the two interfaces, and d cm is the thickness of the membrane. The E's are the electrical potentials and overbars refer to the membrane phase. (′) and (″) are the two aqueous phases on either side of the liquid membrane.

where K_{AX} is the association constant. The electroneutrality condition gives

$$\bar{C}_X = \sum_i \bar{C}_i, \qquad i = A, B, \ldots, n \tag{124}$$

The fluxes of various species can be written as

$$J_i^t = J_i + J_{iX} \tag{125}$$

$$J_X^t = J_X + \sum_i J_{iX} \tag{126}$$

where J_i^t and J_X^t, the total fluxes of counterion and site, are given by the sum of the partial fluxes of i, X, and iX species. Similarly, the total concentrations are given by

$$\bar{C}_i^t = \bar{C}_i + \bar{C}_{iX} \tag{127}$$

$$\bar{C}_X^t = \bar{C}_X + \sum_i \bar{C}_{iX} \tag{128}$$

The fluxes of i, X, and iX are given by the Nernst–Planck flux equation

(47). Thus

$$J_i = -\bar{u}_i \bar{C}_i \frac{\partial}{\partial x}\left(RT \ln \bar{C}_i + z_i F\bar{E}\right) \tag{129}$$

$$J_X = -\bar{u}_X \bar{C}_X \frac{\partial}{\partial x}\left(RT \ln \bar{C}_X + z_x F\bar{E}\right) \tag{130}$$

$$J_{iX} = -\bar{u}_{iX} \bar{C}_{iX} \frac{\partial}{\partial x}\left(RT \ln \bar{C}_{iX}\right) \tag{131}$$

In the steady state, the total fluxes are constant although the partial fluxes are not. The continuity equation applies to total fluxes as

$$\operatorname{div} J_i^{t} = \frac{\partial J_i^{t}}{\partial x} = \frac{\partial \bar{C}_i^{t}}{\partial t} \tag{132}$$

$$\operatorname{div} J_X^{t} = \frac{\partial J_X^{t}}{\partial x} = \frac{\partial \bar{C}_X^{t}}{\partial t} \tag{133}$$

At each of the two boundaries $x = 0$ and $x = d$, the electrochemical potentials of i are equal. So

$$\mu_i^\circ + RT \ln a_i' + z_i FE' = \bar{\mu}_i^\circ + RT \ln \bar{C}_i' + z_i F\bar{E}' \tag{134}$$

$$\mu_i^\circ + RT \ln a_i'' + zFE'' = \bar{\mu}_i^\circ + RT \ln \bar{C}_i'' + z_i F\bar{E}'' \tag{135}$$

Equations (134) and (135) apply to any number of n ions. Applying Eq. (134) to two ions i and j and subtracting them gives

$$\frac{a_i'}{\bar{C}_i'} k_i = \frac{a_j'}{\bar{C}_j'} k_j \tag{136}$$

where

$$k_i = \exp(\mu_i^\circ - \bar{\mu}_i^\circ)/RT \quad \text{and} \quad k_j = \exp(\mu_j^\circ - \bar{\mu}_j^\circ)/RT \tag{137}$$

Thus Eq. (136) applied to all ions may be written as

$$\frac{a_i'}{\bar{C}_i'} k_i = \frac{a_j'}{\bar{C}_j'} k_j = \cdots = \frac{\sum_{i=n} a_i' k_i}{\bar{C}_X'} \tag{138}$$

where according to Eq. (124), $\Sigma \bar{C}_i' = \bar{C}_X'$. Thus

$$\bar{C}_i' = \bar{C}_X' \frac{a_i' k_i}{\sum_i a_i' k_i} \tag{139}$$

A similar relation

$$\bar{C}_i'' = \bar{C}_X'' \frac{a_i'' k_i}{\sum_i a_i'' k_i}$$

for the boundary (″) at $x = d$ exists. The total potential across the membrane is given by

$$E = E'' - E' = (\bar{E}' - E') + (\bar{E}'' - \bar{E}') + (\bar{E}'' - E'')$$

Subtraction of Eq. (134) from Eq. (135) gives on rearrangement

$$E = \psi + \frac{RT}{z_i F} \ln \frac{a_i'}{\bar{C}_i'} - \frac{RT}{z_i F} \ln \frac{a_i''}{\bar{C}_i''} \tag{140}$$

where $\psi = \bar{E}'' - \bar{E}'$. Substitution of Eq. (139) into Eq. (140) gives

$$E = \psi + \frac{RT}{z_i F} \ln \frac{\sum a_i' k_i}{\sum a_i'' k_i} + \frac{RT}{z_i F} \ln \frac{\bar{C}_X''}{\bar{C}_X'} \tag{141}$$

The properties of the sites in the membrane are assumed to be such that the concentration of sites is reflected at the membrane boundaries and their total concentration in the membrane is constant. Further, in the steady state, $J_X^t = 0$, and in the nonsteady state

$$J_X^t(t, ') = J_X^t(t, '') = 0$$

The current I carried by the charged species is given by

$$I = F \sum z_i J_i + F z_X J_X$$

Since $z_i = -z_X$,

$$I = F z_i \left(\sum J_i - J_X \right) \tag{142}$$

Substituting from Eqs. (129) and (130) into Eq. (142) gives, on rearrangement for the condition $I = 0$, the expression

$$\frac{\partial \bar{E}}{\partial x} = -\frac{RT}{z_i F} \frac{1}{\sum \bar{u}_i \bar{C}_i + \bar{u}_X \bar{C}_X} \frac{\partial}{\partial x} \left(\sum \bar{u}_i \bar{C}_i - \bar{u}_X \bar{C}_X \right) \tag{143}$$

$\partial \bar{C}_X / \partial x$ can be evaluated by combining Eqs. (126), (130), (131), and (123). Thus

$$-\frac{\partial \bar{C}_X}{\partial x} = \frac{(J_X^t / RT) + \bar{u}_X \bar{C}_X z_X (F/RT)(\partial \bar{E}/\partial x) + \sum \bar{u}_{iX} K_{iX} \bar{C}_X (\partial \bar{C}_i / \partial x)}{\left(\bar{u}_X + \sum \bar{u}_{iX} K_{iX} \bar{C}_i \right)} \tag{144}$$

Substitution of Eq. (144) into Eq. (143), on rearrangement, gives

$$\frac{z_iF}{RT}\frac{\partial \bar{E}}{\partial x} = -\frac{\sum \bar{u}_i(\partial \bar{C}_i/\partial x)}{\sum \bar{u}_i\bar{C}_i} - t\left\{\frac{\sum \bar{u}_{i\mathrm{X}}K_{i\mathrm{X}}(\partial \bar{C}_i/\partial x)}{\sum \bar{u}_{i\mathrm{X}}K_{i\mathrm{X}}\bar{C}_i} - \frac{\sum \bar{u}_i(\partial \bar{C}_i/\partial x)}{\sum \bar{u}_i\bar{C}_i}\right\} - \frac{(\bar{u}_\mathrm{X}J_\mathrm{X}^\mathrm{t}/RT)}{\left(\bar{u}_\mathrm{X} + \sum \bar{u}_{i\mathrm{X}}K_{i\mathrm{X}}\bar{C}_i\right)\sum \bar{u}_i\bar{C}_i + \bar{u}_\mathrm{X}\bar{C}_\mathrm{X}\sum \bar{u}_{i\mathrm{X}}K_{i\mathrm{X}}\bar{C}_i} \tag{145}$$

where

$$t = \frac{\bar{u}_\mathrm{X}\bar{C}_\mathrm{X}}{\left(\left(\bar{u}_\mathrm{X}\bar{C}_\mathrm{X}/\sum \bar{u}_{i\mathrm{X}}\bar{C}_{i\mathrm{X}}\right) + 1\right)\sum \bar{u}_i\bar{C}_i + \bar{u}_\mathrm{X}\bar{C}_\mathrm{X}} \tag{146}$$

Integration of Eq. (145) gives the internal potential ψ as

$$\frac{z_iF}{RT}\psi = -\ln\frac{\sum \bar{u}_i\bar{C}_i''}{\sum \bar{u}_i\bar{C}_i'} - \int_{'}^{''} t\, d\ln\left[\frac{\sum \bar{u}_{i\mathrm{X}}K_{i\mathrm{X}}\bar{C}_i}{\sum \bar{u}_i\bar{C}_i}\right] - \int_{'}^{''} \frac{(\bar{u}_\mathrm{X}J_\mathrm{X}^\mathrm{t}/RT)\, dx}{\left(\bar{u}_\mathrm{x} + \sum \bar{u}_{i\mathrm{X}}K_{i\mathrm{X}}\bar{C}_i\right)\sum \bar{u}_i\bar{C}_i + \bar{u}_\mathrm{X}\bar{C}_\mathrm{X}\sum \bar{u}_{i\mathrm{X}}K_{i\mathrm{X}}\bar{C}_i} \tag{147}$$

Substitution of Eq. (147) into Eq. (141) and elimination of the boundary concentrations with the help of Eq. (139) gives

$$E = -\frac{RT}{z_iF}\left[\ln\frac{\sum \bar{u}_i a_i'' k_i}{\sum \bar{u}_i a_i' k_i} + \int_1 + \int_2\right] \tag{148}$$

where $\int_1$ and $\int_2$ are the integrals appearing in Eq. (147).

In the steady state when $J_\mathrm{X}^\mathrm{t} = 0$, i.e., $\int_2 = 0$, two cases may be considered:

(1) When there is complete dissociation, $t = 0$ [Eq. (146)], and so $\int_1 = 0$. Therefore Eq. (148) reduces to

$$E = -\frac{RT}{z_iF}\ln\left[\frac{\sum \bar{u}_i k_i a_i''}{\sum \bar{u}_i k_i a_i'}\right] \tag{149}$$

(2) When there is strong association, for the case of two counterions A and B such that $\bar{C}_X = \bar{C}_A + \bar{C}_B$, $\int_1$ can be written as

$$\int_1 = \int_{'}^{''} \frac{\bar{u}_X\left[\left(\bar{C}_A/\bar{C}_B\right) + 1\right]}{(\bar{u}_A + \bar{u}_X)\left(\bar{C}_A/\bar{C}_B\right) + \bar{u}_B + \bar{u}_X} \times d\ln\left[\frac{\left(\bar{u}_{AX}\bar{C}_A K_{AX}/\bar{C}_B\right) + \bar{u}_{BX}K_{BX}}{\bar{u}_A\left(\bar{C}_A/\bar{C}_B\right) + \bar{u}_B}\right] \tag{150}$$

Equation (150) has been integrated[97–99] and substituted into Eq. (148) to give

$$E = \frac{RT}{z_i F}\left[(1-\tau)\ln\frac{(\bar{u}_A + \bar{u}_X)k_A a'_A + (\bar{u}_B + \bar{u}_X)k_B a'_B}{(\bar{u}_A + \bar{u}_X)k_A a''_A + (\bar{u}_B + \bar{u}_X)k_B a''_B} + \tau\frac{\bar{u}_{AX}K_{AX}k_A a'_A + \bar{u}_{BX}K_{BX}k_B a'_B}{\bar{u}_{AX}K_{AX}k_A a''_A + \bar{u}_{BX}K_{BX}k_B a''_B}\right] \tag{151}$$

where

$$\tau = \frac{\bar{u}_X(\bar{u}_{BX}K_{BX} - \bar{u}_{AX}K_{AX})}{(\bar{u}_A + \bar{u}_X)\bar{u}_{BX}K_{BX} - (\bar{u}_B + \bar{u}_X)\bar{u}_{AX}K_{AX}} \tag{152}$$

The validity of these equations has been discussed elsewhere.[98, 100–102]

7. Liquid Membrane with Electrically Neutral Ligands

In these systems, the liquid membrane immiscible with the aqueous phases (′) and (″) contains a dissolved carrier X, usually an electroneutral macrocylic compound, which forms a complex with the ions determining the membrane potential. Eisenman and co-workers[99, 103–105] have developed expressions to describe the behavior of the membrane system shown in Fig. 4.

It is assumed that the following reactions occur in both the aqueous and membrane phases. Ideal behavior is assumed (i.e., $\gamma_i = 1$) for all species in both the membrane and aqueous phases except the species A and Y in the aqueous phase. Thus

$$A^+ + X \rightleftarrows AX^+; \qquad K_{AX}^+ = \frac{C_{AX}}{a_A C_X} \tag{153}$$

and

$$AX^+ + Y^- \rightleftarrows AXY; \qquad K_{AXY} = \frac{C_{AXY}}{C_{AX} a_Y} \tag{154}$$

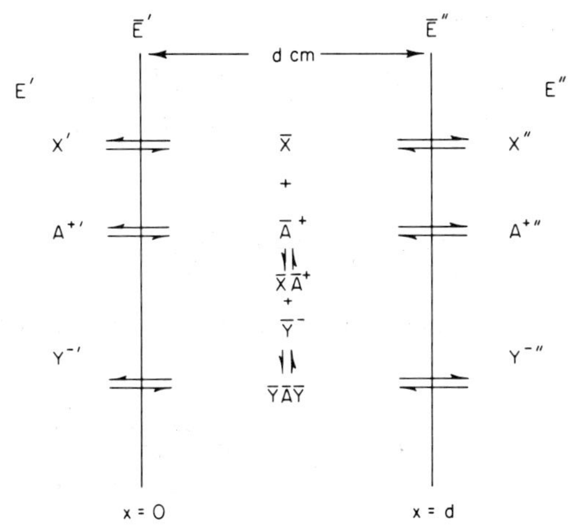

Fig. 4. Schematic diagram of liquid membrane system containing a neutral ligand. Three independent types of species, A^+ (cation), Y^- (anion), and X (electroneutral ligand), are present in all the phases. (′) and (″) are the two aqueous phases on either side of the membrane. The E's are the electrical potentials and overbars refer to the membrane phase. $\overline{XA^+}$ and $\overline{XAY}$ are the complexed cation and the neutral complex.

The state of equilibrium between the membrane and the aqueous phase for neutral species (e.g., X, AXY) may be described by the partition equilibria as

$$X \rightleftarrows \overline{X}; \qquad k_X = \frac{\bar{C}_X}{C_X} \tag{155}$$

The equilibria for charged species (e.g., $R^{\pm} = A^+, AX^+, Y^-$) are

$$R^{\pm} \rightleftarrows \overline{R}^{\pm}; \qquad k_r = \frac{\bar{a}_r \exp[z_r F\bar{E}/RT]}{a_r \exp[z_r FE/RT]} \tag{156}$$

The electric current density I is given by

$$\frac{I}{F} = \sum_{i=1}^{n} z_{iX} J_{iX} + \sum_{i=1}^{n} z_i J_i + \sum_{i=1}^{n} z_Y J_Y \tag{157}$$

It is assumed for simplicity that $z_{iX} = z_i = -z_Y = 1$ and that in the membrane phase, the concentrations of $\overline{A}^+$ and $\overline{Y}^-$ are negligible compared to the concentration of $\overline{AX}^+$. Consequently, Eq. (157) becomes

$$I \cong F \sum_{i=1}^{n} J_{iX} \tag{158}$$

Thus the Nernst–Planck flux equation applied to all the iX species for the condition $I = 0$ becomes

$$\sum_{i=1}^{n} \bar{u}_{i\mathrm{X}} \left[\frac{d\bar{C}_{i\mathrm{X}}}{dx} + \bar{C}_{i\mathrm{X}} \frac{F}{RT} \frac{d\bar{E}}{dx} \right] = 0$$

i.e.,

$$-\frac{d\bar{E}}{dx} = \frac{RT}{F} \frac{d}{dx} \ln \left[\sum_{i=1}^{n} \bar{u}_{i\mathrm{X}} \bar{C}_{i\mathrm{X}} \right] \tag{159}$$

Integration of Eq. (159) gives the diffusion potential ψ within the membrane. Hence

$$\psi = \bar{E}'' - \bar{E}' = -\frac{RT}{F} \ln \frac{\sum_{i=1}^{n} \bar{u}_{i\mathrm{X}} \bar{C}''_{i\mathrm{X}}}{\sum_{i=1}^{n} \bar{u}_{i\mathrm{X}} \bar{C}'_{i\mathrm{X}}} \tag{160}$$

Equation (156) may be used to express $\bar{C}'_{i\mathrm{X}}$ and $\bar{C}''_{i\mathrm{X}}$ as

$$\bar{C}'_{i\mathrm{X}} = C'_{i\mathrm{X}} k_{i\mathrm{X}} \exp\left[-(F/RT)(\bar{E}' - E') \right] \tag{161}$$

$$\bar{C}''_{i\mathrm{X}} = C''_{i\mathrm{X}} k_{i\mathrm{X}} \exp\left[-(F/RT)(\bar{E}'' - E'') \right] \tag{162}$$

Substitution of Eqs. (161) and (162) into Eq. (160) gives

$$\bar{E}'' - \bar{E}' = -\frac{RT}{F} \ln \frac{\sum_{i=1}^{n} \bar{u}_{i\mathrm{X}} k_{i\mathrm{X}} \bar{C}''_{i\mathrm{X}}}{\sum_{i=1}^{n} \bar{u}_{i\mathrm{X}} k_{i\mathrm{X}} \bar{C}'_{i\mathrm{X}}} + (\bar{E}'' - E'') + (E' - \bar{E}')$$

or

$$E = E'' - E' = \frac{RT}{F} \ln \frac{\sum_{i=1}^{n} \bar{u}_{i\mathrm{X}} k_{i\mathrm{X}} \bar{C}'_{i\mathrm{X}}}{\sum_{i=1}^{n} \bar{u}_{i\mathrm{X}} k_{i\mathrm{X}} \bar{C}''_{i\mathrm{X}}} \tag{163}$$

As the concentration of $C_{i\mathrm{X}}$ is unknown, it is desirable to relate it to known ionic activities and the total concentration of the carrier X. Eqs. (153) and (154) can be written for a general case as

$$C_{\mathrm{X}} = \frac{C_{i\mathrm{X}}}{K^{+}_{i\mathrm{X}} a_i} = \cdots = \frac{C_{n\mathrm{X}}}{K^{+}_{n\mathrm{X}} a_n} = \cdots = \frac{C_{i\mathrm{XY}}}{K^{+}_{i\mathrm{X}} K_{i\mathrm{XY}} a_i a_{\mathrm{Y}}}$$

$$= \cdots = \frac{C_{n\mathrm{X}m}}{K^{+}_{n\mathrm{X}} K_{n\mathrm{X}m} a_n a_m} \tag{164}$$

Adding the numerators and the denominators gives

$$C_{\mathrm{X}} = \frac{\sum_{i=1}^{n} C_{i\mathrm{X}} + \sum_{i=1}^{n} \sum_{y=1}^{m} C_{i\mathrm{XY}}}{\sum_{i=1}^{n} K^{+}_{i\mathrm{X}} a_i + \sum_{i=1}^{n} \sum_{y=1}^{m} K^{+}_{i\mathrm{X}} K_{i\mathrm{XY}} a_i a_{\mathrm{Y}}} \tag{165}$$

But the total concentration of the carrier X is given by

$$C_X^t = C_X + \sum_{i=1}^{n} C_{iX} + \sum_{i=1}^{n} \sum_{y=1}^{m} C_{iXY} \tag{166}$$

Substitution of Eq. (166) into Eq. (165) gives, on rearrangement,

$$C_X = \frac{C_X^t}{1 + \sum_{i=1}^{n} K_{iX}^+ a_i + \sum_{i=1}^{n} \sum_{y=1}^{m} K_{iX}^+ K_{iXY} a_i a_Y} \tag{167}$$

which can be combined with Eq. (164) to give

$$C_{iX} = \frac{C_X^t K_{iX}^+ a_i}{1 + \sum_{i=1}^{n} K_{iX}^+ a_i + \sum_{i=1}^{n} \sum_{y=1}^{m} K_{iX}^+ K_{iXY} a_i a_Y} \tag{168}$$

Now substitution of Eq. (168) into Eq. (163) gives, on rearrangement,

$$E = \frac{RT}{F} \ln \frac{\sum_{i=1}^{n} \bar{u}_{iX} k_{iX} K_{iX}^+ a_i'}{\sum_{i=1}^{n} \bar{u}_{iX} k_{iX} K_{iX}^+ a_i''} + \frac{RT}{F} \ln \frac{C_X^{t\prime}}{C_X^{t\prime\prime}}$$

$$+ \frac{RT}{F} \ln \frac{1 + \sum_{i=1}^{n} K_{iX}^+ a_i'' + \sum_{i=1}^{n} \sum_{y=1}^{m} K_{iX}^+ K_{iXY} a_i'' a_Y''}{1 + \sum_{i=1}^{n} K_{iX}^+ a_i' + \sum_{i=1}^{n} \sum_{y=1}^{m} K_{iX}^+ K_{iXY} a_i' a_Y'} \tag{169}$$

Under the conditions of (a) equal total concentrations of the carrier on both sides of the membrane, (b) the presence of only two monovalent ions i and j, and (c) negligible concentration of the neutral complexes iXY and jXY, Eq. (169) simplifies to

$$E = \frac{RT}{F} \ln \frac{a_i' + [\bar{u}_{jX} k_{jX} K_{jX}^+ / \bar{u}_{iX} k_{iX} K_{iX}^+] a_j'}{a_i'' + [\bar{u}_{jX} k_{jX} K_{jX}^+ / \bar{u}_{iX} k_{iX} K_{iX}^+] a_j''} + \frac{RT}{F} \ln \frac{1 + K_{iX}^+ a_i'' + K_{jX}^+ a_j''}{1 + K_{iX}^+ a_i' + K_{jX}^+ a_j'} \tag{170}$$

When dilute solutions are used, negligible complex ion formation takes place and the second term in Eq. (170) becomes insignificant; thus Eq. (170) simplifies further to

$$E = \frac{RT}{F} \ln \frac{a_i' + [\bar{u}_{jX} k_{jX} K_{jX}^+ / \bar{u}_{iX} k_{iX} K_{iX}^+] a_j'}{a_i'' + [\bar{u}_{jX} k_{jX} K_{jX}^+ / \bar{u}_{iX} k_{iX} K_{iX}^+] a_j''} \tag{171}$$

The product $k_{iX} K_{iX}^+$ can be expressed in terms of a so-called bulk extraction (or partition) constant K_i, which is for the reaction

$$i^+ + Y^- + X^* \rightleftharpoons iX^{*+} + Y^{*-}$$

where * indicates the organic phase. Thus

$$K_i = \frac{a_{iX}^* a_Y^*}{a_i a_Y a_X^*} \tag{172}$$

Substituting from Eqs. (153), (155), and (156), Eq. (172) becomes

$$K_i = \frac{k_{iX} K_{iX}^{+} k_Y}{k_X} \tag{173}$$

The potential difference disappears because of the oppositely charged ions involved. A similar equation may be written for K_j and since the species X and Y are common for the system, the relation

$$\frac{K_j}{K_i} = \frac{k_{jX} K_{jX}^{+}}{k_{iX} K_{iX}^{+}} \tag{174}$$

is obtained.

Experimental verification of some of these relations has been presented by Eisenman and co-workers.[99, 103, 105–107]

8. Selectivity of Membrane Electrodes

The relations of the selectivity of the membrane electrodes to the various physicochemical parameters as derived in the theoretical equations given above are summarized[88, 89, 99, 108] in Table 1, from which it is seen that the selectivity of the membrane electrode is governed by both the mobility of the ions in the membrane and the equilibrium that exists at the membrane–solution interfaces (e.g., partition coefficients, ion exchange equilibrium constants). The factors responsible for the selectivity of the membrane arising from differences in the mobilities of the ions in the membrane have not yet been investigated. The factors controlling or responsible for the membrane selectivity due to physicochemical equilibrium conditions have been analyzed by a number of investigators. The work concerned with ion exchange membranes has been reviewed by Reichenberg[109] and intensively pursued by Eisenman,[70, 77, 78, 82, 83, 99, 110] the principal features of whose work are outlined below.

For the ion exchange reaction

$$A^+ \text{ (membrane)} + B^+ \text{ (aqueous)} \rightleftharpoons B^+ \text{ (membrane)} + A^+ \text{ (aqueous)}$$

the standard free energy change $(-\Delta G_{ij}^\circ)$ is dependent on the partial molal free energies (μ's) of the interactions of the species with water and the membrane site (X^-). Thus

$$\Delta G_{ij}^\circ = (\mu_A - \mu_B) + (\bar{\mu}_B - \bar{\mu}_A) \tag{175}$$

The term $\mu_A - \mu_B$ can be evaluated from the known hydration energies of ions. The term $\bar{\mu}_B - \bar{\mu}_A$ was evaluated by Eisenman by using a model in which the anionic sites (X^-), devoid of water molecules, are contained in the membrane at various distances of separation. Using Coulomb's law the electrostatic interaction energies $\bar{\mu}_{BX}$ and $\bar{\mu}_{AX}$ in kilocalories per mole

TABLE 1

Expressions for Selectivity between a Primary Ion i and an Interfering Ion j for Different Types of Membrane Electrodes[a]

Type of membrane	Species within the membrane	$K_{ij}^{pot} = 1/K_{ji}^{pot}$	Equation	Ref.
Solid membrane				
Ion exchanger	X^-, i^{z+}, j^{z+}	$\frac{\bar{u}_j}{\bar{u}_i} K_{ij}$	(117)	82, 83
Solid state of the silver halide type	M^+, i^{z-}, j^{z-}	$\frac{S_{Mi}}{S_{Mj}}$	(179)	85–87, 108, 117
Liquid membrane				
Ion exchanger: dissociated	X^-, i^{z+}, j^{z+}	$\frac{\bar{u}_j}{\bar{u}_i}\frac{k_j}{k_i}$	(149)	97–99, 102
Ion exchanger: associated				
(1) $\tau = 0$				
(poorly mobile site)	iX_z, jX_z	$\frac{(\bar{u}_j + \bar{u}_X)k_j}{(\bar{u}_i + \bar{u}_X)k_i}$	(151)	
When $\bar{u}_X \ll \bar{u}_i, \bar{u}_j$		$\frac{\bar{u}_j k_j}{\bar{u}_i k_i}$		
(2) $\tau = 1$				
(highly mobile site)		$\frac{\bar{u}_{jX} k_j K_{jX}}{\bar{u}_{iX} k_i K_{iX}}$ or $= \frac{\bar{u}_{jX}}{\bar{u}_{iX}} K_{ij}$	(151)	
With electrically neutral ligand	$X, i^{z+}, j^{z+}, iX_n^{z+}, jX_n^{z+}, Y^-$	$\frac{\bar{u}_{jX} k_{jX} K_{jX}^+}{\bar{u}_{iX} k_{iX} K_{iX}^+} = \frac{\bar{u}_{jX} K_j}{\bar{u}_{iX} K_i}$	(171), (174)	89, 103–108
When $\bar{u}_{jX} = \bar{u}_{iX}$		$\frac{K_j}{K_i}$		

[a] k_i, k_j: partition coefficients; K_{iX}, K_{jX}: association constants; K_i, K_j: bulk partition coefficients; S_{Mi}, S_{Mj}: solubility products of sparingly soluble precipitates of metal M.

(distance in angstroms) were calculated from

$$\bar{\mu}_{BX} = -332/(r_{B^+} + r_-); \qquad \bar{\mu}_{AX} = -332/(r_{A^+} + r_-)$$

where the r's are the radii. Thus from Eq. (175) ΔG_{ij}° could be evaluated as a function of r_- (i.e., the anionic field strength). A plot of ΔG_{ij}° vs. r_- gave an isotherm, and isotherms were constructed for a number of monovalent ions. Taking the Cs ion as the reference, these isotherms gave for an anionic site of largest radius (lowest electrostatic field strength) the selectivity sequence for alkali metal ions as Cs > Rb > K > Na > Li. The sequence was reversed for an anionic site of high field strength (small r_-). Other sequences exist between these extremes. Out of a possible 120, the following sequences were found:

I. Cs > Rb > K > Na > Li
II. Rb > Cs > K > Na > Li, or
IIa. Cs > K > Rb > Na > Li
III. Rb > K > Cs > Na > Li, or
IIIa. K > Cs > Rb > Na > Li
IV. K > Rb > Cs > Na > Li
V. K > Rb > Na > Cs > Li
VI. K > Na > Rb > CS > Li
VII. Na > K > Rb > Cs > Li, or
VIIa. K > Na > Rb > Li > Cs
VIII. Na > K > Rb > Li > Cs
IX. Na > K > Li > Rb > Cs
X. Na > Li > K > Rb > Cs
XI. Li > Na > K > Rb > Cs

Even if water molecules are admitted into the vicinity of the anionic site, the same order of selectivity is found, although the magnitude of the selectivity among ions is decreased. The important facts emerging from these studies are that (i) the pattern of selectivity is predominantly governed by r_- (i.e., field strength), (ii) the number of water molecules in the neighborhood of the cationic and anionic sites influences only the magnitude and not the sequence of selectivity, and (iii) the spatial distribution of anionic sites influences the selectivity through overlap of electrostatic forces of the sites. These sequences of selectivity have been observed in a number of artificial and biological systems.[70]

The same type of analysis has been applied to anions and to divalent cations.[70, 110] Among the halide ions, the following sequences have been predicted and some of them have been observed experimentally[70]:

I. I > Br > Cl > F
II. Br > I > Cl > F
III. Br > Cl > I > F
IV. Cl > Br > I > F
V. Cl > Br > F > I
VI. Cl > F > Br > I
VII. F > Cl > Br > I

The effect of field strength on divalent cations is the same as it is on monovalent cations. However, when there is a mixture of mono- and divalent cations, the sites at low field strengths prefer monovalent to divalent ions, but with an increase in field strength, the divalent ions are preferred. In these situations, it was found that the number of sites per counterion also influenced the selectivity.

An analogous type of analysis has been provided by different schools of investigators[99, 111–116] to explain the selectivity of membranes containing electroneutral ligands. In these treatments, the homopolar binding site is replaced by a neutral dipole. The partial molar free energy of the complexed ion in such a case is given by

$$\mu_{i(\text{complex})} = \left[\frac{-332}{r_+ + r_\text{n}} + \frac{332}{r_+ + r_\text{p}} \right](qN)$$

where q is the fractional value of electronic charge and N is the coordination number of the ligands, r_+ is the cationic radius, and r_n and r_p are the distances from the surface of the dipole of the negative and positive charges, respectively. Selectivity isotherms have been constructed for such a model. When the dipole separation is large, i.e., $r_\text{p} - r_\text{n} \gg r_n$, the selectivities approach those calculated for the simple homopolar site. The selectivity sequences have been found to be the same as those given above for monopolar binding sites, for a wide range of dipolar charge distributions.

In the case of solid ion exchangers, the two factors, i.e., K_{ij} and $\bar{u}_j/\bar{u}_i$, determining K_{ij}^pot act in such a way as to oppose each other. If the ion is preferred more strongly, its mobility within the membrane will be reduced. Because of this poor mobility, there are few solid ion exchangers that are selective to divalent ions. This limitation will be obviated if the sites are made to move. As a result, there are a number of liquid ion exchange membranes that act selectively to divalent ions.[89] Furthermore, in the case of solid state membranes (e.g., silver halide precipitate membrane) in which there is no charge transport through the solid phase (i.e., no diffusion potential) the selectivity, as given in Table 1, is determined entirely by the solubility products $S_{\text{M}i}$ and $S_{\text{M}j}$ of the metal precipitates of composition $\text{M}_a i_b$ and $\text{M}_n j_m$ where a, b, m, and n are constants determining the stoichiometry of the two sparingly soluble metal precipitates and i and j are the anions.

In the absence of a diffusion potential, the selectivity of a solid state membrane K_{ij}^pot becomes equivalent to the equilibrium constant which for the general reaction[85–87]

$$\frac{1}{a}(\text{M}_a i_b) + \frac{m}{n} j \rightleftharpoons \frac{1}{n}(\text{M}_n j_m) + \frac{b}{a} i$$

becomes

$$K_{ij} = \frac{\bar{a}_j^{m/n} a_i^{b/a}}{\bar{a}_i^{b/a} a_j^{m/n}} \tag{176}$$

Multiplying and dividing by a_{M}, the activity of the metal ion, and substituting for the solubility products $S_{\text{M}i} = a_{\text{M}}^a a_i^b$ and $S_{\text{M}j} = a_{\text{M}}^n a_j^m$, Eq. (176) becomes

$$K_{ij} = \frac{S_{\text{M}i}^{1/a}}{S_{\text{M}j}^{1/n}} \left(\bar{a}_j^{m/n} \bar{a}_i^{-b/a}\right) \tag{177}$$

When the membrane is in contact with a solution containing equivalent concentrations of i and j, Eq. (177) may be written as[85]

$$K_{ij} = \frac{S_{\text{M}i}^{1/a}}{S_{\text{M}j}^{1/n}} \left[\bar{a}_j^{(m/n)-(b/a)}\right] \tag{178a}$$

When the valences of i and j are unity, Eq. (178a) becomes

$$K_{ij} = S_{\text{M}i}/S_{\text{M}j} \tag{178b}$$

Thus Eq. (116), which determines the boundary potential for $n = 1$, becomes[108, 117]

$$E' = \text{const} + \frac{RT}{F} \ln\left(a_i + \frac{S_{\text{M}i}}{S_{\text{M}j}} a_j\right) \tag{179}$$

In the case of liquid membranes in which there is strong dissociation, the selectivity (see Table 1) depends entirely on the properties of the solvent. There is little dependence on the properties of the lipophilic exchanger molecule, other than what is due to the sign of the charge on the molecule. On the other hand, in the case of a mixed valence system (i.e., response to monovalent cations in the presence of a divalent cation) it has been shown by Buck and Sandifer[118] that the selectivity is dependent on membrane loading, i.e., the concentration of sites. In the case of other liquid membranes in which there is an association of counterions with the sites so as to make the sites less mobile than the counterions, the selectivity again depends on the properties of the solvent. But if the sites become more mobile, the selectivity becomes dependent on the properties of both the solvent and the sites. As there are no ionexchange sites but only neutral molecules forming stoichiometric complexes with ions in liquid membranes containing neutral ligands, the selectivity in these membranes is determined entirely by the equilibrium parameters such as partition and equilibrium constants.

REFERENCES

1. S. Glasstone, "Thermodynamics for Chemists." Van Nostrand-Reinhold, Princeton, New Jersey, 1947.
2. S. Glasstone, "An Introduction to Electrochemistry." Van Nostrand-Reinhold, Princeton, New Jersey, 1942.

3. R. A. Robinson and R. H. Stokes, "Electrolyte Solutions." Academic Press, New York, 1959.
4. E. A. Guggenheim, *J. Amer. Chem. Soc.* **52**, 1315 (1930).
5. P. Henderson, *Z. Phys. Chem.* **59**, 118 (1907); **63**, 325 (1908).
6. M. Planck, *Ann. Phys.* **39**, 161 (1890); **40**, 561 (1890).
7. N. Lakshminarayanaiah, "Transport Phenomena in Membranes." Academic Press, New York, 1969.
8. D. A. MacInnes, "The Principles of Electrochemistry." Dover, New York, 1961.
9. G. N. Lewis and L. W. Sargent, *J. Amer. Chem. Soc.* **31**, 363 (1909).
10. K. R. Johnson, *Ann. Phys.* (*Leipzig*) **14**, 995 (1904).
11. H. Pleijel, *Z. Phys. Chem.* **72**, 1 (1910).
12. U. Behn, *Ann. Phys.* **62**, 54 (1897).
13. N. Lakshminarayanaiah, *in* "Electrochemistry" (Specialist Periodical Reports), (G. J. Hills, ed.), Vol. 2. The Chemical Society, London, 1972.
14. W. Simon, H. R. Wuhrmann, M. Vasak, L. A. R. Pioda, R. Dohner, and Z. Stefanac, *Angew. Chem. Int. Ed.* **9**, 445 (1970).
15. E. J. Harris, "Transport and Accumulation in Biological Systems," p. 62. Academic Press, New York, 1960.
16. E. A. Moelwyn-Hughes, "Physical Chemistry," p. 1106. Pergamon, Oxford, 1961.
17. R. Schlogl, *Z. Phys. Chem. (Frankfurt)* **1**, 305 (1954).
18. T. Teorell, *Proc. Soc. Exp. Biol. Med.* **33**, 282 (1935); *Proc. Nat. Acad. Sci. U.S.* **21**, 152 (1935); *Z. Elektrochem.* **55**, 460 (1951).
19. K. H. Meyer and J. F. Sievers, *Helv. Chim. Acta* **19**, 649, 665, 987 (1936).
20. K. S. Spiegler and M. R. J. Wyllie, *in* "Physical Techniques in Biological Research" (G. Oster and A. M. Pollister, eds.), Vol. 2, p. 301. Academic Press, New York, 1956.
21. N. Lakshminarayanaiah, *in* "Electrochemistry" (Specialist Periodical Reports), (H. R. Thirsk, ed.), Vols. 4 and 5. The Chemical Society, London, 1974 and 1975.
22. R. Schlogl, *Discuss. Faraday Soc.* **21**, 46 (1956); *Ber. Bunsenges. Phys. Chem.* **71**, 755 (1967).
23. R. Schlogl, "Stofftransport durch Membranen." Steinkopff, Darmstadt, 1964; *Ber. Bunsenges. Phys. Chem.* **70**, 400 (1966).
24. G. Scatchard, *in* "Ion Transport across Membranes" (H. T. Clarke and D. Nachmanshon, eds.) p. 128. Academic Press, New York, 1954; *in* "Electrochemistry in Biology and Medicine" (T. Shedlovsky, ed.) p. 18. Wiley, New York, 1955.
25. A. J. Staverman, *Trans. Faraday Soc.* **48**, 176 (1952).
26. J. G. Kirkwood, *in* "Ion Transport across Membranes" (H. T. Clarke and D. Nachmanshon, eds.), p. 119. Academic Press, New York, 1954.
27. J. W. Lorimer, E. I. Boterenbrood, and J. J. Hermans, *Discuss. Faraday Soc.* **21**, 141 (1956).
28. Y. Kobatake, *J. Chem. Phys.* **28**, 146 (1958).
29. K. S. Spiegler, *Trans. Faraday Soc.* **54**, 1408 (1958).
30. P. Meares, *Trans. Faraday Soc.* **55**, 1970 (1959).
31. D. Mackay and P. Meares, *Trans. Faraday Soc.* **55**, 1221 (1959).
32. G. J. Hills, P. W. M. Jacobs, and N. Lakshminarayanaiah, *Proc. Roy. Soc.* **A262**, 246 (1961).
33. O. Kedem and A. Katchalsky, *Trans. Faraday Soc.* **59**, 1918, 1931, 1941 (1963).
34. T. Hoshiko and B. D. Lindley, *Biochim. Biophys. Acta* **79**, 301 (1964).
35. G. Scatchard, *J. Amer. Chem. Soc.* **75**, 2883 (1953).
36. S. R. de Groot, "Thermodynamics of Irreversible Processes." North-Holland Publ., Amsterdam, 1963.
37. A. Katchalsky and P. F. Curran, "Nonequilibrium Thermodynamics in Biophysics." Harvard Univ. Press, Cambridge, Massachusetts, 1965.

38. H. Eyring, R. Lumbray, and J. W. Woodbury, *Rec. Chem. Progr.*, **10**, 100 (1949).
39. B. J. Zwolinski, H. Eyring, and C. E. Reese, *J. Phys. Colloid Chem.* **53**, 1426 (1949).
40. K. J. Laidler and K. E. Shuler, *J. Chem. Phys.* **17**, 851, 856 (1949).
41. K. E. Shuler, C. A. Dames, and K. J. Laidler, *J. Chem. Phys.* **17**, 860 (1949).
42. M. Nagasawa and Y. Kobatake, *J. Phys. Chem.* **56**, 1017 (1952).
43. F. H. Johnson, H. Eyring, and M. J. Polissar, "The Kinetic Basis of Molecular Biology," p. 754. Wiley, New York, 1954.
44. R. B. Parlin and H. Eyring, *in* "Ion Transport across Membranes" (H. T. Clarke and D. Nachmanshon, eds.), p. 106. Academic Press, New York, 1954.
45. M. Nagasawa and I. Kagawa, *Discuss. Faraday Soc.* **21**, 52 (1956).
46. S. Glasstone, K. J. Laidler, and H. Eyring, "The Theory of Rate Processes." McGraw-Hill, New York, 1941.
47. Y. Kobatake, N. Takeguchi, Y. Toyoshima, and H. Fujita, *J. Phys. Chem.* **69**, 3981 (1965).
48. Y. Kobatake, Y. Toyoshima, and N. Takeguchi, *J. Phys. Chem.* **70**, 1187 (1966).
49. Y. Toyoshima, Y. Kobatake, and H. Fujita, *Trans. Faraday Soc.* **63**, 2814 (1967).
50. Y. Kobatake and N. Kamo, *Progr. Polym. Sci. Japan* **5**, 257 (1973).
51. J. S. Mackie and P. Meares, *Proc. Roy. Soc.* **A232**, 485 (1955).
52. G. E. Boyd and K. Bunzl, *J. Amer. Chem. Soc.* **89**, 1776 (1967).
53. T. Teorell, *Progr. Biophys. Biophys. Chem.* **3**, 305 (1953).
54. N. Lakshminarayanaiah and V. Subrahmanyan, *J. Polym. Sci. Part. A* **2**, 4491 (1964).
55. N. Lakshminarayanaiah, *J. Phys. Chem.* **70**, 1588 (1966).
56. J. R. Wilson, "Demineralization by Electrodialysis," p. 84. Butterworths, London and Washington, D. C., 1960.
57. M. R. J. Wyllie, *J. Phys. Chem.* **58**, 67 (1954).
58. F. Helfferich and R. Schlogl, *Discuss. Faraday Soc.* **21**, 133 (1956).
59. F. Helfferich, *Discuss. Faraday Soc.* **21**, 83 (1956).
60. F. Helfferich, "Ion Exchange." McGraw-Hill, New York, 1962.
61. M. R. J. Wyllie and S. L. Kanaan, *J. Phys. Chem.* **58**, 73 (1954).
62. F. Helfferich and H. D. Ocker, *Z. Phys. Chem. (Frankfurt)* **10**, 213 (1957).
63. J. B. Andelman and H. P. Gregor, *Electrochim. Acta* **11**, 869 (1966).
64. A. Ilani, *Biophys. J.* **6**, 329 (1966).
65. N. Lakshminarayanaiah, *Chem. Rev.* **65**, 491 (1965).
66. D. Woermann, K. F. Bonhoeffer, and F. Helfferich, *Z. Phys. Chem.* (*Frankfurt*) **8**, 265 (1956).
67. D. E. Goldman, *J. Gen. Physiol.* **27**, 37 (1943).
68. A. L. Hodgkin and B. Katz, *J. Physiol.* **108**, 37 (1949).
69. W. D. Stein, "The Movement of Molecules across Membranes." Academic Press, New York, 1967.
70. G. Eisenman, *Bol. Inst. Estud. Med. Biol.* (*Univ. Nac. Auton. Mex.*) **21**, 155 (1963); *Proc. Int. Congr. Physiol. Sci., 23rd, 1965, Tokyo* p. 489.
71. G. Karreman and G. Eisenman, *Bull. Math. Biophys.* **24**, 413 (1962).
72. F. Conti and G. Eisenman, *Biophys. J.* **5**, 247, 511 (1965); **6**, 227 (1966).
73. G. Eisenman and F. Conti, *J. Gen. Physiol.* **48**, 65 (1965).
74. G. Eisenman, J. P. Sandblom, and J. L. Walker, Jr., *Science* **155**, 965 (1967).
75. D. Mackay and P. Meares, *Kolloid-Z.* **171**, 139 (1960).
76. E. Hogfeldt, *Acta Chem. Scand.* **9**, 151 (1955).
77. G. Eisenman, D. O. Rudin, and J. U. Casby, *Science* **126**, 831 (1957).
78. G. Eisenman, *Biophys. J. Suppl.* **2**, 314 (1962).
79. V. Rothmund and G. Kornfeld, *Z. Anorg. Allgem. Chem.* **103**, 129 (1918).

80. M. Dole, *J. Chem. Phys.* **2**, 862 (1934); "The Glass Electrode, Methods, Applications and Theory." Wiley, New York, 1941.
81. B. P. Nicolsky, *Acta Physicochim URSS* **7**, 597 (1937).
82. G. Eisenman, (ed.), "Glass Electrodes for Hydrogen and other Cations, Principles and Practice," Chapters 3–6. Dekker, New York, 1967.
83. G. Eisenman, *Advan. Anal. Chem. Instrum.* **4**, 213 (1965); also reprinted in G. Eisenman, R. Bates, G. Mattock, and S. M. Friedman, "The Glass Electrode." Wiley (Interscience), New York, 1966.
84. R. M. Garrels, M. Sato, M. E. Thompson, and A. H. Truesdell, *Science* **135**, 1045 (1962); also see G. Eisenman (ed.), "Glass Electrodes for Hydrogen and other Cations, Principles and Practice," Chapter 11. Dekker, New York, 1967.
85. E. Pungor, *Anal. Chem.* **39**, 28A (1967).
86. E. Pungor and K. Toth, *Anal. Chim. Acta* **47**, 291 (1969).
87. E. Pungor and K. Toth, *Analyst* **95**, 625 (1970).
88. H. R. Wuhrmann, W. E. Morf, and W. Simon, *Helv. Chim. Acta* **56**, 1011 (1973).
89. W. E. Morf, D. Ammann, E. Pretsch, and W. Simon, *Pure Appl. Chem.* **36**, 421 (1973).
90. W. J. V. Osterhout, *Cold Spring Harbor Symp. Quant. Biol.* **8**, 51 (1940).
91. R. Beutner, "Physical Chemistry of Living Tissues and Life Processes." Williams and Wilkins, Baltimore, Maryland, 1933.
92. A. Germant, "Ions in Hydrocarbons." Wiley (Interscience), New York, 1962.
93. K. Sollner and G. M. Shean, *J. Amer. Chem. Soc.* **86**, 1901 (1964).
94. G. M. Shean and K. Sollner, *Ann. N. Y. Acad. Sci.* **137**, 759 (1966).
95. K. Sollner, *in* "Diffusion Processes." (G. N. Sherwood, A. V. Chadwick, W. M. Muir, and F. L. Swinton, eds.), p. 655. Gordon and Breach, New York, 1971.
96. F. Conti and G. Eisenman, *Biophys. J.* **6**, 227 (1966).
97. J. Sandblom, G. Eisenman, and J. L. Walker, Jr., *J. Phys. Chem.* **71**, 3862, 3871 (1967).
98. G. Eisenman, *Anal. Chem.* **40**, 310 (1968).
99. G. Eisenman, *in* "Ion Selective Electrodes" (R. A. Durst, ed.), p. 1. Nat. Bur. of Std. Spec. Publ. 314, Washington, D. C., 1969.
100. J. Sandblom, *J. Phys. Chem.* **73**, 249, 257 (1969).
101. J. Sandblom and F. Orme, *in* "Membranes" (G. Eisenman, ed.), Vol. 1, p. 125. Dekker, New York, 1972.
102. J. L. Walker, Jr., G. Eisenman, and J. Sandblom, *J. Phys. Chem.* **72**, 978 (1968).
103. G. Eisenman, S. M. Ciani, and G. Szabo, *Fed. Proc.* **27**, 1289 (1968).
104. S. Ciani, G. Eisenman, and G. Szabo, *J. Membrane Biol.* **1**, 1 (1969).
105. G. Szabo, G. Eisenman, and S. M. Ciani, *in* "Physical Principles of Biological Membranes" (F. Snell, J. Wolken, G. Iverson, and J. Lam, eds.), p. 79. Gordon and Breach, New York, 1970.
106. G. Eisenman, S. Ciani, and G. Szabo, *J. Membrane Biol.* **1**, 294 (1969).
107. G. Szabo, G. Eisenman, and S. Ciani, *J. Membrane Biol.* **1**, 346 (1969).
108. J. Koryta, *Anal. Chim. Acta* **61**, 329 (1972).
109. D. Reichenberg, *in* "Ion-Exchange" (J. A. Marinsky, ed.), Vol. 1, p. 227. Dekker, New York, 1968.
110. G. Eisenman, *in Symp. Membrane Transport Metabolism* (A. Kleinzeller and A. Kotyk, eds.) p. 163. Academic Press, New York, 1961.
111. G. Eisenman, G. Szabo, S. McLaughlin, and S. M. Ciani, *J. Bioenerget.* **4**, 93 (1972).
112. G. Szabo, G. Eisenman, R. Laprade, S. Ciani, and S. Krasne, *in* "Membranes" (G. Eisenman, ed.), Vol. 2, Chapter 3. Dekker, New York, 1972.
113. G. Eisenman, G. Szabo, S. Ciani, S. McLaughlin, and S. Krasne, *Progr. Surface Membrane Sci.* **6**, 139 (1973).

114. W. Simon and W. Morf, *in Symp. Mole. Mech. Antibiotic Action Protein Biosynthesis Membranes* (D. Vasquez, ed.). Springer-Verlag, Berlin and New York, 1971; *in* "Membranes" (G. Eisenman, ed.), Vol. 2, Chapter 4. Dekker, New York, 1972.
115. H. Diebler, M. Eigen, G. Ilgenfritz, G. Maas, and R. Winkler, *Pure Appl. Chem.* **20**, 93 (1969).
116. M. Eigen and R. Winkler, *Neurosci. Res. Program. Bull.* **9**, 330 (1971).
117. R. P. Buck, *Anal. Chem.* **40**, 1432 (1968).
118. R. P. Buck and J. R. Sandifer, *J. Phys. Chem.* **77**, 2122 (1973).

PART II

Solid and Liquid Membrane Electrodes

Chapter 4

ORGANIC ION EXCHANGERS

In recent years, various solid and liquid organic ion exchangers have been marketed by a number of commercial firms. The liquid ion exchangers, in which anionic or cationic sites have a considerable degree of freedom to move about compared to solid ion exchangers, show selectivity either to cations or anions without displaying any special preference to any one particular cation or anion. This is also true of solid ion exchangers which, because of the ease with which they can be manipulated, are finding increasing usage in a number of important industrial operations. Some of the organic liquid ion exchangers that are used in the construction of ion-selective electrodes are discussed in Chapter 8, which deals with liquid membrane electrodes. The solid membrane electrodes discussed in the various chapters use organic polymers to hold special compounds which exhibit selectivity to particular ions. The typical solid organic ion exchangers, although very useful in a number of separation processes, have to be processed into a form suitable for use as membrane electrodes even though none of them by themselves have proved particularly useful as ion-selective electrodes. However, in fundamental studies involving ion exchange and ion transport across barriers, solid ion exchangers in the form of membranes have proved very valuable and so have been used as models for biological membranes. A number of authoritative books[1–7] and review articles[8–12] describe the properties and applications of ion exchange membranes. This chapter presents a summary of the work concerned with some of the basic properties and applications of ion exchange membranes where they serve as sensing and/or separating devices.

The first membrane to be tried as an electrode was collodion which can be easily formed by spreading and evaporating a solution of it on a suitable surface. A systematic study of these porous membranes of high selectivity dates back to the time of Michaelis[13] whose work on dried collodion molecular sieve membranes is classic in the sense that it laid the foundations for further work by Sollner and his co-workers. All of this early work on the preparation and behavior of collodion-based membranes is reviewed by Sollner.[11]

It was in 1950, with the advent of ion exchange resins, that heterogeneous membranes incorporating these commercially available ion exchangers into thermoplastic polymers were prepared by Wyllie and Patnode[14] by hot pressing a mixture of ion exchange granules and polystyrene. Since then, both homogeneous and heterogeneous membranes have been prepared by using a number of very interesting methods, a summary of which can be found elsewhere.[6, 8, 9] Some heterogeneous membranes are marketed by a number of commercial firms. The homogeneous membranes that have proved very useful in fundamental studies involving membrane phenomena are cross-linked polymethacrylic acid, sulfonated phenol–formaldehyde, dimethyl-2-hydroxybenzylamine phenol–formaldehyde, and sulfonated polystyrene. A number of homogeneous but nonionic membranes have also been prepared and used as separation devices in processes such as demineralization, desalination, and so on.

A. PROPERTIES OF ION EXCHANGE MEMBRANES

Generally the ion exchange membranes contain ionogenic groups fixed to the resin or polymer matrix—negative groups such as $-SO_3^-$, $-COO^-$, *et seq.*, in the case of cation exchange membranes, and positive groups such as $-NH_3^+$, NH_2^+, $-N^+-$, *et seq.*, in the case of anion exchange membranes. The ions of the same charge as the groups on the membrane (coions) are excluded from the membrane phase by electrostatic repulsion. The degree of repulsion is dependent on the concentration of the electrolyte solution with which the membrane is in equilibrium. At low concentrations, the coions are almost absent in the membrane phase but as the concentration is increased the coions, with counterions (ions opposite in charge to the groups in the membrane) to maintain electroneutrality, enter the membrane phase. The number of coions in the membrane phase will be less than the number of counterions by an amount equal to the number of ionogenic groups in the membrane. In the case of nonion exchange membranes, the distinction between counter- and coions being absent, there will be an equal number of positive and negative ions in the membrane phase. The ability of ion exchange membranes to exclude coions makes them permselective.

In general, permselectivity is defined by

$$P_s = \frac{\bar{t}_+ - t_+}{1 - t_+} \tag{1}$$

where t_i is the transport number of the species i and the overbar refers to the membrane phase. The transport number of the counterion referred to the membrane phase can be determined directly in the usual way[6] by passing a known quantity of current through the membrane cell:

Solution (C)/Membrane/Solution (C)

and estimating the concentration change; or it can be derived by measurement of the membrane potential across the membrane when it separates two solutions of different concentrations C_1 and C_2. The relationship between the two transport numbers, i.e., $\bar{t}_+$ when $i \neq 0$ and when $i = 0$, is given by

$$\bar{t}_+ = \bar{t}_{+(\mathrm{app})} + 10^{-3} M m \bar{t}_w \tag{2}$$

where $\bar{t}_+$ is measured directly by passing current i, $\bar{t}_{+(\mathrm{app})}$ is derived from membrane potential measurement, M is the molecular weight of the solvent, m is the molality of the solution, and $\bar{t}_w$ is the transport number of water (moles of water transported for the passage of 1 faraday of current).

Equation (2) follows from equating the integrated form of Eq. (82) of Chapter 3 with the equation

$$E = 2\bar{t}_{+(\mathrm{app})} \frac{RT}{F} \ln \frac{a'}{a''} \tag{3}$$

where Eq. (3) gives the emf E of the membrane cell (57) of Chapter 3 in which anion-reversible Ag–AgCl electrodes are directly used to measure the membrane potential.

Ion exchange membranes, unlike glass membranes or other membranes used in the construction of electrodes, have low electrical resistance. This low resistance or high conductance is due to two factors: (1) the porosity of the membrane and (2) the high density of ionogenic groups. Membranes of low porosity and high charge density are well suited for the construction of electrodes.

B. SOME EXPERIMENTAL RESULTS

The theoretical aspects of concentration and bi-ionic potentials arising across ionexchange membranes have been outlined in Chapter 3. For further details, the monograph by Lakshminarayanaiah[6] may be consulted.

In a preliminary note published nearly three decades ago, Sollner[15] used collodion and protamine-doped collodion membranes as electrodes to estimate the concentrations of cations and anions, respectively. Later Gregor and Sollner[16] extended this work using better membranes. About the same time, Wyllie and Kanaan[17] used a variety of ion exchange membranes and measured the bi-ionic potentials (BIP) arising across them. The BIP was expressed as

$$E_{\mathrm{BIP}} = \frac{RT}{F} \ln \frac{a_i \bar{u}_i}{a_j \bar{u}_j} \tag{4}$$

This equation follows from Eq. (86) of Chapter 3, provided $\bar{\gamma}_i = \bar{\gamma}_j$ and the diffusion coefficients are replaced by mobilities. Wyllie[18] expressed the intramembrane mobility ratio as

$$\frac{\bar{u}_i}{\bar{u}_j} = \frac{\bar{t}_i}{\bar{t}_j} = \frac{\bar{m}_i}{\bar{m}_j} \frac{\bar{k}_i}{\bar{k}_j} \tag{5}$$

where $\bar{t}_i/\bar{t}_j$ is the intramembrane transference ratio, the $\bar{m}$'s are the steady state equilibrium concentrations of i and j in the junction zone, and $\bar{k}_i$ is the conductivity of the membrane when it is wholly in i form and $\bar{k}_j$ is the conductivity of the membrane when it is wholly in j-form. Furthermore, it was shown that the selectivity $K_{ji} \approx \bar{m}_i/\bar{m}_j$. Substituting this relation into Eq. (5) gives

$$\bar{u}_i/\bar{u}_j = K_{ji}\left(\bar{k}_i/\bar{k}_j\right) \tag{6}$$

Thus the ratio of the mobilities is related to the chemical and electrical properties of the membrane.

These equations were tested by Wyllie and Kanaan[17] who derived values for the intramembrane mobility ratio $\bar{u}_i/\bar{u}_j$ by BIP measurements, using a graphical procedure. The potentials were measured keeping the concentration of the electrolyte iX constant at 0.01m and by varying the molality of the electrolyte jX from 0.01 to 4.0m, and again by keeping jX constant at 0.01m and varying iX from 0.01 to 4.0m. The BIPs for these two sets of measurements were plotted against the logarithm of the mean molal activities. Excellent straight lines were obtained, which were extended to cut the activity axis at zero potential. The two sets of a_i, a_j values for which E was zero were thus obtained. According to Eq. (4), the ratio a_i/a_j was equal to $\bar{u}_j/\bar{u}_i$. For the membranes used, K_{ji} values were known and thus $\bar{k}_i/\bar{k}_j$ could be calculated from Eq. (6). These calculated values agreed with those derived by direct measurements of $\bar{k}_i$ and $\bar{k}_j$.

Similarly, Bergsma and Staverman[19] used a variety of membranes and four combinations of ions, H^+–Na^+, K^+–Na^+, H^+–Ag^+, and Na^+–Ag^+, and measured both BIPs and transport numbers. The BIP equation (86) of

Chapter 3 can be written as

$$E_{\text{BIP}} = \frac{RT}{F} \ln \frac{\bar{t}_i}{\bar{t}_j} \tag{7}$$

where

$$\frac{\bar{t}_i}{\bar{t}_j} = \frac{\bar{D}_i \bar{C}_i}{\bar{D}_j \bar{C}_j} = \frac{\bar{D}_i \bar{a}_i \bar{\gamma}_j}{\bar{D}_j \bar{a}_j \bar{\gamma}_i}$$

But according to the Donnan relation, $\bar{a}_i/\bar{a}_j = a_i/a_j$. Thus we have the relation

$$\frac{\bar{t}_i}{\bar{t}_j} = \frac{\bar{D}_i a_i \bar{\gamma}_j}{\bar{D}_j a_j \bar{\gamma}_i} \tag{8}$$

The ratio $\bar{t}_i/\bar{t}_j$ determined by BIP measurement and by direct measurement gave in some cases discrepancies which were attributed to the transport of coions and water.

Andelman and Gregor[20] found in the case of an anion-selective membrane (Nalfilm, Nalco Chemical Co.) that the values of $K_{ji}(i = \text{I}^-, j = \text{Cl}^-)$ varied with the composition of the membrane phase: When the mole fraction of $\text{I}(\bar{x}_{\text{I}})$ was between 0.6 and 1.0, $\ln K_{\text{Cl, I}} = 2.63$; when $\bar{x}_{\text{I}} < 0.6$, $\ln K_{\text{Cl, I}} = 1.5(\bar{x}_{\text{Cl}})^2 + 2.23$. Accordingly, Eq. (86) of Chapter 3 has to be modified to correct for the variation of K_{ji}. The modified equation is

$$\begin{aligned}
\frac{FE}{RT} &= \ln \frac{\bar{D}_j a_j'}{\bar{D}_i a_i''} K_{ji} - \int_{\text{I}}^{\text{II}} \left(\bar{t}_j \, d \ln \bar{\gamma}_j + \bar{t}_i \, d \ln \bar{\gamma}_i\right) \\
&= \ln \frac{\bar{D}_j a_j'}{\bar{D}_i a_i''} K_{ji} + \int_{\text{I}}^{\text{II}} \bar{t}_j \, d \ln K_{ji}
\end{aligned} \tag{9}$$

When the integral is ignored, values of E corresponding to $\bar{x}_{\text{Cl}} = 0$, 0.5, and 1.0 are calculated to be 27.8, 27.8, and 56.3 mV, respectively whereas the observed value is 33.5 mV. The integral term may be written for Cl and I interchange as

$$\int_{\text{I}}^{\text{II}} \frac{\bar{x}_{\text{Cl}}}{\bar{x}_{\text{Cl}} + (\bar{u}_{\text{I}}/\bar{u}_{\text{Cl}})(1 - \bar{x}_{\text{Cl}})} \, d \ln K_{\text{Cl, I}}$$

This analytical function relating $\bar{x}_{\text{Cl}}$ to $K_{\text{Cl, I}}$ was used to evaluate the integral; the value obtained was -0.83. Accordingly, Eq. (9) predicts a value of 34.9 mV for the BIP, i.e., BIP $= (RT/F)(2.16 - 0.83)$. This value agreed with the measured value of 33.5.

For bromobenzene liquid membrane (i.e., bromobenzene saturating a Millipore filter), Ilani[21] found good agreement between observed and calculated BIPs when the variation of the selectivity constant with the membrane composition was taken into account.

The concentration dependence of BIPs in collodion and modified collodion membranes has been studied by Takeguchi and Nakagaki.[22] On the assumption that the common ion concentration in the membrane is constant, they derived the values for the ratio of ionic mobilities of the two counterions or two coions in the membrane for 1 : 1 electrolytes. These were similar to the values found in bulk solutions. On the other hand, the ratios of ionic mobilities of counter- and coions were different from the corresponding values observed in bulk solutions. The mobility ratios of ion pairs Cl–Br, Br–I, and I–NO_3 and individual ion activities in aqueous and aqueous propanol media have been measured using ion exchange membranes.[23]

Both concentration and bi-ionic potentials have been measured by a number of investigators using a variety of membrane systems. Some unusual systems using stearate have been constructed and used in the study of bi-ionic potentials[24–26] to follow the effects of Na and K ions on the lipoid aggregates of the membrane structure. Similarly, the selective behavior of synthetic sulfonic and phosphonic acid membranes toward Na and K ions has been deduced from measurements of BIPs.[27] The potentials have been correlated with the relative mobilities of ions in the membrane and their selectivity coefficients, which were chemically determined. The poly(ethylene–styrene) graft copolymer membrane containing sulfonic acid groups showed selectivity to K over Na at pH 5 and 13, whereas the graft copolymer membrane containing phosphonic acid groups showed selectivity to Na over K. However, the sulfonic acid membrane preferred Na ions over H^+ ions.

The electrolyte pairs KCl–NaCl, KCl–LiCl, and NaCl–LiCl have been used with oxidized collodion membranes by Toyoshima and Nozaki[28] to study bi-ionic potentials. Of the three oxidized collodion membranes used, only one membrane was nonselective; the other two membranes were selective to K^+ ions in the order $K^+ > Na^+ > Li^+$. Similarly, Tombalakian[29] has made measurements of both bi-ionic potentials and the interchange of fluxes across a polystyrene sulfonic acid membrane using combinations of K, Na, Li, and H ions. From the experimental data, values for ionic mobility ratios and for single ion diffusion and interdiffusion coefficients of interchanging cations have been derived. These studies have been extended to another cation exchange membrane, polyethylene membrane containing a sulfonic acid polyelectrolyte, and other divalent ion pairs such as Ca–Ba, Ni–Ba, Cd–Ba, Co–Ba, and Cu–Ba.[30] Other ion exchange membranes containing strong acid or stong base groups have been used in bi-ionic potential measurements.[31, 32] The potentials have been related to selectivity coefficients and mobility ratios of the ions concerned. The relative transport of ions was found to follow the order Li < Na < NH_4 < K in cation exchange membranes and acetate < iodate < benzene sulfonate < Cl < NO_3 < Br < CNS < I in anion exchange membranes. Bipolar membranes (cation exchange and anion exchange membranes cemented together) also have been used in these potential measurements.[33]

A review of the various theories of membrane potential and an assessment of the Scatchard equation (82) (see Chapter 3) as applied to the behavior of three membranes of widely differing fixed charge density $\bar{X}$, namely cross-linked polymethacrylic acid ($\bar{X} \approx 3m$), sulfonated phenol–formaldehyde ($\bar{X} \approx 1m$), and untreated collodion ($\bar{X} \approx 10^{-3}m$), have been given by Lakshminarayanaiah.[34] Numerical integration of Eq. (82) of Chapter 3 using experimentally determined, unambiguous values of $\bar{t}_+$ and $\bar{t}_w$ at various external molalities gave values for E which agreed within 1 mV with the measured values. Similar agreement has been noted by Dawson and Meares[35] and by Gunn and Curran[36] who used commercially available ion exchange membranes. An integrated form of Eq. (82) of Chapter 3 has been used by Botre *et al.*[26] to derive values for $\bar{t}_+$ in gelatin-supported stearate membranes.

The mobilities and activity coefficients of small ions have been determined for collodion-based polystyrene sulfonic acid membranes in aqueous KCl solutions of different concentrations.[37] The activity coefficients were calculated from the Donnan relation by analyzing the amounts of co- and counterions present in the membrane phase. These data in combination with those of membrane potential and ion permeabilities were used to determine the mobilities through the use of a function which interrelated them. This type of study was extended to other salt solutions.[38] In every case, the coion mobility was identical with that in the bulk solution in the whole range of concentrations studied, while that of the counterions decreased very much with decrease of the external salt solution.

A very interesting study of membrane potential using uni- and multivalent ions has been presented by Yamauchi and Kimizuka.[39] For a z_M–z_X electrolyte whose solutions bound an ion exchange membrane, the potential E is given by

$$E = \frac{RT}{F}\left(\frac{t_M}{z_M}\ln\frac{a_{Mo}}{a_{Mi}} + \frac{t_X}{z_X}\ln\frac{a_{Xo}}{a_{Xi}}\right) \tag{10}$$

where the a's are the activities of ions concerned in outside (o) and inside

(i) solutions. Using activities of the electrolyte, Eq. (10) becomes

$$E = \frac{RT}{F}\left(\frac{t_M}{z_M} + \frac{t_X}{z_X}\right)\ln\frac{a_{(M-X)o}}{a_{(M-X)i}} \tag{11}$$

In the case of cation exchange membranes for which $t_X = 0$ and using solutions of $MgSO_4$ or $CaCl_2$, it was shown that an equation E (mV) $= 29 \log(a_{\pm(o)}/a_{\pm(i)})$ was followed. For bi-ionic systems, i.e., external solution z_M–z_X electrolyte and internal solution z_N–z_Y electrolyte, it was shown that the membrane potentials were given by

$$E_{XY} = \frac{2RT}{(z_X + z_Y)F}\ln\frac{(z_Y P_Y a_Y)_o}{(z_X P_X a_X)_i} \quad \text{for an anion exchanger} \tag{12}$$

$$E_{MN} = \frac{2RT}{(z_M + z_N)F}\ln\frac{(z_N P_N a_N)_o}{(z_M P_M a_M)_i} \quad \text{for a cation exchanger} \tag{13}$$

For the case where the concentration of the univalent ion N was varied holding the concentration of the multivalent ion M constant, the experimental results followed the equation

$$E_{MN} = A + \frac{118}{z_M + z_N}\log a_{\pm(o)} \tag{14}$$

where

$$A = \frac{118}{z_M + z_N}\log\frac{z_N P_N (f_N/f_\pm)}{(z_M P_M a_M)_i} \tag{15}$$

In Eq. (13) P_N/P_M is assumed to be constant, f_N is the activity coefficient of ion N, and $f_\pm$ is the mean activity coefficient of a 1 : 1 electrolyte. According to Eq. (14) slopes of 40 mV/p*a* for 2 : 1 systems and of 29 mV/p*a* for 3 : 1 systems were found.

A solution of tetraheptyl ammonium bromide in ethyl bromide used as a liquid membrane (M) in the cell:

$$H_2(1\text{ atm})|HBr(a_1)||M||HBr(a_2)|H_2(1\text{ atm})$$

gave Nernstian behavior.[40] The membrane, as expected of an anion exchanger, was permselective to Br^- ions. Electromotive forces of similar electrochemical cells in which cation exchange membranes and solutions of $MgCl_2$ and $BaCl_2$ are used have been measured.[41]

Danesi *et al.*[42] used benzene solutions of NO_3^-, Cl^-, and Br^- salts of tetraheptyl ammonium interposed between two aqueous electrolyte solutions. Bi-ionic potentials measured with NO_3–Cl, NO_3–Br, and Cl–Br couples have been quantitatively correlated with ion exchange constants, ion pair formation constants, and ionic mobilities. Mono-ionic concentration potentials followed the usual Nernst relation. Also the dependence of the selectivity constant on the concentration of a liquid membrane (benzene solution of tetraheptyl ammonium nitrate) has been evaluated.[43] Similarly, Shean and Sollner[44] have measured bi-ionic potentials arising in cells of type

$$A_+ \, L_- \, (C_1) \| \text{Liquid membrane of high selectivity} \| A_+ \, M_- \, (C_1)$$

where the liquid membrane was a solution of trioctylpropyl ammonium salt in *o*-dichlorobenzene. The BIP arising with any pair of critical ions (e.g., L_- and M_-) is independent not only of the activity of the electrolyte solution but also of the degree of loading of the membrane with the ion exchanger compound. The BIPs followed in a series of cells were algebraically additive in the sense that $BIP_{L_-/M_-} + BIP_{M_-/N_-} = BIP_{L_-/N_-}$.

Membrane potentials arising across a parchment-supported AgI precipitate membrane separating different concentrations of the same electrolyte ($BaCl_2$, $CaCl_2$, or $MgCl_2$) have been measured.[45] Cobalt ferrocyanide membrane has also been used in the measurements.[46] A new type of flow cell has been used to measure the membrane potentials.[47] In order to follow the formation of an ion barrier by precipitation of $BaSO_4$ in porous cellophane membrane, Hirsch-Ayalon[48] monitored the development of the membrane potential with time when the porous membrane separated solutions of $Ba(OH)_2$ and H_2SO_4. A sudden rise in potential indicated the creation of an ion barrier. The factors that control counterdiffusion of ions leading to the development of the barrier have been delineated. Similarly, the preparation and characterization of a number of membrane electrodes ($BaSO_4$ incorporated in parchment paper, polyvinyl chloride–tricresyl phosphate membrane, Ag_2S ceramic membrane) subject to an imposed electric field have been discussed.[49] The responses of parchment-supported $BaSO_4$ membrane to solutions of different ionic strength and pH have been evaluated and found to be selective to H^+ ions without interference from Na^+ ions.[50]

C. PROPERTIES AND APPLICATIONS OF ASYMMETRIC MEMBRANE ELECTRODES

Membranes that have a gradient of fixed charge running through different layers of thickness (asymmetric membrane) have been prepared and their properties have been studied.[51] When an asymmetric membrane made from collodion–polystyrene sulfonic acid and containing a gradient of fixed charge density (i.e., its two faces f_i and f_o have a fixed charge density of 5×10^{-4} and 5×10^{-1} equiv/kg) separated the same KCl or NaCl solution (10^{-3} *M*), potentials of the order of 50–70 mV were observed. These decreased as the concentration was increased. Under similar conditions, uniformly charged membranes gave zero potential. The asymmetry potential was explained in terms of a model in which two membranes, one of high charge density (M_H—highly selective to cations) and the other of low charge density (M_L—low selectivity to cations), were considered to hold between them a high concentration of counterions associated with the polyelectrolyte (polystyrene sulfonic acid). According to this model,[52] which can be represented by the membrane cell

Saturated calomel electrode	Reference saline solution a_1	Membrane of high charge M_H	Solution of speci-men a_M	Membrane of low charge M_L	Reference saline solution a_2	Saturated calomel electrode

(16)

the asymmetry potential arose as a result of the differences in the selectivities of the two membranes. Expressing selectivities in terms of counterion transport numbers, Lakshminarayanaiah and Siddiqi[53] showed that the emf E across the composite membrane system was given by

$$E = \frac{RT}{F}\left[(2\bar{t}_{+(H)} - 1)\ln\frac{a_M}{a_1} - (2\bar{t}_{+(L)} - 1)\ln\frac{a_M}{a_2}\right] \tag{17}$$

where $\bar{t}_{+(H)}$ and $\bar{t}_{+(L)}$ are the transport numbers of counterions in high and low charge density membranes, respectively, a_M is the activity of the counterions existing between the membranes, and a_1 and a_2 are the activities of the two solutions contacting the membranes M_H and M_L. When $a_1 = a_2 = a$, the asymmetry potential given by Eq. (17) becomes

$$E = \frac{2RT}{F}(\bar{t}_{+(H)} - \bar{t}_{+(L)})\ln\frac{a_M}{a} \tag{18}$$

The magnitude of E has been shown[53] to be controlled more by the value of a_M than by the factor $\bar{t}_{+(H)} - \bar{t}_{+(L)}$. If a_M changed, the value of E would change. As a result Lakshminarayanaiah and Siddiqi[54] were unable to notice steady asymmetry potentials when the composite membrane system was subject to long-term (about 30 hr) equilibration due to the fact that the value of a_M changed with time. Despite this the three-chambered cell (16) has been used as an indicator electrode[55] to follow the course of action of biologically important compounds and biopolymers.[56] The reference solution (NaCl or KCl, 1×10^{-2} M) of known and constant concentration ($a_1 = a_2 = a$) was placed in the two external half-cells. The central compartment contained the solution of a colloidal electrolyte or a biopolymer that could be stepwise diluted, concentrated, heated, or titrated with another solution.

A precise calibration curve was obtained by maintaining the two external solutions constant and equal at 0.01 M and varying the concentration of NaCl or KCl in the central compartment of the cell. The emf was recorded at each step. The values agreed with those calculated according to Eq. (18). If the solution in the central compartment was replaced by a solution of sodium lauryl sulfate and its concentration was progressively changed, it was possible to follow the transitions. The emf of the cell was followed with different concentrations of sodium lauryl sulfate. On plotting the emf versus the logarithm of the molarity of sodium lauryl sulfate, a break in the straight line marked the critical micelle concentration (CMC). The value so derived agreed with the value obtained by the conductance method. Structural changes following temperature changes can also be studied by this method. Botre and co-workers used this technique to follow the mechanism of action of antihistaminics and cortisol on bovine serum albumin[57] and interactions of *E. coli* cells with poly-DL-ornithine and other antibiotics.[58, 59]

Using other anion and/or cation exchange membranes, concentration cells of the type[60, 61]

Calomel	3 *M* NH_4NO_3 – agar bridge	SDS C	Cation or anion exchange membrane	SDS 1 m*M*	3 *M* NH_4NO_3 – agar bridge	Calomel

where SDS is sodium dodecyl sulfate solution, or of the type[62]

Reference electrode	Reference solution I	A_m	Test solution II SDS	C_m	Reference solution I	Reference electrode

where test solution II contains SDS or cetylpyridinium bromide bounded by an anion exchanger membrane A_m and a cation exchange membrane C_m, have been used to determine the CMC of surfactants. In the first type of cell, SDS (1 m*M*) acted as its own reference solution and the potential was given by $E = (RT/F)\ln(1/C)$; whereas in the second type of cell, the potential was given by $E = (2RT/F)\ln(a_I/a_{II})$. At concentrations below CMC levels, a Nernstian response was observed. When the concentrations exceeded CMC levels, E deviated from Nernstian behavior. However, these procedures, which could be used to estimate the activity of surfactants, have been questioned by Birch and Clarke[63] who, using the second type of cell, found that the membranes lost their permselectivity on prolonged soaking and thus gave inconsistent values for the surfactant activity. Unless these results are confirmed with other membranes of different degrees of cross-linking and water content, the findings of Birch and Clarke[63] cannot be considered to be generally true for all ion exchange membranes.

The problems of specificity and the mechanism of ion transfer in ion exchange membranes have been considered by Shults.[64] In a number of studies, simple membrane electrodes made from organic or inorganic ion exchangers and showing little specificity to any particular ion have been used in the estimation of ion activities[65–75] and in the determination of a second dissociation constant of H_2SO_4.[76] They have been used to follow the mobility ratios of bi- and univalent ions.[77, 78] Ion exchange membranes have been used as a preconcentration step[79] in the analysis of certain trace metal ions by electrochemical and neutron activation techniques. They have also been used in the study of solvation of cations.[80] Here the change in membrane potential following solvent change has been correlated with the solvation of cations.

Potentiometric titrations involving two univalent ions[81] or a uni- and a bivalent or a uni- and a tervalent ion[82] have been carried out by using ion exchange membrane electrodes. Similarly, they have been used to follow acid–base titrations.[83, 84] Polystyrene-based cation and anion exchange membranes have been used in electrometric titrations of Li, Na, K, Cu(II), and Mn(II) in the case of cation exchangers and of Cl, NO_3, and SO_4 in the case of anion exchangers.[85] The various experimental parameters involved in these titrations and their effects on the shapes of potentiometric curves have been discussed by Ijsseling and Van Dalen.[86] Interfering ions, precipitation reactions., etc., generally lead to errors in the detection of end points. Different aspects of this problem related to use of ion-selective membrane electrodes have been considered by a number of investigators.[87–90] Estimations using titration procedures employ the technique of either standard addition to a sample or sample addition to a standard. To hasten these operations a nomograph has been presented.[91]

In acid–base titrations, following changes in conductance of the solutions become very difficult when the solutions are highly conducting. This problem has been solved by interposing a polyvinyl chloride membrane (containing Alassion CS and dioctylphthalate) between the two membrane electrodes. The end point is marked by a sudden decrease in conductance.[92]

A strong acid cation exchange membrane (Permaplex) has been used[93] to measure the concentration of HCl, H_2SO_4, and HNO_3 over the concentration range 10^{-3}–7.0 *M*. HF interfered when its concentration approached that of the other strong acid. The electrode showed selectivity to H^+ ions over cations such as Cr^{3+}, Fe^{3+}, Ni^{2+}, Ag^+, and K^+. The membrane electrode could be used for direct measurement of acidity in stainless steel pickling baths.

REFERENCES

1. H. T. Clarke and D. Nachmansohn (eds.), "Ion Transport across Membranes." Academic Press, New York, 1954.
2. T. Shedlovsky (ed.), "Electrochemistry in Biology and Medicine." Wiley, New York, 1955.

3. J. R. Wilson (ed.), "Demineralization by Electrodialysis." Butterworths, London and Washington, D.C., 1960.
4. F. Helfferich, "Ion Exchange." McGraw-Hill, New York, 1962.
5. S. B. Tuwiner, "Diffusion and Membrane Technology." Van Nostrand-Reinhold, Princeton, New Jersey, 1962.
6. N. Lakshminarayanaiah, "Transport Phenomena in Membranes." Academic Press, New York, 1969.
7. G. Eisenman (ed.), "Membranes," Vol. 1. Dekker, New York, 1972.
8. F. Bergsma and C. A. Kruissink, *Fortschr. Hochpolym. Forsch.* **2**, 307 (1961).
9. N. Lakshminarayanaiah, *Chem. Rev.* **65**, 491 (1965).
10. S. R. Caplan and D. C. Mikulecky, *in* "Ion Exchange" (J. A. Marinsky, ed.), Vol. 1, p. 1. Dekker, New York, 1966.
11. K. Sollner, *J. Macromol. Sci.* **A3**, 1 (1969); *in* "Intestinal Absorption Metal Ions, Trace Elements and Radionuclides" (S. C. Skoryna and D. Waldron-Edward, eds.), p. 21. Pergamon, Oxford, 1970; *in* "Diffusion Processes" (J. N. Sherwood, A. V. Chadwick, W. M. Muir, and F. L. Swinton, eds.), p. 655. Gordon and Breach, New York, 1971.
12. N. Lakshminarayanaiah, *in* "Electrochemistry." The Chemical Society, London. (G. J. Hills, ed.), Vol. 2, p. 203, 1972; (H. R. Thirsk, ed.), Vol. 4, p. 167, 1974.
13. L. Michaelis, *J. Gen. Physiol.* **8**, 33 (1925); *Kolloid-Z.* **62**, 2 (1933).
14. M. R. J. Wyllie and H. W. Patnode, *J. Phys. Chem.* **54**, 204 (1950).
15. K. Sollner, *J. Amer. Chem. Soc.* **65**, 2260 (1943).
16. H. P. Gregor and K. Sollner, *J. Phys. Chem.* **58**, 409 (1954).
17. M. R. J. Wyllie and S. L. Kanaan, *J. Phys. Chem.* **58**, 73 (1954).
18. M. R. J. Wyllie, *J. Phys. Chem.* **58**, 67 (1954).
19. F. Bergsma and A. J. Staverman, *Discuss. Faraday Soc.* **21**, 61 (1956).
20. J. B. Andelman and H. P. Gregor, *Electrochim. Acta* **11**, 869 (1966).
21. A. Ilani, *Biophys. J.* **6**, 329 (1966).
22. N. Takeguchi and M. Nakagaki, *Biochim. Biophys. Acta* **233**, 753 (1971).
23. M. Adhikari and G. G. Biswas, *Indian J. Chem.* **10**, 209 (1972).
24. C. Botre and W. Dorst, *Farmaco. Ed. Sci.* **24**, 373 (1969).
25. C. Botre, M. Mascini, A. Memoli, and M. Marchetti, *Farmaco. Ed. Sci.* **24**, 873 (1969).
26. C. Botre, W. Dorst, M. Marchetti, and A. Memoli, *Biochim. Biophys. Acta* **193**, 333 (1969).
27. D. K. Hale and K. P. Govindan, *J. Electrochem. Soc.* **116**, 1373 (1969).
28. Y. Toyoshima and H. Nozaki, *J. Phys. Chem.* **74**, 2704 (1970).
29. A. S. Tombalakian, *Can. J. Chem. Eng.* **50**, 203 (1972).
30. A. S. Tombalakian and G. K. Markarian, *Can. J. Chem. Eng.* **51**, 124 (1973).
31. R. Natarajan and M. S. Rajawat, *Indian J. Technol.* **8**, 76B (1970).
32. K. P. Govindan, *Indian J. Technol.* **7**, 274 (1969).
33. H. Kawabe and M. Yanagita, *Bull. Chem. Soc. Japan* **42**, 1029 (1969).
34. N. Lakshminarayanaiah, *in Proc. Conf. Natur. Synthetic Membranes* (C. Saravis, K. Gershengorn, and M. E. Brown, eds.), p. 125. NIH, Bethesda, Maryland, 1969.
35. D. G. Dawson and P. Meares, *J. Colloid Interface Sci.* **33**, 117 (1970).
36. R. B. Gunn and P. F. Curran, *Biophys. J.* **11**, 559 (1971).
37. N. Kamo, T. Toyashima, H. Nozaki, and Y. Kobatake, *Kolloid-Z.* **248**, 914 (1971).
38. T. Ueda, N. Kamo, N. Ishida, and Y. Kobatake, *J. Phys. Chem.* **76**, 2447 (1972).
39. A. Yamauchi and H. Kimizuka, *J. Theoret. Biol.* **30**, 285 (1971).
40. R. Galli and T. Mussini, *Nature (London)* **223**, 179 (1969).
41. Z. S. Alagova, O. K. Stefanova, and E. A. Materova, *Electrokhimiya* **5**, 1116 (1969).
42. P. R. Danesi, F. Salvemini, G. Scibona, and B. Scuppa, *J. Phys. Chem.* **75**, 554 (1971).
43. P. R. Danesi, G. Scibona, and B. Scuppa, *Anal. Chem.* **43**, 1892 (1971).

44. G. Shean and K. Sollner, *J. Membrane Biol.* **9**, 297 (1972).
45. M. A. Beg and S. Pratap, *J. Electroanal. Chem. Interfacial Electrochem.* **36**, 349 (1972).
46. K. K. Panday and V. K. Agrawal, *J. Indian Chem. Soc.* **48**, 775 (1971).
47. J. C. T. Kwak, *Desalination* **11**, 61 (1972).
48. P. Hirsch-Ayalon, *J. Membrane Biol.* **12**, 349 (1973).
49. C. Liteanu, I. C. Popescu, and E. Hopirtean, *in* "Ion Selective Electrodes" (E. Pungor, ed.), p. 51. Akadémiai Kiadó, Budapest, 1973.
50. C. Liteanu and I. C. Popescu, *Talanta* **19**, 974 (1972).
51. A. M. Liquori and C. Botre, *Ric. Sci.* **34** (6), 71 (1964).
52. A. M. Liquori and C. Botre, *J. Phys. Chem.* **71**, 3765 (1967).
53. N. Lakshminarayanaiah and F. A. Siddiqi, *Biophys. J.* **11**, 617 (1971).
54. N. Lakshminarayanaiah and F. A. Siddiqi, *in* "Membrane Processes in Industry and Biomedicine" (M. Bier, ed.), p. 301. Plenum Press, New York, 1971.
55. C. Botre, S. Borghi, and M. Marchetti, *Biochim. Biophys. Acta* **135**, 208 (1967).
56. C. Botre, S. Borghi, M. Marchetti, and M. Baumann, *Biopolymers* **4**, 1046 (1966).
57. C. Botre, M. Marchetti, C. D. Vechio, G. Liometti, and A. Memoli, *J. Medicin. Chem.* **12**, 832 (1969).
58. C. Botre, M. Marchetti, S. Borghi, and A. Memoli, *Biochim. Biophys. Acta* **183**, 249 (1969).
59. C. Botre, A. Memoli, S. Borghi, and M. Benignetti, *Experentia Supple.* No. 18, 173 (1971).
60. K. Shirahama, *Kolloid-Z.* **250**, 620 (1972).
61. W. U. Malik, S. K. Srivastava, and D. Gupta, *J. Electroanal. Chem. Interfacial Electrochem.* **35**, 247 (1972).
62. C. Botre, D. G. Hall, and R. V. Scowen, *Kolloid-Z.* **250**, 900 (1972).
63. B. J. Birch and D. E. Clarke, *Anal. Chim. Acta* **69**, 473 (1974).
64. M. M. Shults, *Dokl. Akad. Nauk SSSR.* **194**, 377 (1970).
65. B. K. Pain and S. K. Mukherjee, *J. Indian Chem. Soc.* **46**, 341 (1969).
66. S. C. Ghosh, A. Sarkar, and S. K. Mukherjee, *J. Indian Chem. Soc.* **46**, 784 (1969).
67. M. Adhikari and D. Ghosh, *J. Indian Chem. Soc.* **47**, 384 (1970).
68. M. Adhikari and G. G. Biswas, *J. Indian Chem. Soc.* **47**, 399 (1970).
69. S. C. Ghosh and S. K. Mukherjee, *J. Indian Chem. Soc.* **47**, 467 (1970).
70. G. K. Pillai and D. R. Pandit, *J. Indian Chem. Soc.* **47**, 669 (1970).
71. R. Tamamushi and S. Sato, *Null. Chem. Soc. Japan* **43**, 3420 (1970).
72. V. S. Shterman, A. V. Gordievskii, E. L. Filippov, and S. V. Bruk, *Zh. Fiz. Khim.* **44**, 2059 (1970).
73. M. Adhikari and M. Paul, *J. Appl. Polym. Sci.* **14**, 2675 (1970).
74. M. Adhikari, M. Paul, and D. Gangopadhyay, *Indian J. Chem.* **8**, 79 (1970).
75. C. Liteanu and E. Hopirtean, *Rev. Roumaine Chim.* **15**, 749 (1970).
76. M. Adhikari, D. Ganguli, G. G. Biswas, and D. Ghosh, *J. Indian Chem. Soc.* **46**, 1131 (1969).
77. B. K. Pain and S. K. Mukherjee, *J. Indian Soc. Soil Sci.* **17**, 407 (1969).
78. S. C. Ghosh and S. K. Mukherjee, *J. Indian Chem. Soc.* **47**, 162 (1970).
79. U. Eisner and H. B. Mark, Jr., *Talanta* **16**, 27 (1969).
80. M. Adhikari and S. K. Mukherjee, *J. Indian Chem. Soc.* **47**, 109 (1970).
81. V. S. Shterman, N. A. Rozenkevich, A. V. Gordievskii, and E. L. Filippov, *Zh. Fiz. Khim.* **43**, 1552 (1969).
82. F. P. Ijsseling and E. Van Dalen, *Anal. Chim. Acta* **45**, 493 (1969).
83. R. Geyer and W. Erxlenben, *Z. Chem.* **9**, 237 (1967).
84. C. Liteanu and E. Hopirtean, *Talanta* **17**, 1067 (1970); *Rev. Roumaine Chim.* **15**, 1331 (1970).

85. R. Nagarajan and M. S. Rajawat, *Indian J. Technol.* **10**, 139 (1972).
86. F. P. Ijsseling and E. Van Dalen, *Anal. Chim. Acta* **45**, 121 (1969).
87. F. A. Schultz, *Anal. Chem.* **43**, 502, 1523 (1971).
88. P. W. Carr, *Anal. Chem.* **43**, 425 (1971); **44**, 452 (1972).
89. J. Buffle, N. Parthasarathy, and D. Monnier, *Anal. Chim. Acta* **59**, 427 (1972).
90. N. Parthasarathy, J. Buffle, and D. Monnier, *Anal. Chim. Acta* **59**, 447 (1972).
91. B. Karlberg, *Anal. Chem.* **43**, 1911 (1971).
92. C. Liteanu and L. G. Mirza, *Talanta* **19**, 980 (1972).
93. T. Eriksson and G. Johansen, *Anal. Chim. Acta* **63**, 445 (1973).

Chapter 5

ELECTRODES SELECTIVE TO HALIDE IONS

Membrane electrodes selective to halide ions contain either a heterogeneous membrane or a homogeneous membrane. In the case of the latter, the active material is a reasonably conducting single crystal whose conductance, if it is low, is increased by doping it with other compatible material. An example is the case of lanthanum trifluoride whose conductance is increased by doping it with europium. In the case of a heterogeneous membrane, a halide of silver is usually the active material; it is held in an inactive matrix such as paraffin, thermoplastic polymer material, or silicone rubber. Silver halides behave like solid electrolytes and the silver ion acts as the charge carrier moving through the crystal lattice by the Frenkel mechanism,[1] which is shown for silver bromide in Fig. 1.

The crystal, mechanically stable, chemically inert in the sample solution, and of low solubility, can be used in a thin section as the membrane electrode. The crystal membrane could be highly selective since conduction (see Fig. 1) occurs by a lattice defect mechanism and an ion of the right size, shape, and charge can only fit the vacancy in the crystal lattice. All other ions, since they cannot move through the lattice, do not contribute to the conduction process. Unlike other membranes, such as liquid membranes, the solid state membranes derive their selectivity by preventing ions other than the one to which they are selective from entering the membrane phase. Thus Nernstian behavior is always noted. Interferences arise only from chemical reactions occurring at the crystal surface. In these systems, the current is not carried by the ions that take part in the ion exchange reaction at the membrane surface. Thus no diffusion potential is generated within the membrane and the membrane potential is given by Eq. (179) of Chapter 3.

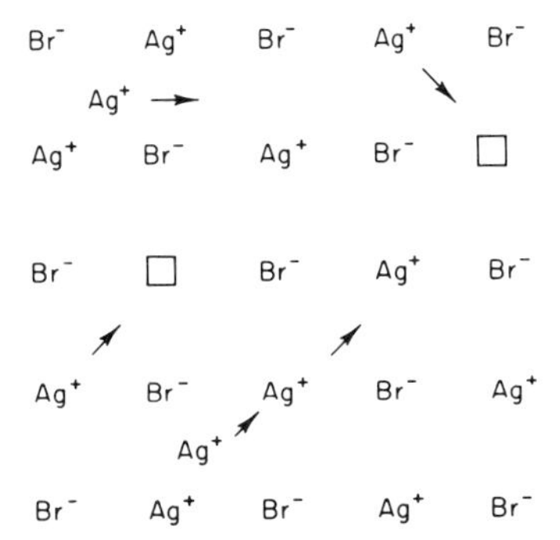

Fig. 1. A scheme for the transport of Ag^+ ions in the AgBr crystal lattice (Frenkel mechanism). The number of interstitial silver ions in the crystal lattice is the same as the number of unoccupied lattice positions.

A. PREPARATION OF THE MEMBRANE ELECTRODE AND FORMATION OF THE MEMBRANE CELL

Homogeneous solid state electrodes for fluoride, chloride, bromide, and iodide have been prepared and supplied by commercial firms.

The lanthanum trifluoride electrode has a membrane, as already mentioned, made from lanthanum fluoride single crystal and doped with europium.[2] Doping is employed to increase the membrane conductance. Fluorides of lanthanide series crystallize with a hexagonal structure (LaF_3-type lattice; e.g., La, Ce, Pr, Nd, Sm) or with an orthorhombic structure (YF_3-type lattice; e.g., Sm after heating at 700°C for 2 hr, Lu).[3–7] In the hexagonal lattice, each metal ion is surrounded by five fluoride ions and there are six other fluoride ions as the closest neighbors. The lattice consists of La(III) and F^- ions with a layer of fluoride ions on either side of the LaF^{2+} network. In this structure, F^- ions are mobile. In the orthorhombic structure, which is not suitable for use as an electrode, each metal ion is surrounded by eight F^- ions.

The search for a fluoride electrode started with the testing of insoluble fluorides such as calcium and barium fluorides. Bismuth trifluoride pressed at 50,000 psi and 550°C gave a disk which, sealed to a tube, acted as an electrode and gave results at various F^- concentrations following the Nernst equation up to pF 4. Cl^- ion interference was found.[8] Similarly, electrodes made from rare earth fluorides such as niobium, praseodymium, and cerium displayed Nernstian behavior up to pF 3.5. As opposed to these results, Russian workers[9] found the electrode behavior to be ideal in the range pF 0–3.

The lanthanum fluoride ion-selective electrodes are manufactured by the firms of Beckman,[10] Coleman,[11] Corning Instruments,[12] Foxboro,[13] Orion,[14] Philips,[15] and the Research Institute of Single Crystals, Turnov, Czechoslavakia (Crytur).[16]

Single crystals of AgCl and AgBr have been used by the "Monokrystaly" Research Institute of Single Crystals in Turnov, Czechoslovakia, in the production of homogeneous ion-selective membrane electrodes. Also membranes made of pure AgCl single crystals, annealed at 320°C in evacuated glass ampules, have been used in the preparation of selective chloride electrodes. Annealing improved the electrode properties in that the membrane resistance decreased and the potential response was Nernstian up to 10^{-5} *M* chloride.[17] Pressed polycrystalline iodide materials have been used in the construction of iodide electrodes.[18] Similar crystalline materials have been used by the firm of Philips[19] in the construction of AgCl, AgBr, and AgI electrodes, also.

The disadvantages of these compact AgCl, AgBr membranes are their resistance[20] and the need to work in constant light because of the development of photoelectric potentials. Although AgI is a good conductor, the material in pure form is difficult to form into a pellet since it undergoes several phase transitions on heating or application of pressure. A pellet formed by fusion or by pressing, on returning to room conditions, reverts to its original state. In so doing, it is fractured. To obviate these difficulties, the precipitates of silver halide (Cl, Br, and I) are incorporated into a silver sulfide matrix that is less soluble than the silver halides.[14] Coprecipitated silver halide and sulfide could be pressed in a die at 100,000 psi to form a membrane.[21] Similarly, HgS and Hg_2Cl_2 in the optimal composition range 30–70 M % (HgS) to 70–30 M % (Hg_2Cl_2) have been pressed and used as electrodes selective to Cl^- ions.[21a] These electrodes conduct well and generate potentials typical of pure silver halides. Besides Orion, other firms[10, 12, 22, 23] also produce these silver halide electrodes.

The so-called Selectrodes also use the precipitates of silver halides by themselves or as a mixture with silver sulfide. The silver halide by itself or mixed with silver sulfide is melted and a graphite rod is immersed into the melt for impregnation. After cooling to room temperature, the rod is rendered hydrophobic by treatment with carbon tetrachloride, benzene, or mesitylene.[24] In an improved method,[25] the halide is precipitated with the sulfide by treating a mixture of sodium halide and sodium sulfide with excess $AgNO_3$. The precipitate is washed and dried overnight. An aqueous suspension of the precipitate is rubbed on the surface of a porous graphite rod and dried at 200°C overnight. Finally it is rendered hydrophobic by treatment with a solvent or with Teflon. In these types of electrodes (see Fig. 2) graphite is directly connected to the potentiometer so that no inner reference solution and electrode are required.

An interesting variation of this method is to fill a plastic body with carbon paste (graphite powder) and Nujol or paraffin wax (5 : 1 w/v in the case of Nujol or 3 : 1 w/w in the case of paraffin) containing a mixture of silver halide–silver sulfide (1–30%)[26] or a liquid ion exchanger[27] containing the ion of interest (Aliquat in the ionic form desired).

In the preparation of heterogeneous solid state membrane electrodes, the silver halide precipitate is held in the matrix of a polymer. Paraffin has been used to hold the precipitate of AgI to form a membrane.[28] In later work, silicone rubber was used by Pungor and co-workers.[28–31] In the preparation of the membrane, a precipitate of silver halide is dispersed in polysiloxane using a laboratory mixer. Later a roller mill is used to homogenize and at this stage a cross-linking agent (a silane derivative) and a catalyst are added. The final shape of the membrane is determined by calendering. The quality of the membrane with respect to its function depends on the physical nature and the quantity of the precipitate incorporated into the membrane. The latter determines the manner in which the particles are held at the membrane surface. These selective membrane electrodes of AgCl, AgBr, and AgI are produced by the firm of Radelkis.[32] These electrodes require soaking for 1–2 hr before use in a dilute solution of the appropriate potassium halide.

The preparation and functioning of a heterogeneous fluoride membrane electrode have been described by Macdonald and Toth.[33] Thorium fluoride was incorporated into silicone rubber material (1 : 1 weight ratio) with cold polymerization. This membrane was not selective to fluoride ion. Thorium fluoride precipitated from 25–35% excess of thorium ion in the presence of *p*-ethoxy chrysoidine (which causes an increase in the specific volume of AgI precipitate[34]) was selective to F^- ions; however, the sensitivity to F^- ions was not good and the potentials observed were erratic. On the other hand, LaF_3 precipitated from NaF and 30% excess lanthanum acetate in the presence of *p*-ethoxy chrysoidine and incorporated into silicone rubber gave a membrane that was selective to F^- ions in the concentration range 10^{-2}–10^{-4} *M*. Below this range the sensitivity was poor. Similarly, calcium fluoride-containing membranes were also prepared and their response to F^- ions was better than the response of the thorium fluoride membranes. So far no heterogeneous membrane electrode selective to F^- ions and equal or superior in performance to the homogeneous membrane has been found.

The functional characteristics of these membranes are determined by the physical nature (e.g., specific volume of the precipitate incorporated) and the quality of the precipitate incorporated into the membrane. To ensure contact between particles to facilitate conduction, the correct ratio of active material to binder must be achieved. Other factors include particle size, adhesion, surface tension, and cross-linking of the polymeric material,

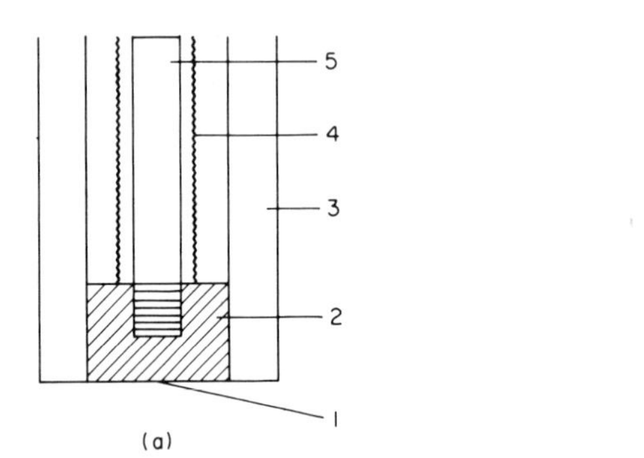

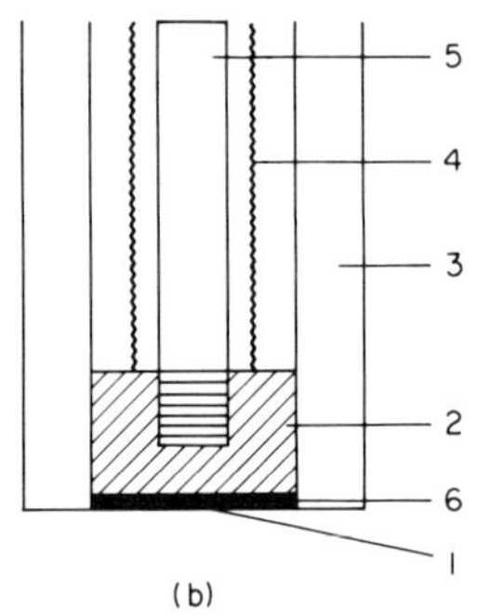

Fig. 2. The construction of a solid state Selectrode: (a) Basic shape: 1, sensitive surface; 2, cylinder pressed from graphite hydrophobized by Teflon; 3, Teflon tubing; 4, screening; and 5, stainless steel contact. (b) Activated solid state Selectrode: 1, surface enriched by the electroactive material 6.

especially its resistance to swelling in water.[8] Cured membranes (thickness 0.3–0.5 mm) are cut into circles and fixed on the end of a glass tube with some silicone rubber glue.

An alternative procedure is to use the method of Buchanan and Seago.[35] According to this method, powdered silver halide may be mixed with vulcanized silicone rubber (General Electric "Clear Seal") and pressed between a heavy polyethylene plate and a polyvinyl chloride foil to give a sheet 0.5 mm thick.[35] Also membrane electrodes of silver halide with the halide precipitate mixed with a thermoplastic polymer membrane have been prepared by molding.[36, 37] In this process any thermoplastic polymer can be used. Mascini and Liberti[36] used methacrylic esters and polythene. In these electrodes, conditioning of the electrode is very important[35] and soaking the electrode for a few hours in an appropriate solution is suggested. Electrodes made of silver halides in a polymeric matrix are produced by the firm of Coleman.[38]

The possible arrangement of a membrane cell using a membrane electrode is shown in Fig. 1 of Chapter 3. The reference electrode could be a standard calomel electrode (SCE) or an electrode of the second kind used as shown in the schemes:

$$\mathrm{Hg{-}Hg_2Cl_2}\ \Big|\ \text{Satd. KCl}\ \underset{E_L}{\Big\|}\ \text{Solution 1}\ \Big|\ \begin{matrix}\text{Membrane}\\ \mathrm{LaF_3}\end{matrix}\ \Big|\ 0.1\ M\ \mathrm{NaF}\ \Big\|\ \text{Satd. KCl}\ \Big|\ \mathrm{Hg_2Cl_2{-}Hg} \qquad (1)$$

$$\mathrm{Hg{-}Hg_2Cl_2}\ \Big|\ \text{Satd. KCl}\ \underset{E_L}{\Big\|}\ \text{Solution 1}\ \Big|\ \begin{matrix}\text{Membrane}\\ \mathrm{LaF_3}\end{matrix}\ \Big|\ 0.1\ M\ \mathrm{NaF};\ 0.1\ M\ \mathrm{NaCl}\ \Big|\ \mathrm{AgCl{-}Ag} \qquad (2)$$

where E_L is the liquid junction potential. Treating the whole complex of an ion-selective electrode as a single unit, schemes (1) and (2) may be written as

$$\mathrm{Hg{-}Hg_2Cl_2}\,\Big|\,\mathrm{Satd.\ KCl}\,\Big\|\,\mathrm{Solution\ 1}\,\Big|\,\mathrm{LaF_3\ electrode} \tag{3}$$

The emf of this cell is given by [see Eq. (121), Chapter 3]

$$E = E^\circ - (RT/\mathrm{F})\ln\left[a_\mathrm{F}(1) + K^\mathrm{pot}_{\mathrm{F,OH}}a_\mathrm{OH}\right] + E_\mathrm{L} - E_{\mathrm{Hg_2Cl_2{-}Hg\ (Satd.\ KCl)}} \tag{4}$$

A metal contact may be used in place of the internal electrolyte solution and the reference electrode. This type of solid contact with glass electrodes was established many years ago. In this case the internal wall is plated with metal and a metallic connection is soldered. Such glass electrodes give a Nernstian response to H^+ ions.[39, 40] Thermoplastic-based AgCl electrodes with an internal solid contact (silver foil) respond well to Cl^- ions.[37] Such electrodes with internal contacts made of silver are produced by the firm of Coleman.[41] The Selectrodes[24, 25] already referred to also have an internal solid contact.

B. ION ACTIVITY STANDARDS

Equation (4) contains single ion activity coefficients and the terms E_L and E°. If the values of the latter are known accurately, values for the single ion activity for a number of standard solutions could be determined. However, precise values for E_L and E° are difficult to assign[42–44] and so it becomes necessary to use various approximations or modifications.

In a number of experiments using ion-selective electrodes, solution 1 in cells (1)–(3) generally contains an indifferent electrolyte in concentrations that are probably higher than those of the other ions present in the solution. In such a case the activity coefficients of the ions, for example F^- and OH^-, to which the electrode is sensitive (LaF_3 membrane) become dependent on the concentration of the indifferent electrolyte and not on the concentrations of F^- or OH^- ions.[45] The liquid junction potential E_L depends mainly on the indifferent electrolyte concentration in solution 1. Equation (4) therefore can be written for a certain concentration of the indifferent electrolyte as

$$E = E^\circ - (RT/F)\ln\gamma_\mathrm{F} + E_\mathrm{L} - E_{\mathrm{Hg_2Cl_2{-}Hg(Satd.\ KCl)}} - (RT/F)\ln\left(m_\mathrm{F} + K^\mathrm{pot}_{m_\mathrm{F,OH}}m_\mathrm{OH}\right) \tag{5}$$

where

$$K^{\mathrm{pot}}_{m_{\mathrm{F,OH}}} = K^{\mathrm{pot}}_{\mathrm{F,OH}}(\gamma_{\mathrm{OH}}/\gamma_{\mathrm{F}})$$

The first four terms on the right-hand side of Eq. (5) are constants at a given indifferent electrolyte concentration and can be determined for any given ion-selective electrode using a standard solution of known fluoride concentration such that $m_{\mathrm{F}} \gg K^{\mathrm{pot}}_{m_{\mathrm{F,OH}}} m_{\mathrm{OH}}$. Equation (4) is frequently used in cases where there is no excess of indifferent electrolyte. Then the data obtained are subject to errors due to variation of the activity coefficient and the liquid junction potential. A solution to the problem of single ion activity coefficients is to use some convention with regard to the scale of ion activities as done in the case of pH measurements.[44, 46–48]

There are three conventions[49] that could be used:

(1) The pH convention of Bates and Guggenheim[50] who used the Debye–Hückel relation

$$-\log \gamma_{\mathrm{Cl}} = \frac{AI^{1/2}}{1 + 1.5I^{1/2}} \tag{6}$$

where $A = 0.512$ mole$^{-1/2}$ kg$^{1/2}$ at 25°C, to obtain the activity coefficient of the Cl^- ion. Consequently, γ_{Na} is defined since the mean activity coefficient $\gamma_\pm$ of NaCl is obtained by a number of thermodynamic methods. In a similar way, values for γ_{Ca}, γ_{K}, etc., in their chloride solutions can be obtained.

(2) The MacInnes convention,[51] according to which the activity coefficients of K and Cl ions are equated to the mean activity coefficient of KCl in a solution of equivalent ionic strength. This convention has been advocated by Garrels[52] for establishing the activity standards for ion-selective electrodes.

(3) The valence relations of Debye–Hückel theory (see Chapter 2) are applied to separate the mean activity coefficient of the electrolyte into the individual ionic contributions. This leads to the relation $\gamma_+^4 = \gamma_-^4 = \gamma_{2+} = \gamma_{2-}$ where the subscripts indicate the ionic charge.

Following these conventions, Bates and Alfenaar[44] derived values for pM ($-\log a_{\mathrm{M}}$) or pX, where M is a cation and X is an anion, for electrolytes of the same charge type. The values so derived are very close to one another at ionic strengths of 0.1 and less, but large divergences are found in concentrated solutions.[44, 49] In order to find close agreement between pM or pX values derived by these conventions in concentrated solutions, the hydration theory of strong electrolytes developed by Stokes and Robinson[53] (see Chapter 2) taking the ion–solvent interaction into account has been used by Bates *et al.*[54] According to the hydration theory, the activity coefficient $\gamma_\pm$ of an electrolyte in a solution of molality m is given by Eq.

(118) of Chapter 2, i.e.,

$$\log \gamma_{\pm} = |z_+ z_-| \log f_{DH} - (h/\nu)\log a_1 - \log[1 + 0.018m(\nu - h)] \quad (7)$$

where $\log f_{DH}$ is the Debye–Hückel rational activity coefficient and is given by

$$\log f_{DH} = -\frac{AI^{1/2}}{1 + BaI^{1/2}} \quad (8)$$

In the case of an electrolyte M^{z+} Cl of molality m ($z_+ = 1$), the Gibbs–Duhem equation [Eqs. (23) and (71) of Chapter 2] gives

$$-\frac{55.51}{m} d \log a_1 = d \log \gamma_{M^{z+}} m + d \log a_{Cl} \quad (9)$$

The hydration of the Cl ion is assumed to be zero and that of M is h; then m', the molality in terms of unbound water, is related to m by

$$m' = \frac{1000m}{1000 - hm18} = \frac{m}{1 - 0.018hm} \quad (10)$$

The activity of Cl is the same on the two scales, i.e., $m'\gamma'_{Cl} = m\gamma_{Cl}$. So Eq. (10) can be transformed into

$$-\frac{55.51}{m'} d \log a_1 = -\frac{55.51}{m} d \log a_1 + h\, d \log a_1$$
$$= d \log \gamma'_{M^{z+}} m' + d \log a_{Cl} \quad (11)$$

Substituting Eq. (9) into Eq. (11) gives

$$d \log \gamma_{M^{z+}} m + h\, d \log a_1 = d \log \gamma'_{M^{z+}} m'$$

or using Eq. (10) gives

$$\log \gamma_{M^{z+}} = \log \gamma'_{M^{z+}} - h \log a_1 - \log(1 - 0.018hm) \quad (12)$$

According to the hydration theory,[53] $\log f_i$ of the hydrated species is equal to $z_i^2 \log f_{DH}$. Hence converting $\log f_i$ to the molal scale [see Eq. (40) of Chapter 2] gives

$$\log \gamma'_{M^{z+}} = z_+^2 \log f_{DH} - \log(1 + 0.018\nu m') \quad (13)$$

Equations (13) and (10) substituted into Eq. (12) give

$$\log \gamma_{M^{z+}} = z_+^2 \log f_{DH} - h \log a_1 - \log[1 + 0.018m(\nu - h)] \quad (14)$$

Eliminating $\log f_{DH}$ between Eqs. (7) and (14) and using the definition of molal osmotic coefficient ϕ [see Eq. (67) of Chapter 2] gives, on rearrangement,

$$|z_+ z_-| \log \gamma_{M^{z+}} = z_+^2 \log \gamma_{\pm} - [z_+^2 - |z_+ z_-| \nu] 0.00782hm\phi$$
$$+ [z_+^2 - |z_+ z_-|] \log[1 + 0.018m(\nu - h)] \quad (15)$$

Equation (15) applied to a 1 : 1 electrolyte ($z_+ = z_- = 1$ and $\nu = 2$)

becomes

$$\log \gamma_{M^+} = \log \gamma_\pm + 0.00782hm\phi \qquad \text{for the cation} \tag{16}$$

$$\log \gamma_{Cl^-} = \log \gamma_\pm - 0.00782hm\phi \qquad \text{for the anion} \tag{17}$$

since $2 \log \gamma_\pm = \log \gamma_{M^+} + \log \gamma_-$.

Equation (15) applied to alkaline earth chlorides MCl_2 ($z_+ = 2, z_- = 1; \nu = 3$) becomes

$$\log \gamma_{M^{2+}} = 2 \log \gamma_\pm + 0.00782hm\phi + \log[1 + 0.018m(3 - h)] \tag{18}$$

For the anion, since $3 \log \gamma_+ = \log \gamma_{M^{2+}} + 2 \log \gamma_{Cl^-}$, Eq. (18) becomes

$$2 \log \gamma_{Cl^-} = \log \gamma_\pm - 0.00782hm\phi - \log[1 + 0.018m(3 - h)] \tag{19}$$

The mean activity coefficients and the osmotic coefficients for a number of alkali and alkaline earth chlorides have been determined at high ionic strength and tabulated.[55] Using these data, single ion activity coefficients have been calculated[8, 40, 54] by using Eqs. (16)–(19). Since these values are useful for the calibration of ion-selective electrodes, they are given in Tables 1–3.

The hydration equation (7) has been used to analyze the activity data for potassium fluoride. This analysis[56] has given a value of 1.87 for the fluoride ion hydration, which is almost identical to 1.9 for the K^+ ion on the basis of $h = 0$ for the Cl^- ion.

If both cations and anions are hydrated, Eqs. (16) and (17) become

$$\log \gamma_+ = \log \gamma_\pm + 0.00782(h_+ - h_-)m\phi \tag{20}$$

$$\log \gamma_- = \log \gamma_\pm + 0.00782(h_- - h_+)m\phi \tag{21}$$

As $h_+ \approx h_-$ in the case of KF, it follows that $\gamma_\pm = \gamma_+ = \gamma_-$, and this is the basis of the values for KF given in Table 4. These considerations have been extended[49, 57, 58] to binary mixtures of electrolytes with a common unhydrated anion or a common hydrated cation, one hydrated and one

TABLE 1

Single Ion Activity Coefficients at 25°C Based on the Hydration Theory[a]

Reference substance	0.1 m		1.0 m	
	γ_+	γ_-	γ_+	γ_-
KF	0.775	0.775	0.645	0.645
NaCl	0.783	0.773	0.697	0.620
KCl	0.773	0.768	0.623	0.586
$CaCl_2$	0.269	0.719	0.263	0.690
NaBr	0.788	0.776	0.739	0.639
KBr	0.769	0.775	0.639	0.596
$CaBr_2$	0.283	0.729	0.378	0.748

[a] Compiled by Covington.[8]

TABLE 2

Single Ion Activity Coefficients of Uni–Univalent Chlorides at 25°C Based on the Hydration Theory of Bates *et al.*[54]

Molality	HCl		LiCl		NaCl		KCl		RbCl		CsCl		NH_4Cl	
(m)	γ_+	γ_-	γ_+	γ_-	γ_+	γ_-	γ_+	γ_-	γ_+	γ_-	γ_+	γ_-	γ_+	γ_-
0.01	—	—	0.906[a]	—	0.904[a]	—	0.906[a]	—	0.904[a]	—	0.902[a]	—	0.906[a]	—
0.1	0.807	0.785	0.799	0.781	0.783	0.773	0.773	0.768	0.766	0.762	0.756	0.756	0.772	0.768
0.2	0.788	0.746	0.775	0.739	0.744	0.726	0.722	0.714	0.712	0.706	0.694	0.694	0.722	0.714
0.5	0.812	0.706	0.786	0.695	0.701	0.661	0.659	0.639	0.640	0.628	0.606	0.606	0.657	0.641
1.0	0.940	0.697	0.882	0.680	0.697	0.620	0.623	0.586	0.594	0.572	0.544	0.544	0.619	0.588
2.0	1.421	0.717	1.233	0.688	0.756	0.590	0.610	0.538	0.568	0.525	0.496	0.496	0.601	0.541
3.0	2.357	0.735	1.893	0.706	0.870	0.586	0.626	0.517	0.569	0.505	0.479	0.479	0.608	0.518
4.0					1.038	0.591	0.659	0.506	0.584	0.496	0.474	0.474	0.624	0.502
5.0					1.272	0.600			0.606	0.492	0.475	0.475	0.645	0.490
6.0					1.594	0.610							0.667	0.477

[a] According to Covington.[8]

TABLE 3

Single Ion Activity Coefficients for Alkaline Earth Chlorides at 25°C Based on the Hydration Theory of Bates *et al.*[54]

Molality	$MgCl_2$		$CaCl_2$		$SrCl_2$		$BaCl_2$	
(m)	γ_+	γ_-	γ_+	γ_-	γ_+	γ_-	γ_+	γ_-
0.0333	0.392[a]	—	0.378	0.784	0.377[a]	—	0.374[a]	—
0.1	0.279	0.726	0.269	0.719	0.266	—	0.259	0.712
0.2	0.239	0.697	0.224	0.685	0.218	—	0.204	0.668
0.333	0.226[a]	—	0.205[a]	—	0.197[a]	—	0.176[a]	—
0.5	0.234	0.688	0.204	0.665	0.190	—	0.165	0.630
1.0	0.344	0.732	0.263	0.690	0.226	—	0.167	0.620
1.8	—	—	—	—	—	—	0.229	0.642
2.0	1.439	0.898	0.768	0.804	0.542	—	—	—

[a]According to Covington.[8]

TABLE 4

Single Ion Activity Coefficients of Potassium and Fluoride Ions in Solutions of KF at 25°C Based on Hydration Theory[56]

Molality (m)	Molarity (C)	$\gamma_\pm = \gamma_K = \gamma_F$	$pF = -\log a_F\,(m)$
0.01	0.00997	0.903	2.044
0.05	0.04983	0.820	1.387
0.1	0.09961	0.775	1.111
0.2	0.1990	0.727	0.837
0.5	0.4961	0.670	0.475
1.0	0.9868	0.645	0.190
2.0	1.951	0.658	− 0.119
3.0	2.888	0.705	− 0.325
4.0	3.794	0.779	− 0.494

unhydrated anion. These can be further extended to salts containing Br^- and I^- assuming that these anions are also unhydrated.

For a mixture of total molality m composed of $x_{MX}m$ moles of MX and $x_{NX}m$ moles of NX ($x_{MX} + x_{NX} = 1$), it has been shown[57] that the hydration theory gives the following expression for the activity coefficient of the anion X^-:

$$\log \gamma_X = x_{MX} \log \gamma_{MX} + x_{NX} \log \gamma_{NX} - 0.00782hm\phi \tag{22}$$

where γ_{MX} and γ_{NX} are the mean activity coefficients of MX and NX in the mixture and $h = x_{MX}h_M + x_{NX}h_N$. The activity coefficients of the cations are given by

$$\log \gamma_{M^+} = 2 \log \gamma_{MX} - \log \gamma_X \tag{23}$$

$$\log \gamma_{N^+} = 2 \log \gamma_{NX} - \log \gamma_X \tag{24}$$

In order to use these equations the value of ϕ as a function of composition of the electrolyte mixture must be known. This can be derived from the Harned rule coefficients.[59]

In the case of a mixture of MX and MY where M^+ and X^- are hydrated and Y^- is unhydrated, expressions for individual ionic activity coefficients, although more complex, have been derived.[58] Similarly, Leyendekkers[60] has developed expressions for single ion activity coefficients in multicomponent systems. The predictions of the theory have been verified by using the fluoride-selective electrode for NaCl–NaF and KCl–NaF systems.

Butler and Huston[61] carried out measurements on NaCl–NaF mixtures using the cell

$$\text{Ag, AgCl} \left| \, \text{Na}^+ \, \text{Cl}^- \, \text{F}^-, \text{H}_2\text{O} \, \right| \text{LaF}_3 \text{ membrane electrode}$$

The potential of this cell is given by

$$E = E^\circ + (RT/F)\ln(m_{\text{Cl}}/m_{\text{F}}) + (2RT/F)\ln(\gamma_{12}/\gamma_{21}) \tag{25}$$

where m_{Cl} and m_{F} are the molal concentrations of Cl and F in a mixed electrolyte, and γ_{12} is the mean activity coefficient of NaCl (component 1), and γ_{21} is that of NaF (component 2) in the mixed electrolyte. The test solutions are compared with a calibration solution with $m_{\text{Cl}} = m_{\text{F}}$. In all the solutions, $m = m_{\text{Cl}} + m_{\text{F}} = 1$. Equation (25) applied to the measurements of potentials with both reference (r) and test (t) solutions whose values of m are known gives

$$R_{21} = \log \frac{\gamma_{21}^{\text{r}} \gamma_{12}^{\text{t}}}{\gamma_{12}^{\text{r}} \gamma_{21}^{\text{t}}} \tag{26}$$

Equation (26) in combination with Harned's rule coefficients α_{12} and α_{21} and osmotic coefficients [see Eq. (148) of Chapter 2] has been used to determine γ_{12}^{t} and γ_{21}^{t}. The values obtained for the Harned coefficients α_{12} and α_{21} were $+0.028$ and -0.027, respectively. These values indicate that the interactions of Na^+ and F^- ions are stronger than those of Na^+ and Cl^-. A value for the equilibrium constant K_f for the formation of the Na^+–F^- pair has been derived ($\log K_f = -0.79$). Robinson *et al.*[56] have calculated a value of -0.27 for $\log K_f$. In view of this ion association existing in solutions of NaF and because KF is more soluble than NaF, solutions of KF have been recommended[56] for purposes of calibration of the fluoride-selective electrode.

Concentrated solutions of KF up to 3 m have been used by Bagg and Rechnitz[62] in activity measurements with the fluoride membrane electrode. The electrode was used in a cell with transport containing a hetero-ionic liquid junction whose potential was evaluated using the Henderson equation. This together with the single ion activity convention of Bates *et al.*[54]

gave emf data that agreed with the measurements. Similar agreement was obtained for other halide-selective electrodes used in solutions up to 6 *m* NaCl, 4 *m* KCl, and 1 *m* LiCl (Cl electrode) and 4 *m* KBr (Br electrode). In the case of the iodide electrode, deterioration of the electrode occurred at concentrations less than 0.5 *m*. Similarly, activity coefficients of fluoride ion in mixtures of trace concentrations of KF in KCl, KBr, and KI solutions at concentrations up to 4 *m* and trace concentrations of NaF in NaCl solutions at concentrations up to 1 *m* have been determined.[63]

Ion-selective membrane electrodes used in cells of type (3) can be calibrated with the help of electrolyte solutions whose single ion activity coefficients are given in Tables 1–4. Equation (4), in which $K^{pot}_{F,OH} a_{OH} \approx 0$, will be the basis for calibration of the cell. If the liquid junction potential E_L is assumed to remain unchanged when a standard reference solution is used in place of the unknown solution X, then considering the analogy with the operational definition of pH,[44, 50] Eq. (4) may be written as

$$\mathrm{pM}(X) = \mathrm{pM}(S) + \frac{n(E_X - E_S)F}{2.303RT} \tag{27}$$

for measurements with cation (M)-selective electrodes and

$$\mathrm{pY}(X) = \mathrm{pY}(S) - \frac{n(E_X - E_S)F}{2.303RT} \tag{28}$$

for measurements with anion Y-selective electrodes.

From Eq. (4) it is seen that the single ion activity and the liquid junction potential are interdependent. Complete equality of liquid junction potentials during measurements of ion activity using unknown (X) and standard (S) solutions is seldom achieved. A correction can, however, be made simply by using the Henderson equation [see Eq. (40) of Chapter 3]. If δE_L is the difference in these measurements, i.e., $\delta E_L = E_L(X) - E_L(S)$, then Eqs. (27) and (28) become

$$\mathrm{pM}(X) = \mathrm{pM}(S) + \frac{n(E_X - E_S - \delta E_L)F}{2.303RT} \tag{27a}$$

$$\mathrm{pY}(X) = \mathrm{pY}(S) - \frac{n(E_X - E_S - \delta E_L)F}{2.303RT} \tag{28a}$$

C. DETERMINATION OF SELECTIVITY CONSTANTS (K^{pot}_{ij})

The practical aspects of selectivity and sensitivity of electrodes have been discussed by Moody and Thomas.[64] There are a number of methods[30, 65, 66] that can be used to determine the values for K^{pot}_{ij}.

Method 1: For the primary ion i ($i = 1$) only in solution (i.e., $a_j = 0$), Eq. (12) of Chapter 3 becomes

$$E_1 = E^\circ + (RT/F)\ln a_1 \tag{29}$$

If the solution is without i (i.e., $a_i = 0$) and contains only ion j ($j = 2$), then Eq. (121) of Chapter 3 becomes

$$E_2 = E^\circ + (RT/F)\ln K_{ij}^{\text{pot}} a_2 \tag{30}$$

For the condition $a_1 = a_2$, Eqs. (29) and (30) give, at 25°C, the relation

$$\log K_{ij}^{\text{pot}} = \frac{E_2 - E_1}{59.2} \tag{31}$$

where E_1 and E_2 are in millivolts. This method has been used by Eisenman *et al.*,[67] Rechnitz and co-workers,[66, 68, 69] and others.[70]

Method 2: If the concentrations of the solution of ion i and of the solution of ion j are so chosen that $E_1 = E_2$, then Eqs. (29) and (30) give

$$K_{ij}^{\text{pot}} = \frac{a_1}{a_2} \tag{32}$$

This method has been used only by a limited number of investigators.[66, 68, 71]

Method 3: In methods 1 and 2, only one ion is used in any test solution. In method 3, both the ions are utilized. Equation (121) of Chapter 3 becomes

$$E^* = E^\circ + (RT/F)\ln(a_1 + K_{ij}^{\text{pot}} a_2) \tag{33}$$

Equation (33) can be combined with Eq. (29) to give

$$E^* - E_1 = \frac{RT}{F}\ln\left(\frac{a_1 + K_{ij}^{\text{pot}} a_2}{a_1}\right)$$

which on rearrangement becomes

$$K_{ij}^{\text{pot}} = \left[\exp\left\{\frac{(E^* - E_1)F}{RT}\right\}a_1 - a_1\right] \Big/ a_2 \tag{34}$$

This method was used by Light and Swartz[70] directly, whereas Srinivasan and Rechnitz[66] used a graphical procedure to evaluate K_{ij}^{pot}. This is done by adding increasing quantities of ion 2 to a solution containing ion 1. After each addition of ion 2, the potential E^{**} is measured. E^{**} can also be measured for different solutions containing ions 1 and 2 in different proportions. Thus E^{**} according to Eq. (33) can be written as

$$E^{**} = E^\circ + (RT/F)\ln(a_1^* + K_{ij}^{\text{pot}} a_2^*) \tag{35}$$

Combining Eqs. (29) and (35) gives

$$\exp\left\{\frac{(E^{**} - E_1)F}{RT}\right\}a_1 - a_1^* = K_{ij}^{\text{pot}}a_2^* \tag{36}$$

The left-hand side of Eq. (36) may be plotted against a_2^* and the slope gives the value for K_{ij}^{pot}, but this procedure is not sensitive at low values of K_{ij}^{pot}. Under such conditions, the concentration of ion 2 is kept high and ion 1 is added gradually. For these conditions, Eqs. (33) and (35) may be combined to yield the relation

$$-\exp\left\{\frac{(E^{**} - E^*)F}{RT}\right\}a_1 + a_1^* = K_{ij}^{\text{pot}}\left[\exp\left\{\frac{(E^{**} - E^*)F}{RT}\right\}a_2 - a_2^*\right] \tag{37}$$

Since all the quantities are known, K_{ij}^{pot} can be evaluated graphically or numerically.

Pungor and Toth[30, 65] argue that making measurements using only solution 1 or 2 as in methods 1 and 2 does not give the real value of K_{ij}^{pot} since the conditions for potential measurements are not well defined probably due to electrode surface contamination by solution 1 while making measurements in solution 2. They favor performing potential measurements in solutions containing both the primary ion i and the interfering ion j. Instead of the procedures described above, they recommend determination of K_{ij}^{pot} from the dependence of the ion-selective membrane electrode potential on the logarithm of the concentration of ion j in the presence of a constant concentration of i (C_i). The other procedure of keeping the concentration of j constant and varying the concentration of i is also recommended. The plot of cell emf versus $-\log C_j$ gives two linear portions whose point of intersection (the "break point" of the curve) gives the concentration of C_j. The ratio C_i/C_j is equal to K_{ij}^{pot}. This is called the direct method as opposed to the other procedure of keeping the concentration of j constant and varying the concentration of i (indirect method). In this procedure also, there are theoretically two possibilities: (i) to follow the direct method by preparing solutions of different proportions of j and i keeping j always constant; (ii) to follow a titration procedure by starting with a mixture of i and j in equal concentrations and varying i by titrating it with a suitable reagent and following the potential. At the "coprecipitation point" the concentration of i can be estimated by using the Nernst equation. Since the concentration of j is known, K_{ij} can be calculated.

In order to compare the values of K_{ij}^{pot} determined by different procedures some data collected for the bromide- and iodide-selective membrane electrodes are given in Table 5 together with values calculated by using Eq. (178b) of Chapter 3.

The differences in the values of K_{ij}^{pot} derived by the various methods (Table 5) are probably due to the different procedures employed in the determinations.

TABLE 5

Comparison of Values Determined for the Selectivity Constant K_{ij}^{pot} by Different Methods

Heterogeneous solid membrane	Calculated[a] Eq. (178b), Chapter 3	Pungor *et al.*[30, 65] graphical[b]	Rechnitz *et al.*[68, 69, 71] Eq. (32)[c]
AgBr electrode			
$K_{Br,Cl}^{pot}$	4.9×10^{-3}	1.5×10^{-3}	1.0×10^{-2}
AgI electrode			
$K_{I,Br}^{pot}$	1.9×10^{-4}	2.0×10^{-4}	4.8×10^{-3}
$K_{I,Cl}^{pot}$	9.6×10^{-7}	1.0×10^{-6}	5.9×10^{-6}

[a]Values for the solubility product taken from "Handbook of Chemistry,"[72] measurements at 25°C.

[b]Measurements at 25°C. Reciprocal of the values given in the references.

[c]Measurements at 30°C. Reciprocal of the values given in the references.

D. PROPERTIES AND APPLICATIONS OF FLUORIDE-SELECTIVE ELECTRODES

Besides the properties described in the foregoing, some other properties and applications have been reviewed by Ross,[20] Butler,[73] and others.[8, 30, 40, 74] A summary of these is presented in the following.

The conductivity of lanthanum fluoride crystal is 10^{-7} ohm^{-1}cm^{-1} at 25°C.[75] The charge carrier is the F^- ion and so the mechanism involved is considered to be[75, 76]

$$LaF_3(+\text{ molecular hole}) \rightleftharpoons LaF_2^+ + F^-$$

Using Orion's 94-09 and 94-09A fluoride electrodes, Stahr and Clardy[77] studied the mechanism of charge transfer in the LaF_3 crystal. The electrode was kept in 15% sodium acetate solution (pH 6.3) in the presence and absence of ^{18}F and voltage was applied. ^{18}F accumulated on the electrode surface very rapidly and also diffused rapidly into the body of the crystal, which has about 0.1% neutral Schotty defects at 27°C. Since the Debye temperature is 87°C for LaF_3, F^- ion is very mobile. This mobility is further increased by doping the crystal with europium. Thus the potential of the electrode seems to be determined by the ion exchange mechanism and diffusion. In view of this, it may be considered an analog of the glass membrane electrode.[78]

The impedance characteristics of the lanthanum fluoride membrane–electrolyte system has been studied by Brand and Rechnitz.[79] Similarly, Vesely[80] studied the resistance characteristics of the LaF_3 single crystal electrode doped with five rare earths. The cell resistance decreased with increasing concentration of the dopant and the potential response of the electrode was unaffected.

An interesting discussion concerning the estimation of the lower limit of detectability of the F^- ion with the help of the electrode has been given by Butler.[73] A number of workers[81–84] have shown that the electrode gave a Nernstian response in the fluoride ion concentration range from 0.1 *M* to below 10^{-5} *M*. The estimation of the lower limit is subject to some uncertainties due to the variable liquid junction potential and the activity coefficient of the F^- ion. These factors and others influencing the behavior of the electrode have been considered in detail by Parthasarathy *et al.*[85]

The divergence from a linear relationship between the cell potential and log C_F is generally attributed to (a) solubility of the membrane, (b) the presence of impurities in the supporting electrolyte, or (c) adsorption of test ions on the walls of the container. A detailed study[85] has shown that in the case of the fluoride electrode, the solubility of the membrane did not interfere with the proper functioning of the electrode whereas the ions of the supporting electrolyte and the chloride impurities did. The two factors, namely solubility of the crystal membrane and adsorption of F^- ions at the membrane–solution interface, since they affect the sensitivity of the LaF_3 electrode, have been investigated in depth by Buffle *et al.*[86] They found that adsorption of the F^- ions rather than the extent of solubility of the crystal itself determined the lower limit of detection of the electrode. Similarly, Vesely and Stulik[87] studying the effect of solution acidity on the response of the electrode found that the electrode behavior was determined by the competitive adsorption of OH^- and F^- ions and of the various fluoride-containing species in the hydrophilic film formed on the electrode. The crystal itself was not attacked by the ions in solution.

Buffle *et al.*[86] in their study of the solubility of the LaF_3 crystal estimated a value of 10^{-30} or less for the solubility product S_p, whereas a value of 10^{-29} was given by Frant and Ross.[2] This implies that the fluoride ion concentration in a saturated solution of LaF_3 is approximately 10^{-7} *M*.

From the plots of E vs. log a_F given by Frant and Ross,[2] Butler[73] estimates a value of 3.2×10^{-25} for S_p and a value of 1.0^{-6} M as the lower limit of detectability of the F^- ion in solution. As Frant and Ross used no supporting electrolyte, this value for S_p should correspond to a compact lanthanum fluoride crystal. On the other hand, Lingane[88] measured a value of 1.2×10^{-18} by titration of F^- with lanthanum nitrate in a solution of ionic strength of about 0.08 M. Similarly, a value of 2.2×10^{-17} for the S_p of europium trifluoride has been determined.[88] This value is about 10 times larger than the value of LaF_3. In the case of freshly precipitated LaF_3, the value of S_p is nearly 10^7–10^8 times larger than that of the compact or massive crystal of LaF_3. This large discrepancy is atrributed to the differences in the crystal surface energy which is larger with a fresh hydrated polycrystalline material than with a massive single crystal. This may be the reason it is difficult to produce a successful heterogeneous fluoride membrane electrode.[33]

The fluoride membrane electrode responds in both neutral and moderately acidic media within the fluoride ion concentration range 1 to 10^{-5} or 10^{-6} M.[20, 73, 89] In pure fluoride solutions, the response even reaches the limit of 10^{-7} M.[90]

Interference with the ideal functioning of the F^- ion-selective electrode by a number of compounds and/or ions has been studied by a number of investigators. Bock and Strecker[83] showed that the Nernstian slope of pure solutions of KF or NaF was not affected when 0.1 M KNO_3 or $MgSO_4$ or 1.0 M NaCl was added to them. This result showed that K^+, Na^+, Mg^{2+}, NO_3^-, Cl^-, and SO_4^{2-} did not interfere with the response of the electrode to F^- ions. The responses of the electrode to 10^{-4} M NaF in a buffer consisting of 0.18 M sodium potassium tartrate, 0.012 M disodium citrate, and 0.008 M sodium hydroxide to which various substances are added (pH 5.0–6.3) are shown in Table 6. It is expected that the interfering ion makes the potential of the electrode more negative. Only silicate solution does this to a very small extent (0.3 mV for a 65-fold excess of silicate over fluoride). This small change could have been brought about by a number of factors such as fluoride impurity in the silicate, adsorption of silicate on the membrane, or even a change in liquid junction potential. The other additions in Table 6 give potentials that are more positive. Possible explanations are changes in ionic strength (NaCl, $MgSO_4$, H_3BO_3, $TiOSO_4$) or formation of complexes. The effects of nonionic compounds such as glucose, urea, and H_2O_2 on the electrode response were also studied and the changes in potential in the case of glucose and urea (more negative) were attributed to changes in the activity coefficient of NaF and in the case of H_2O_2 (more positive) to formation of weak complexes between fluoride ions and protons of H_2O_2.

TABLE 6

Lanthanum Fluoride Membrane Electrode Response to Various Added Substances[83a]

Added substance	Concentration (M)	pH	E (mV)
None (control)	—	5.9	68.2
NaCl	0.1	5.9	68.9
$MgSO_4$	0.01	5.8	69.0
	0.01	5.4	69.0
H_3BO_3	0.10	6.3	70.0
H_2SiO_3	0.0065	5.9	67.9
$TiOSO_4 + Na_2SO_4$	0.01 + 0.10	6.2	69.2
$BeSO_4$ + NaOH	0.01	5.7	100.4
$ZrOCl_2$ + NaOH	0.01	6.2	93.2
$Fe_2(SO_4)_3$ + NaOH	0.01	5.4	69.0
$NH_4Al(SO_4)_2$ + NaOH	0.01	5.2	131.0

[a]Control solution: sodium fluoride (10^{-4} M) + sodium potassium tartrate (0.18 M) + disodium citrate (0.012 M) + NaOH (0.008 M).

Similarly, Baumann[91] showed that 1, 5, and 10 M phosphoric acid added to NaF solution did not change the slope of the curves of potential versus log concentration of NaF although the potentials shifted to more positive values as the concentration of acid was increased. In 10 M phosphoric acid Fe(III), Al(III), UO_2^{2+}, or Th(IV) had little effect on the potential since they complexed with the acid and were unavailable to complex with the fluoride ion. Also in this study, the noninterference of NO_3^- and Cl^- with the function of the electrode was confirmed. However, monofluorophosphate ion interfered with the response of the electrode.[92]

In the work described above, Baumann[91] used the standard addition technique[93] which is a very convenient method for the determination of the total concentration of a species in very complex systems. In using this technique, the response of the electrode (slope of potential versus log concentration) is evaluated to as high a degree of precision as possible with standard solutions. Then addition of one or more aliquots of the species of interest to the unknown sample solution is carried out under conditions of constant ionic strength.

The observed initial potential of the sample solution of unknown concentration (C_x) is given by the Nernst equation. Thus

$$E_1 = E^\circ + \frac{RT}{nF} \ln(C_x \gamma_x) + E_L \tag{38}$$

On addition of a known amount of test ion (V_s ml of known concentration

C_s to initial volume V_x), the new potential measured is given by

$$E_2 = E^\circ + \frac{RT}{nF} \ln \frac{C_x V_x + C_s V_s}{V_x + V_s} \gamma'_x + E_L \tag{39}$$

Assuming the constancy of E_L and $\gamma_x = \gamma'_x$, subtracting Eq. (38) from Eq. (39) gives

$$\Delta E = E_2 - E_1 = \frac{RT}{nF} \ln \frac{C_x V_x + C_s V_s}{(V_x + V_s) C_x} \tag{40}$$

On rearrangement Eq. (40) gives

$$\frac{\Delta E}{S} = \log \frac{C_x V_x + C_s V_s}{(V_x + V_s) C_x} \tag{41}$$

where S is the Nernst slope ($= 2.303 RT/nF$) experimentally determined by using a series of known standard solutions. Equation (41) may be rearranged to give

$$C_x = \frac{C_s V_s}{V_x + V_s} \left[10^{\Delta E/S} - \frac{V_x}{V_x + V_s} \right]^{-1} \tag{42}$$

Thus C_x can be evaluated. If there is no change in volume (V_x) due to the addition of $C_s V_s$ quantity of the standard, i.e., $V_x \gg V_s$, then Eq. (42) becomes

$$C_x = C_s (V_s / V_x)(10^{\Delta E/S} - 1)^{-1} \tag{43}$$

where $C_s(V_s/V_x) = \Delta C$, the increase in concentration. Thus Eq. (43) becomes

$$\frac{C_x}{\Delta C} = (10^{\Delta E/S} - 1)^{-1} \tag{44}$$

This standard addition technique has been used by Bruton[94] to determine simultaneously the quantity of F^- and Cl^- present in calcium halagenophosphate.

Two variations of this procedure have also been used. In the first variation, which is called Gran's plot,[95, 96] a graphical procedure is used. Equation (39) can be rearranged to give

$$(V_x + V_s) 10^{E_2 F/2.303RT} = 10^{(E^\circ + E_L)F/2.303RT} \gamma'_x (C_x V_x + C_s V_s) \tag{45}$$

A plot of $(V_x + V_s)10^{E_2 F/2.303RT}$ vs. V_s gives a straight line which intercepts the abscissa for a value of V_s called V_e where $C_x V_x = -C_s V_e$. Thus C_x can be evaluated since V_e, V_x, and C_s are known. This method has been illustrated by Liberti and Mascini[97] using the fluoride-selective electrode, and has also been used by Selig[98] to estimate F^- ion concentration using the TISAB buffer (total ionic strength adjustment buffer). Since the constancy of E_L and of the activity coefficient is dependent on the maintenance of high ionic strength, TISAB has been recommended by Frant and Ross[99] who used it in the estimation of F^- in water derived from various sources. This buffer had 0.25 M acetic acid, 0.75 M sodium acetate, 1.0 M sodium chloride, and 10^{-3} M sodium citrate. The citrate complexed with the interfering ions such as Al(III) and Fe(III).

The second variation, which is the inverse of standard additions, also called "analate additions" potentiometry by Durst[100] is based on the addition of aliquots of the unknown solution to a known volume of the standard solution of the same species. Using C_s in place of C_x in Eq. (38) and subtracting it from Eq. (39) gives, on rearrangement, the expression for the unknown concentration C_x as

$$C_x = C_s\left[\left(\frac{V_s + V_x}{V_x}\right)10^{\Delta E/S} - \frac{V_s}{V_x}\right] \tag{46}$$

This technique has been applied to the determination of F in solutions containing electrochemically generated F^- ions.[101]

In another study, Baumann[102] showed with the help of complexing ions such as Th^{4+} and Zr^{4+} that the fluoride electrode responded to F concentration to pF 9. The fluoride ions liberated by the solution from the crystal probably complexed. A very extensive test of the specificity of the fluoride electrode for the F^- ion in the presence of 1.0 M Cl^- has been carried out by Mesmer.[84] The response of the electrode was Nernstian to F concentrations below 10^{-4} M. When correction for the presence of fluoride impurity in the NaCl or KCl samples was applied, Nernst behavior was obtained even below 10^{-5} M. This brings out the importance of the purity of compounds in doing experiments at constant ionic strength. Addition of complexing agents such as HCl or beryllium established the fact that chloride interference came about only at concentrations less than 2×10^{-8} M in the presence of 1.0 M chloride. Similarly the response of the electrode in water and in NaCl solutions has been tested by Warner.[103] Despite this superb selectivity of the electrode to F^- ions in the presence of Cl^- ions, considerable interference comes from the presence of OH^- ions.[2, 73] Frant and Ross[2] estimate the selectivity constant $K^{pot}_{F,OH}$ to be 0.1. However, Butler[73] has shown that it could vary from zero in 10^{-1} M fluoride ion to a value greater than unity in 10^{-5} M fluoride ion. In addition, Zentner[104] finds that the nonionic compound 1-fluoro-2, 4-dinitrobenzene (FDNB) also interferes with the functioning of the fluoride electrode. This has been attributed to the high polarizing power of the F atoms which imparted appreciable dipole moment to the organic compound leading to its adsorption on the electrode surface. Calcium ion interference in the range 10^{-3}–1.0 *M* Ca has been also recorded. Below 10^{-3} *M* Ca, the fluoride electrode response to Ca was absent and so it is likely that fluoride was present as an impurity in the solution of Ca used in this investigation.

The kinetic response of the electrode in higher concentrations (> 1 m*M*) of F^- ion is limited by the recorder response time of 0.5 sec.[105, 106] In very dilute solutions, the response time could be longer. In 10^{-6} *M* NaF solution, Bock and Strecker[83] found that a steady potential was reached in about 1 hr, whereas a response time of less than 3 min was found by Raby and Sunderland.[107] The ultimate speed of response of the electrode has not been determined and seems to depend on the diffusion of the test solution into the electrode to reach equilibrium.

A microelectrode from europium-doped LaF_3 crystal has been constructed and its response characteristics have been described.[108] The lanthanum fluoride electrode has been used in a number of studies because of its versatile behavior:

(1) The fluoride in rocks and minerals[109–117] and in exploration ores, fluorspar, opal glass, phosphate rock, other phosphates, various geological and production samples has been estimated.[118–123]

(2) The fluoride in micro- and submicroamounts in organic and/or some biological compounds[124–130] has been determined. In the case of some organic and organometallic compounds,[131] closed flask combustion and direct measurement with TISAB buffer have been employed.

(3) The electrode has been used in the determination of fluoride in seawater,[132, 133] potable or natural water and waste waters,[134–145] air and smoke gases,[146–149] atmospheric precipitation,[150] fuming nitric acid,[151] chromium plating baths,[152] fluorosilicic acid,[153] welding fluxes and coatings,[154] nuclear fuel processing solutions,[155] tungsten,[156] selenium fluoride,[157] the analysis of the composition of alloys of Th–U and U–Zr,[158] aluminum reduction materials,[159] and coal.[160]

The electrode has been used in the following analytical processes:

(a) The determination of fluoride in sugarcane.[161] This, however, requires prior removal of Si, Al, and Fe from sugarcane by leaching from a solution of sodium carbonate–zinc oxide fusion, followed by complexing the residual trace elements with citrate.

(b) Determination of soluble fluoride from phosphate fertilizer and aluminum reduction plants.[162] The results obtained with the electrode are equivalent to those obtained by SPADNS zirconium lake procedure.

(c) The estimation of soluble fluoride in rain, snow, fog, or aerosols in samples (10 ml) containing microgram per kilogram quantities (0.28 ppb).[163]

(d) The assay of fluoride ion in glass[164] that contains permissible amounts of aluminum using 0.5 *M* citrate buffer at pH 6.0. If the concentration of aluminum exceeds permissible limits, a successive dilution step at constant ionic strength (citrate added) is recommended.

(e) The determination of the solubility product of cryolite.[165]

(f) The determination of fluorine in petroleum and petroleum process catalysts.[166] After alkali fusion or treatment with sodium biphenyl, fluoride is directly measured in aqueous extracts. Limits of estimations were found to be $\geqslant 0.01$ ppm F^- (distillates), $\geqslant 0.1$ ppm F^- (crudes), and $\geqslant 2$ ppm F^- (catalysts).

(g) The determination of HF in strong acid solutions such as stainless steel pickling baths.[167] If HF should be mixed with other strong acids such as HNO_3, acidity may also be determined by using a quinhydrone electrode[168] or a Permaplex ion exchange membrane electrode.[167]

(h) Fluoride in biological samples has been determined by a reverse extraction technique.[169] This technique is illustrated using bovine serum, human serum, etc. The fluoride contents of other biological samples[170–178] have also been estimated. In addition, the electrode has been used to determine the amount of fluoride present in bones,[179, 180] teeth,[181, 182] serum,[183–185] saliva,[186, 187] urine,[188–191] toothpastes,[192] enamel,[193] multivitamin preparations,[194] beverages,[195] foodstuffs,[196] vegetation,[148, 197] and plants.[198, 199]

The electrode has been used in flowing systems to monitor the response to fluoride ions under computer control.[200] Direct potentiometry using the fluoride electrode has been modified by a number of investigators[82, 93, 100, 101, 103, 201, 202] to refine and/or clearly demarcate the end point so that estimations of F^- are more accurate.

The fluoride-selective electrode has been used as an indicator electrode in automatic[203, 204] and other potentiometric titrations,[81, 88, 93, 97, 129, 203, 205–209] in the estimation of aluminum[210, 211] and as a reference electrode in the determination of nitrate ions.[212] Similarly, the electrode is used in certain molten fluorides as part of a reference electrode to obtain thermodynamic data of interest to the designers of molten-salt breeder reactors.[213] It has been used in the estimation of F^- ions in polar nonaqueous solvents. Subnanomoles of F^- in methanol, 2-propanol, and 1, 4-dioxane containing not more than 5% water have been determined.[214] Its use as a reference electrode in mixed solvents in electrochemical cells without liquid junction has been evaluated.[215] It was found that the crystal (LaF_3) electrode itself was less soluble in 4 *M* acetonitrile ($S_p = 4.3 \times 10^{-20}$) than in water ($4.8 \times 10^{-18}$). The electrode has also been used as a reference electrode in the study of the formation constants of acetonitrile and allyl alcohol complexes of AgI.[216] Studies of LaF_3 solubility in relation to the applications of the electrode have been reported by Evans *et al.*[217]

In titrations in which precipitations occur, as in the case of fluoride ion and lanthanum or thorium,[81, 88] tetraphenyl arsonium sulfate,[210] tetraphenyl antimony sulfate,[218] or lithium,[219] the shape of the titration curve is very important. It is generally asymmetrical in the vicinity of the equivalence point since the potential dependence, $E = E_1 - (RT/F) \ln a_{F^-}$ is replaced by the dependence $E = E_1 + (RT/3F) \ln a_{La^{3+}}$ in that region. Arsenate has been estimated by Selig[220] by precipitating it with an excess of $La(NO_3)_3$; the excess La^{3+} was estimated by titration with a fluoride solution using the fluoride-selective electrode.

Carboxylate buffers used in potentiometric titrations of fluoride with lanthanum nitrate have been shown to interfere with precipitation reactions.[7, 221] Formate, acetate, propionate, and butyrate (A^-) formed precipitates of formula $LnF_{3-x}A_x$ (Ln = La, Ce, Pr, Nd, or Sm) and caused the fluoride electrode to work sluggishly. The sluggishness could be overcome by polishing the electrode surface with diamond paste and then immersing the electrode in a dilute solution of NaF saturated with LaF_3.

The formation of fluoride complexes with various metal ions and compounds has been studied by using the LaF_3 electrode. The important equilibria that the fluoride ion participates in are those involving the hydrogen ion. With the help of the electrode, the free fluoride ion concentration $(F)_f$ can be determined. A known amount of total fluoride concentration $(F)_t$ is changed by several orders of magnitude to test for the formation of polynuclear complexes. The most popular function used to analyze the data is the ligand number defined by

$$\bar{n} = \frac{(F)_t - (F)_f}{(M)_t} \tag{47}$$

where $(M)_t$ is the total concentration of metal ion taken. Thus $\bar{n}$ is the ratio of bound fluoride to total metal ion and is related to the successive formation constants by

$$\bar{n} = \frac{\beta_1(F)_f + 2\beta_2(F)_f^2 + 3\beta_3(F)_f^3 + \cdots}{1 + \beta_1(F)_f + \beta_2(F)_f^2 + \beta_3(F)_f^3 + \cdots} \tag{48}$$

where

$$\beta_i = \frac{[MF_i^{(x-i)+}]}{[M^{x+}][F^-]^i}$$

which are evaluated by well-established methods.[222] Thus for the equilibria

$$H^+ + F^- \rightleftarrows HF; \qquad \beta_1 = K_1 = (HF)/[(H^+)(F^-)]$$

$$HF + F^- \rightleftarrows HF_2^-; \qquad \beta_2 = K_1K_2 \quad \text{and} \quad K_2 = (HF_2^-)/[(HF)(F^-)]$$

Eq. (48) becomes

$$\bar{n} = \frac{\beta_1(F^-) + 2\beta_2(F^-)^2}{1 + \beta_1(F^-) + \beta_2(F^-)^2} \tag{49}$$

The linear form of this equation can be obtained by rearrangement as follows:

$$\frac{\bar{n}}{(1 - \bar{n})(F^-)} = \beta_1 + \beta_2\left[\frac{2 - \bar{n}}{1 - \bar{n}}\right](F^-) \tag{50}$$

The values of K_1 and K_2 derived for the HF equilibria by a number of investigators using the LaF_3 membrane electrode are given in Table 7.

The equilibrium constant K_1 for the reaction

$$Fe^{3+} + F^- \rightleftharpoons FeF^{2+}$$

TABLE 7

Equilibrium Constant for the H^+-F^- System[a]

Ionic medium	log K_1	log K_2	Ref.	Remarks
—	3.189	—	223	Extrapolated
0.05 *M* $HClO_4 + NaClO_4$	3.044	—	223	
0.10 *M* $HClO_4 + NaClO_4$	3.031	—	223	
0.20 *M* $HClO_4 + NaClO_4$	2.938	—	223	
0.50 *M* $HClO_4 + NaClO_4$	2.751	—	223	
1.0 *M* NaCl	2.89	—	84	
1.0 *M* $NaNO_3$	2.90	0.77		
	2.89	0.86	105	
3.0 *M* KCl	3.31	0.72	73	
1.0 *M* NaCl	2.887	0.983	224	
1.0 *M* $NaClO_4$	2.928	0.861	224	
—	4.164	0.70	225	Extrapolated
0.01 *M* NH_4NO_3	3.08	0.76	225	
0.1 *M* NH_4NO_3	2.94	0.83	225	
0.3 *M* NH_4NO_3	2.87	0.57	225	
0.5 *M* NH_4NO_3	2.84	0.70	225	
0.5 *M* $NaClO_4$	2.91	—	226	
1.0 *M* $NaClO_4$	2.98	0.73	227	
2.0 *M* $NaClO_4$	3.13	0.84	227	
3.0 *M* $NaClO_4$	3.33	0.94	227	
4.0 *M* $NaClO_4$	3.54	1.05	227	
0.008–0.03 *M* (NaF + HCl)	3.233	0.587	228	In the calculation of K_1 a fixed value of 3.86 was used for K_2
3.0 *M* $(HClO_4 + NaClO_4)$	3.28	—	229	

[a] Measurements at 25°C.

has been determined in the course of a kinetic study of complex formation between fluoride and Fe^{3+}.[230] The value for K_1 was estimated to be $10^{5.06}$ and the value for K_2 was found to be less than 10^4. Similarly, the fluoride electrode has been used in the kinetic study of the Fe^{3+}–iodide reaction

$$Fe^{3+} + I^- \rightleftharpoons Fe^{2+} + \tfrac{1}{2}I_2$$

in fluoride media.[231] The results have been explained on the basis of the inactivity of the FeF^{2+} toward iodide ions. Also, the kinetics of the complex formation of AlF^{2+} have been described.[230] On the other hand, Baumann[225] has derived values for the stability constants of various aluminum fluoride complexes as a function of the ionic strength of the medium. The values in ammonium nitrate medium extrapolated to infinite dilution are shown in the following table.

Aluminum–fluoride reaction	Stability constant K	log K
$Al^{3+} + F^- \rightleftharpoons AlF^{2+}$	K_1	6.98
$Al^{3+} + 2F^- \rightleftharpoons AlF_2^+$	K_2	5.62
$Al^{3+} + 3F^- \rightleftharpoons AlF_3$	K_3	4.05
$Al^{3+} + 4F^- \rightleftharpoons AlF_4^-$	K_4	2.38

The following values for the beryllium–fluoride and fluoborate stability constants have been determined.

Ionic strength	log K_1	log K_2	log K_3	log K_4	Ref.
Reaction: $Be^{2+} + nF^- = BeF_n^{2-n}$					
In 1 M NaCl	4.9	3.8	2.8	1.4	224
In 1 M $NaClO_4$	5.0	3.8	2.8	1.4	224
Reaction: $B(OH)_3 + nF^- = BF_n(OH)_{4-n}^- + (n-1)OH^-$, $1 \leqslant n \leqslant 4$					
In 1 M $NaNO_3$	− 0.3	− 6.10	− 7.83	− 7.39	232

The stabilities of the mixed species $BF_2(OH)_2^-$, $BF(OH)_3^-$ have also been reported.[233] Other stability constants for some other metal–fluoride complexes are given in Table 8.

In addition, the kinetics of aquation of fluoropentamine cobalt(III)

$$(NH_3)_5CoOH_2^{3+} + F^- \rightleftharpoons (NH_3)_5CoF^{2+} + H_2O$$

have been studied and a value of $\approx 170\ M^{-1}$ has been derived[243] for the equilibrium constant of this reaction at an ionic strength of 0.1 M, pH 4.5. Similarly, the reaction of sulfur hexafluoride (SF_6) with hydrated electrons has been investigated.[244] SF_6 has been found to be a specific electron scavenger in water and D_2O, producing high yields of detectable F^- ions.

TABLE 8

Stability Constants of Metal–Fluoride Complexes[a]

Metal ion	Ionic medium	β_1	β_2	β_3	Ref.	Remarks
Th^{4+}	3.0 M ($HClO_4$ + $NaClO_4$)	3.3×10^4	1.8×10^7	7.9×10^8	229	S_p of ThF_4 =
Th^{4+}	4.0 M $HClO_4$	4.8×10^4	4.5×10^7	—	234	6.8×10^{-16}
U^{4+}	4.0 M $HClO_4$	3.5×10^4	5.3×10^7	4.8×10^9	234	at 20°C
Gd^{3+}	0.5 M $NaClO_4$	2.46×10^3			226	
Eu^{3+}	0.5 M $NaClO_4$	2.54×10^3			226	
Y^{3+}	0.5 M $NaClO_4$	8.22×10^3	14.4×10^6	—	226	
Sc^{3+}	0.5 M $NaClO_4$	1.7×10^6	2.2×10^{11}	1.7×10^{15}	226	
Fe^{3+}	0.5 M $NaClO_4$	1.5×10^5	1.2×10^9	$\sim 1.1 \times 10^{12}$	226	
Sn^{2+}	0.85 M $NaClO_4$	1.8×10^6	5.8×10^8	1.8×10^9	235	
Mg^{2+}	0.5 M $NaClO_4$	20.7			226	
	1.0 M $NaClO_4$	24.0			236	
	1.0 M NaCl	18.6			237	
	0.1 M NaCl	28.7			237	
	1.0 M $NaNO_3$	20.0			238	
	1.0 M $NaClO_4$	20.8			239	
	0.5 M $NaClO_4$	18.0			240	At 16°C
Ca^{2+}	0.5 M $NaClO_4$	4.97			226	
	1.0 M $NaClO_4$	3.4			236	
	1.0 M NaCl	3.85			237	
	0.4 M NaCl	5.01			237	
	1.0 M $NaClO_4$	4.3			239	

Sr^{2+}	1.0 *M* $NaClO_4$	1.3		236	
	1.0 *M* $NaClO_4$	1.4		239	
Ba^{2+}	1.0 *M* $NaClO_4$	0.7		236	
	1.0 *M* $NaClO_4$	0.6		239	
Cd^{2+}	0.5 *M* $NaClO_4$	3.5 (Cd conc. = 0.1 *M*)		240	At 16°C
	0.05 *M* $NaClO_4$	13 (Cd conc. = 0.01 *M*)		240	At 16°C
Zn^{2+}	0.5 *M* $NaClO_4$	2.8		240	At 16°C
Ni^{2+}	0.5 *M* $NaClO_4$	1.5		240	At 16°C
Pb^{2+}	1.0 *M* $NaClO_4$	42.	β_2 could not be detected	241	At 15°C
	0.1 *M* $NaClO_4$	54.		241	At 15°C
	1.0 *M* $NaClO_4$	25.		242	
Ag^{+}	0.5 *M* $NaClO_4$	0.4		240	At 16°C
Tl^{+}	0.5 *M* $NaClO_4$	No complex detected		240	At 16°C
		Mixed stability constants of Pb(II)			
Pb–F–Cl	1.0 *M* $NaClO_4$	$B_{11} = 600$		242	
Pb–F–Br	1.0 *M* $NaClO_4$	$B_{11} = 800$		242	
Pb–F–I	1.0 *M* $NaClO_4$	Precipitation of PbI_2		242	Not amenable to measurement

[a] Measurements at 25°C.

Fluorosilicic acid like its sodium salt has been found with the help of the fluoride electrode to exist with 95% of it dissociated.[136] The complex formed between fluoride and lanthanum alizarin complexone (H_4A) has been investigated using glass and fluoride electrodes.[245] The composition of the complex has been found to correspond to the formula $La(LaA)_4F_2$.

E. PROPERTIES AND APPLICATIONS OF HALIDE-SELECTIVE MEMBRANE ELECTRODES

Many years ago Kolthoff and Sanders[246] reported that fused AgCl and AgBr could be used to prepare membrane electrodes. Both AgCl and AgBr are ionically conducting, the mobile species being Ag^+. Still the precipitates used in suitable form as membranes sense the halides like the fluoride electrode by a solubility mechanism.[20] In view of the disadvantages the halides suffer because of their sensitivity to light and high electrical resistance, the electrodes as already described contain an "inert" silver sulfide matrix that is less soluble than the silver halide. The silver halide solubility also should be such that the level of anion released into the solution is negligible compared to the level expected in the sample solution itself.

The halide-selective electrodes are resistant to surface poisoning and can be used in the presence of oxidizing agents. The response times of the electrodes have been studied by a number of investigators.[69, 71, 247] Rechnitz and co-workers[69, 71] found it to vary from 8 sec for AgI to 20 sec for AgCl electrodes for a twofold change in the anion concentration. Similar response times have been reported for the iodide electrode used in solutions containing cyanide ions.[248] In this respect, Pungor and co-workers[249, 250] have reached the following conclusions:

(1) The response time t, the time taken by the electrode to register a constant value for the potential, can be described by one exponential equation, i.e., $E = E_{t=0} + k_1(1 - e^{k_2 t})$ where k_1 and k_2 are constants.

(2) t is only slightly dependent on the thickness of the membrane layer.

(3) t decreases with an increase in the concentration of noninterfering ion.

(4) t for iodide electrodes in cyanide solutions and in iodide solutions has the same value.

The emf versus time curves have been interpreted on the assumption that the desolvation of the primary ion is the rate-determining step involved in the sensing of the ion by the electrode.[251] As opposed to the definition of response time given above, the dynamic response time t_{95}, the time required by the elctrode to attain 95% of the equilibrium potential in response to a concentration step change, has been used to evaluate the response times of polycrystalline halide-selective membrane electrodes that have been used in flowing systems.[252]

Pungor and his associates, who developed the silicone rubber-based heterogeneous halide-selective membrane electrodes, have published extensively[29, 30, 250, 253, 254] describing the properties and applications of these electrodes.

The chloride, bromide, and iodide electrodes gave linear responses as follows: pX = 1–4 for chloride and bromide and pX = 1–5 for the iodide with a slope of 56 mV/pX (concentration). These electrodes responded also to Ag^+ ions.[29] A slope of 57 mV/pAg in the concentration range pAg = 1–4 was obtained with a chloride electrode. The effects of cations on the calibration curve appear only by affecting the mean ionic activity coefficients. In the case of the iodide electrode it was found that the slopes of the calibration curves and the individual cell emf values were identical for solution samples prepared from KI, ZnI_2, BaI_2, or CeI_3.[69] Similar behavior was observed with the other halide electrodes.[68] Electrolytically generated iodide ion has been recommended for electrode calibration at low concentrations (10^{-4}–10^{-7} M).[255]

The lower limit of detection of the electrodes for the halide ions depends, as did the fluoride electrode, on the solubility product of the halide precipitate. Pungor and co-workers[29, 30, 65, 256–258] have shown that the potential E arising across the membrane electrode is given by

$$E = \frac{RT}{F} \ln \frac{\sqrt{S_p} + \sqrt{S_p + 4a_1^2}}{\sqrt{S_p} + \sqrt{S_p + 4a_2^2}} \tag{51}$$

where a_1 and a_2 are the activities of ions to be estimated or examined and S_p is the solubility product of the silver halide precipitate from which the membrane electrode is formed. When $a \gg S_p$, Eq. (51) reduces to the Nernst equation for a concentration cell. When $a \approx S_p$, the value of E reaches a limit and this gives the lower limit of detection. This way it was shown[257] that the silver bromide electrode ($S_p = 7.7 \times 10^{-13}$) had a limit of detection of 10^{-6} M KBr and silver chloride ($S_p = 1.5 \times 10^{-10}$) a limit of 10^{-5} MKCl. The iodide electrode had a limit of 10^{-7} M.

Covington[42] has given a simpler method of calculating the lower limit of concentration at which the electrode would respond. If S is the solubility of the precipitate (electrode material) in the halide solution of concentration C, then the additional potential ΔE due to the solubility of the precipitate is given by $\Delta E = (RT/F)\ln[(C + S)/C]$. This corresponds to about 2.4 mV when $S = C/10$. This change in potential could be just detected as a deviation from a linear calibration curve. The solubility product S_p for a silver halide material is given by $S_p = S(S + C)$. Thus the quadratic $1.1\ C^2 = 10\ S_p$ gives $C = (9.1\ S_p)^{1/2}$ from which the limits of detection follow as 3×10^{-8}, 2.7×10^{-6}, and 3.8×10^{-5} M for iodide, bromide, and chloride electrodes, respectively.

A theoretical treatment of the selectivity behavior and the detection limit of membrane electrodes made of silver compounds has been given by Morf *et al.*[259] The detection limit is shown to be determined by either the solubility of the membrane material or the activity of the silver defects in the membrane surface, whichever is larger.

Some of the properties of the halide membrane electrodes are given in Table 9. The values for the selectivity constants determined for these electrodes by different methods are given in Table 10, and in Table 11 are given the values supplied by the manufacturers of these electrodes.

Although quantitative values for the selectivity constant K_{ij}^{pot} have been given, the values vary with the concentration of the interfering ion and that of the background electrolyte,[8] if any, and also probably with the method used in the measurement. The values given in Tables 10 and 11 are to be considered approximate; they are, however, a valuable guide in describing

TABLE 9

Characteristics of Heterogeneous Pungor Type of Halide Electrodes[a]

Electrode	Type	Temperature range (°C)	p(ion) range	Selectivity	Suggested pretreatment
Chloride selective	OP–Cl–711	5–50	1–5 pCl	Conc. not to exceed $1 M$ K_2SO_4 $1 M$ KNO_3	1–2 hr in 10^{-1}–10^{-3} M KCl
Bromide selective	OP-Br-711	5–50	1–6 pBr	Conc. not to exceed 10^{-1} M KCl $1 M$ K_2SO_4 $1 M$ KNO_3	1–2 hr in 10^{-1}–10^{-3} M KBr
Iodide selective	OP–I–711	5–50	1–7 pI 1–5 pAg 1–5 p(CN) (pH > 1)	Conc. range not to exceed 10^{-1} M KCl or KBr $1 M$ K_2SO_4 or KNO_3	1–2 hr in 10^{-1}–10^{-3} M KI

[a]Taken from Covington.[42]

TABLE 10

Selectivity Constants (K_{ij}) of Halide Ion-Selective Electrodes for Some Anions[a]

	Chloride-selective electrode K_{ij}			Bromide-selective electrode K_{ij}			Iodide-selective electrode K_{ij}		
Anion	Calculated[b]	Direct	Indirect	Calculated[b]	Direct	Indirect	Calculated[b]	Direct	Indirect
Cl^-	1	1		2.0×10^{-3}	1.8×10^{-3}	6.0×10^{-3}	9.6×10^{-7}	—	3.7×10^{-7}
Br^-				1	1		2.0×10^{-4}	2.1×10^{-4}	1.8×10^{-4}
I^-							1	1	
CN^-					1			1	
SCN^-				1.5×10^{0}	0.2×10^{0}	—	3.0×10^{-4}	—	2.4×10^{-4}
OH^-							1.0×10^{-8}	—	0.9×10^{-8}
NO_3^-						$< 10^{-7}$			$< 10^{-8}$
CrO_4^{2-}	5.2×10^{-5}		4.5×10^{-5}	2.5×10^{-7}		1.1×10^{-7}	5.0×10^{-11}		6.6×10^{-11}
CO_3^{2-}	6.3×10^{-5}		4.6×10^{-5}	3.1×10^{-7}		1.0×10^{-7}			
$S_2O_3^{2-}$		1			1				
SO_3^{2-}			2.0×10^{-1}						5.5×10^{-7}
SO_4^{2-}			$< 10^{-6}$			$< 10^{-7}$			1.0×10^{-7}
$C_2O_4^{2-}$			4.5×10^{-5}						
PO_4^{3-}	1.3×10^{-4}		0.5×10^{-4}	6.3×10^{-7}		3.1×10^{-7}	1.2×10^{-10}		0.2×10^{-10}
AsO_4^{3-}	3.3×10^{-4}		2.0×10^{-4}	1.6×10^{-6}		1.2×10^{-6}	3.2×10^{-10}		2.6×10^{-10}
$Fe(CN)_6^{4-}$							2.4×10^{-6}		3.5×10^{-6}

[a] Measured by the methods of Pungor and his co-workers.[30, 250, 260, 261]
[b] Calculated using Eq. (178b) of Chapter 3.

TABLE 11

The Selectivity Constants (K_{ij}) for Halide-Selective Electrodes of the Firms of Philips[19, 40] and Orion[14, 30]

	Cl-selective electrode		Br-selective electrode		I-selective electrode	
Ion	Philips	Orion	Philips	Orion	Philips	Orion
Cl^-			6.0×10^{-3}	2.5×10^{-3}	6.6×10^{-6}	—
Br^-	1.2	3.0×10^{2}			6.5×10^{-5}	
I^-	86.5	2.0×10^{6}	20	5.0×10^{3}		
CN^-	400	5.0×10^{6}	25	1.2×10^{4}	0.34	Interferes
OH^-	2.4×10^{-2}	1.2×10^{-2}	1.0×10^{-3}	3.0×10^{-5}	—	—
CO_3^{2-}	3.0×10^{-3}		2.3×10^{-3}		1.2×10^{-4}	
CrO_4^{2-}	1.8×10^{-3}		1.6×10^{-3}		3.7×10^{-3}	
$S_2O_3^{2-}$	60	10^{2}	1.5		7.1×10^{-4}	Interferes
NH_3		8		0.5		
S^{2-}		Must be absent[a]		Must be absent		Interferes

[a] No interference from F^-, NO_3^-, HCO_3^-, SO_4^{2-}, and PO_4^{3-}.

the performance of the electrodes. Strict comparison between electrodes of the same kind but from different sources can only be made by testing them simultaneously using the same method of measurements. Otherwise it may be unwise to attach too much significance to the values of K_{ij}^{pot}.

Scanning electron microscopy of the crystal electrode surface (AgI–Ag_2S) has revealed[262] that upon prolonged use, the electrode surface is attacked and the active material is removed from the membrane.

Selectrodes, another type of halide electrodes, have been tested and found to give Nernstian response over the pH range 3–10 and pX range 3–5 (Cl^- and Br^-) and 3–6 (I^-). The behavior of the electrodes has been formally interpreted[25] by treating them as second-order silver halide electrodes. Thus the potential of the electrode is represented by

$$E = E^\circ_{(Ag)} + (RT/nF)\ln a_{Ag^+} \tag{52}$$

But $S_p = a_{Ag^+} a_{X^-}$, thus Eq. (52) becomes

$$E = E^\circ_{(Ag)} + \frac{RT}{nF}\ln S_p - \frac{RT}{nF}\ln a_{X^-} \tag{53a}$$

or

$$E = E^\circ - 59.1\mathrm{p}S_p + 59.1\mathrm{pX}^- \quad \text{at } 25°\text{C} \tag{53b}$$

The values of E calculated for the electrodes using the values $E^\circ = 558$ mV (vs. SCE), $\mathrm{p}S_p = 10.06$ (AgI), 12.3 (AgBr), and 9.71 (AgCl) agreed very well with the measured values at $pX^- = 3.01$ for Cl and Br electrodes only.

From Tables 10 and 11, it is seen that the halide electrodes respond very well to the cyanide and the thiosulfate ions. Furthermore, Pungor and co-workers[30, 65, 248, 250] showed that the electrodes, particularly the iodide electrode, displayed the same selectivity to various anions in the presence of the cyanide ion as in the presence of the iodide ion itself (see Table 12).

TABLE 12

Selectivity of the Iodide Membrane Electrode to Other Ions in the Presence of the CN^- or I^- Ion[a]

	K_{ij}, primary ion i	
Interfering ion	CN^-	I^-
Cl^-	10^{-5}–10^{-6}	10^{-6}
Br^-	10^{-3}–10^{-4}	10^{-4}
I^-	1	—
CN^-	—	1
NH_4^+	10^{-5}–10^{-6}	10^{-6}
SO_4^{2-}	10^{-5}–10^{-6}	10^{-6}

[a] Data of Toth and Pungor.[248]

The basic reaction is

$$AgX + 2CN^- \rightleftharpoons Ag(CN)_2^- + X^-$$

This reaction is opposed by the liberated halide ion on the electrode surface. The equilibrium constant is given by

$$K_{X,CN} = \frac{a_{Ag(CN)_2}a_X}{a_{CN}^2 a_{AgX}}$$

The composite potential is assumed to be given by

$$E = E^\circ + 0.059 \log\left(a_X + a_X^* + K_{X,CN}a_{CN}^2\right) \tag{54}$$

where a_X is the activity of the halide ion in solution, a_X^* is the activity of the halide ion released by the cyanide in the membrane layer, a_{CN} is the activity of the cyanide ion in solution, and $K_{X,CN}$ is the dissolution constant of the AgX electrode to CN^- ions. If the solution contains only cyanide ions, Eq. (54) becomes

$$E = E^\circ + 0.059 \log\left(\tfrac{1}{2}a_{CN} + K_{X,CN}a_{CN}^2\right) \tag{55}$$

The experimental values for $K_{X,CN}$ were found to lie between 1 and 0.1 and Eq. (54) was followed in the range 10^{-4}–2.0 *M* cyanide concentration.[30, 248] It is doubtful whether this reaction alone determined the electrode response. Fleet and Storp[263] have proposed that the reaction

$$AgX + CN^- \rightleftharpoons AgCN + X^-$$

might also contribute to the potential. When excess cyanide is present, silver halide dissolves completely in addition to responding to the excess ion. This behavior has been found to be true in the case of $S_2O_3^{2-}$ ion. The value of $K_{X,\,S_2O_3}$ has been found to be between 10^{-2} and 10^{-3} [30, 250]

A characteristic potential for the halide-selective membrane electrodes (AgX) has been assigned.[253, 264] This characteristic potential is a common potential pertaining to both the cationic and anionic functions of the membrane electrode when the activities of both components are equal to each other in the membrane phase. This is the average of two potentials at pAg = 0 and pX = 0 (obtained by extrapolation) measured in solutions of $AgNO_3$ and KX using membrane electrodes and the corresponding Ag–AgX electrode of the second kind. It was found that the potential measured with the membrane electrodes was more positive by a constant value than the one measured with the Ag–AgX electrode. The temperature dependence of these standard potentials was also determined. The characteristic potential of the membrane electrode and its temperature dependence were found to depend on a number of factors pertaining to the precipitate used in the preparation of the electrode. The temperature coefficient of the membrane electrode was of opposite sign to that of the corresponding Ag–AgX electrode of the second kind.

The behavior of halide membrane electrodes in nonaqueous solvents has been described by Pungor and associates[250, 265] who used the electrodes in alcohols, ketones, and other solvents such as dimethylformamide (DMF). Experiments with the halide membrane were carried out to determine the solubility products (S_p) of the silver halides in nonaqueous media and the selectivity constants for the AgI electrode in solutions of bromide in various solvent mixtures. Before carrying out an emf measurement it was found necessary to condition the electrode in the appropriate solvent mixture in which it swelled. The results obtained were used to check the applicability of Eq. (178b) of Chapter 3 to these aqueous and nonaqueous systems. The results given in Table 13 show that the concepts underlying Eq. (178b) of Chapter 3 are applicable to describe the behavior of the halide electrodes in these aqueous–nonaqueous media. Similarly, Ficklin and Gofschall[266] used bromide and iodide electrodes to evaluate their performance in several alcohols and glacial acetic acid. The relationship between the dielectric constant of the solvent and the potential for a bromide electrode in LiBr solutions (10^{-4} M) at 23°C is shown in Table 14. There is a change in potential of nearly 114 mV for a change in the dielectric constant of water to that of methanol solution. The corresponding change noted in the case of iodide solutions was about 69 mV.

TABLE 13

Selectivity constant K_{I-Br} for the Iodide-Selective Membrane Electrode in Aqueous–Nonaqueous Media[a]

Nonaqueous liquid	Volume % nonaqueous solvent	Solubility product (S_p) $-\log S_p$			$-\log K_{I-Br}$	
		AgCl	AgBr	AgI	Calculated[b]	Direct
CH_3OH	10	10.10	12.48	16.12	3.64	3.78
CH_3OH	90	10.50	14.40	17.42	3.02	3.02
C_2H_5OH	10	10.20	12.54	16.16	3.62	3.70
C_2H_5OH	90	10.32	14.30	17.30	3.00	2.76
n-C_3H_7OH	10	9.90	12.50	16.20	3.70	3.74
n-C_3H_7OH	40	10.00	12.82	16.40	3.58	3.42
iso-C_3H_7OH	10	9.88	12.30	10.04	3.74	3.78
iso-C_3H_7OH	40	10.28	12.78	16.28	3.50	3.16
$(CH_3)_2CO$	10	9.8	12.40	16.20	3.82	3.75
$(CH_3)_2CO$	40	10.5	12.60	16.20	3.56	3.55
DMF	10	10.04	12.22	15.96	3.74	3.75
DMF	60	10.66	12.70	15.34	2.64	2.60

[a]Data of Pungor and co-workers.[250, 261, 265]
[b]Calculated using Eq. (178b) of Chapter 3.

TABLE 14

Potential of Bromide Electrode in LiBr Solution as a Function of Dielectric Constant[a]

Solvent	Dielectric constant	Potential (mV)
Water	78.54	+ 55
Methanol	32.63	− 59
Ethanol	24.30	− 89
Butanol	17.1	− 125
Pentanol	13.9	− 138
Hexanol	13.3	− 143
Acetic acid	6.15	− 148

[a]Data of Ficklin and Gofschall.[266] Concentration $10^{-4}M$, measurements at 23°C.

The halide electrodes have been used in a number of studies. Their use in automation has been explored by Mallissa and Jellinek,[267] and their sensitivity and selectivity in the presence of interfering ions have been described.[268] Because of the formation of mixed crystals, the electrodes find limited use in the estimation of halide mixtures although they are best suited for individual halide determinations.[269] In titrations involving mixtures of halides (Cl^-, Br^-, I^-) for the estimation of any one halide using the corresponding ion-selective membrane electrode, an incorrect end point would be obtained due to adsorption of the halides on the surface of the precipitate formed during titration. This can be overcome by using a large excess of KNO_3[270] or a lower concentration of $Ba(NO_3)_2$.[30]

Procedures for the determination of the halide contents of water using halide-selective membrane electrodes and for the establishment of water hardness have been worked out.[90] The quantities of Br^-, $S_2O_3^{2-}$, and CN^- in high-purity waters have been determined.[271, 272] There is the possibility of these substances poisoning the electrodes. Also, the halides present in rainfall can be estimated.[273]

These electrodes of the second kind and other solid state electrodes constructed of the same materials have been studied to determine the differences in their responses.[274] The standard potentials of these electrodes have been computed.

The chloride-selective membrane electrode has been used in the following ways:

(1) To determine the Cl^- ion content of biological fluids[275] (urine, blood, serum, without centrifuging the proteins[276]). Dahms *et al.*[277] have determined the Cl^- ion activity in human serum and pointed out that the effect of protein on the chloride activity coefficient should be taken into account.

(2) To determine the single ion activity in pure solutions of NaCl in the concentration range 10^{-5}–6.0 *M*.[278] The relative usefulness of the MacInnes and Bates–Guggenheim conventions (see Section B) about single ion activity coefficients of the Cl^- ion has been discussed.

(3) To determine the amounts of Cl in analytical reagent grade KOH.[257] This required removal of K using a cation exchanger column and use of the effluent to estimate the chloride. Similar procedures have been described for the estimation of chloride in the presence of bromide and iodide ions.

(4) To determine the chloride content of soils,[30, 279, 280] tap water,[30] high-purity waters,[280, 281] natural waters,[272] and milk.[30, 282]

(5) To determine the chloride content of sweat in newborn infants. This determination is useful in cystic fibrosis programs, since the Cl level is elevated in children born with cystic fibrosis. Sweat is stimulated by either heat[283] or pilocarpine iontophoresis.[284]

(6) To determine chloride in cheese,[285] plant tissues,[286] pharmaceutical[287, 288] and other chloride-containing products,[289] organic compounds,[290] pesticides,[291] calcium phosphates,[94, 121] and gaseous mixtures.[292, 293]

Chloride-selective membrane electrodes with liquid–solid–solid connections and only solid–solid connections have been tried in chloride analysis.[294] Electrodes with only solid connections seem to be good.

Microquantities of chloride have been estimated with the help of the electrode.[295] It has been used in a kinetic study involving cyclization of methyl-bis(β-chloroethyl)amine hydrochloride,[296] and to measure the chloride content of aluminum chloride diisopropylate after hydrolysis.[297]

The bromide-selective membrane electrode has been used in the estimation of bromide in organic compounds,[290] serum,[298] plasma,[299] soft drinks,[300] and pharmaceutical products.[288] It is used also in the determination of bromide ion activity in solutions of *n*-decyl-, *n*-dodecyl-, *n*-tetradecyl-, and *n*-hexadecyltrimethyl ammonium bromides.[301] It has been used in the estimation of a variety of thiols by titration with mercuric perchlorate in a medium containing acetone.[302]

The iodide-selective membrane electrode, as already pointed out, responds well to both cyanide and thiosulfate ions. Consequently, the electrode can be used to determine these ions in solution[29, 30, 248] in addition to determining the iodide and iodate ions in other solutions.[303] It can be used to determine iodine in organic substances and in biological material.[304]

In view of the high selectivity of the electrode to iodide ions, it has been used in the direct determination of iodide in mineral waters[29, 30, 305] and to follow chemical reactions in which iodide is released.[29, 30] In this way, the SO_2 content of a gas can be determined by bubbling the gas through an iodine solution. Excess iodine is removed by extracting with CCl_4 and the iodide produced is estimated with the electrode. A method for the determination of ^{131}I and ^{125}I using the iodide-selective electrode in highly radioactive solutions has been worked out.[306] The electrode may be used in the estimation of iodide in serum[298] and to detect Au(III).[307] Its use in following changes in iodide concentration in a system in which both chemical equilibrium and reaction exist has been discussed.[308, 309]

Equilibrium constants for the species existing in aqueous iodine solutions have been computed.[310] The reactions involved are

$$I_2 + H_2O \rightleftharpoons HIO + H^+ + I^- \quad \text{(hydrolysis, } K_1\text{)}$$

$$I_3^- \rightleftharpoons I_2 + I^- \quad \text{(dissociation, } K_2\text{)}$$

$$I_2 + H_2O \rightleftharpoons H_2OI^+ + I^- \quad \text{(hydrolytic dissociation, } K_3\text{)}$$

The concentration of I^- was estimated with the help of the iodide-selective electrode. The concentrations of the other species were determined with the help of a glass electrode (H^+ ion concentration) and spectrophotometer (I_2 and I_3^- concentrations). The values for K_1 and K_2 were determined as a function of termperature. At 50°C, the values were $K_1 = 41.2 \times 10^{-13}$, $K_2 = 2.25 \times 10^{-3}$, and $K_3 = 3 \times 10^{-9}$ (approximate).

The iodide electrode is used to estimate iodide in feeds and plants,[311] in selenium iodide,[157] in milk,[312] and in the detection of the end point in potentiometric titrations involving mercuric ion.[313]

Molybdenum and tungsten catalyze the H_2O_2–iodide reaction whose course has been followed by using the iodide-selective membrane electrode.[314] Conversely, as little as 0.004 $\mu g/ml$ of molybdenum and tungsten can be estimated using this reaction. In a similar way, formaldehyde has been determined by potentiometric titration using the iodide electrode.[315] The reactions involved are

$$HCHO + I_2 + 2\,KOH \rightleftharpoons 2\,KI + HCOOH + H_2O$$
$$KI + AgNO_3 \rightleftharpoons AgI + KNO_3$$
$$HCHO = 2KI = 2AgNO_3$$

An interesting variation of the AgI–Ag_2S membrane has been described.[316] Coprecipitated AgI–Ag_2S powder pressed into a disk (0.7 mm thick) is split into two pieces which are fixed into a plastic body in such a way as to serve as two identical iodide-selective electrodes. Calibration of these electrodes using a reference electrode gave, in the concentration range of I^- from 10^{-5} to 10^{-1} M, values for cell emfs that were fairly close with almost identical slopes of 57.6 mV/decade concentration. This split electrode could be used in a differential mode to estimate iodide concentrations.

REFERENCES

1. C. Kittel, "Introduction to Solid State Physics," 3rd ed. Wiley, New York, 1966.
2. M. S. Frant and J. W. Ross, Jr., *Science* **154**, 1553 (1966).
3. A. Zalkin and D. H. Templeton, *J. Amer. Chem. Soc.* **75**, 2453 (1953).
4. K. Schlyter, *Ark. Kemi* **5**, 61, 73 (1953).
5. R. Solomon, A. Sher, and M. W. Muller, *J. Appl. Phys.* **37**, 3427 (1966).
6. M. Goldman and L. Shen, *Phys. Rev.* **144**, 321 (1966).
7. T. Anfalt and D. Jagner, *Anal. Chim. Acta* **50**, 23 (1970).
8. A. K. Covington, *Crit. Rev. Anal. Chem.* **3**, 355 (1974).
9. R. R. Tarasyants, R. N. Potsepkina, V. P. Roze, and E. A. Bondarevskaya, *Zh. Anal. Chem.* **27**, 808 (1972).
10. Select Ion Electrodes, Bulletin 7145-A. Beckman Instruments Inc., Fullerton, California.
11. Fluoride Ion Selective Electrode 3-803 (1968). Coleman Instruments, Oakbrook, Illinois.
12. pH and Ion Selective Electrodes. Corning–EEL Scientific Instruments, Corning Glass Works, Corning, New York.
13. Model 32 FMS—A Fluoride Measuring System GS 6-1A2A (1968); Fluoride Ion Measuring Electrode, GS 1-3F1H (1970). Foxboro Co., Foxboro, Massachusetts.
14. Analytical Methods Guide (1971). Orion Research, Inc., Cambridge, Massachusetts.
15. Philips Ion Selective Solid State Electrode for Fluoride, Type No. 15 560-F. Philips Electronic Instruments Inc., Mount Vernon, New York.
16. L. Sucha, M. Suchanek, and Z. Urner, *Proc. Conf. Appl. Phys. Chem. 2nd, Veszprem* **1**, 651 (1971).

17. H. Adametzova and R. Vadura, *J. Electroanal. Chem. Interfacial Electrochem.* **55**, 53 (1974).
18. D. Weiss, *Chem. Listy* **65**, 1071 (1971).
19. Philips Ion Selective Solid State Electrodes for Iodide, Bromide, Chloride, and Cyanide, Type IS 550. Philips Electronic Instruments Inc., Mount Vernon, New York.
20. J. W. Ross, Jr., *in* "Ion Selective Electrodes" (R. A. Durst, ed.), Chapter 2. Nat. Bur. Std. Spec. Publ. 314, Washington, D.C., 1969.
21. J. D. Czaban and G. A. Rechnitz, *Anal. Chem.* **45**, 471 (1973).
21a. J. F. Lechner and I. Sekerka, *J. Electroanal. Chem. Interfacial Electrochem.* **57**, 317 (1974).
22. Chloride Ion Measuring Electrode GS 1-3F 1 C. Foxboro Co., Foxboro, Massachusetts.
23. Radiometer A. S. Instructions for Chloride Selectrode, Type F 1012C1; Instructions for Bromide Selectrode, Type F 1022 Br. Marketed by London Co., Cleveland, Ohio.
24. J. Ruzicka and C. Lamm, *Anal. Chim. Acta* **53**, 206 (1971).
25. J. Ruzicka and C. Lamm, *Anal. Chim. Acta* **54**, 1 (1971).
26. S. Mesari and E. A. M. F. Dahm, *Anal. Chim. Acta* **64**, 431 (1973).
27. J. P. Sapio, J. F. Colaruotalo, and J. M. Bobbitt, *Anal. Chim. Acta* **67**, 240 (1973).
28. E. Pungor and E. Hallos-Rokosinyi, *Acta Chim. Acad. Sci. Hung.* **27**, 63 (1961).
29. E. Pungor, *Anal. Chem.* **39**, 28A (1967).
30. E. Pungor and K. Toth, *Analyst* **95**, 625 (1970).
31. A. K. Covington, *in* "Ion Selective Electrodes" (R. A. Durst, ed.), Chapter 3. Nat. Bur. Std. Spec. Publ. 314, Washington, D.C., 1969.
32. Ion Selective Electrodes, 1971. Radelkis Electrochemical Instruments, Budapest, Hungary.
33. A. M. G. Macdonald and K. Toth, *Anal. Chim. Acta* **41**, 99 (1968).
34. E. Pungor, J. Havas, and K. Toth, *Z. Chem.* **5**, 9 (1965).
35. E. B. Buchanan and J. L. Seago, *Anal. Chem.* **40**, 517 (1968).
36. M. Mascini and A. Liberti, *Anal. Chim. Acta* **47**, 339 (1969).
37. J. C. Van Loon, *Anal. Chim. Acta* **54**, 23 (1971).
38. Bromide Ion Selective Electrode 3-801; Chloride Ion Selective Electrode 3-802. Coleman Instruments, Oakbrook, Illinois.
39. M. R. Thompson, *J. Res. Nat. Bur. Std.* **9**, 833 (1932).
40. J. Koryta, *Anal. Chim. Acta* **61**, 329 (1972).
41. The Completely New Coleman Ion Activity Sensing System, Bulletin B-329. Coleman Instruments, Oakbrook, Illinois.
42. A. K. Covington, *in* "Ion Selective Electrodes" (R. A. Durst, ed.), Chapter 4. Nat. Bur. Std. Spec. Publ. 314, Washington, D.C., 1969.
43. D. J. G. Ives and G. J. Janz, "Reference Electrodes." Academic Press, New York, 1961.
44. R. G. Bates and M. Alfenaar, *in* "Ion Selective Electrodes" (R. A. Durst, ed.), Chapter 6. Nat. Bur. Std. Spec. Publ. 314, Washington, D.C., 1969.
45. J. Koryta, J. Dvorak, and V. Bohackova, "Electrochemistry." Methuen, London, 1970.
46. R. G. Bates, *J. Res. Nat. Bur. Std.* **A66**, 179 (1962).
47. R. G. Bates, "Determination of pH," Chapter 3. Wiley, New York, 1964.
48. G. Mattock and D. M. Band, *in* "Glass Electrodes for Hydrogen and other Cations, Principles and Practice" (G. Eisenman, ed.), Chapter 2. Dekker, New York, 1967.
49. R. G. Bates, *Pure Appl. Chem.* **36**, 407 (1973).
50. R. G. Bates and E. A. Guggenheim, *Pure Appl. Chem.* **1**, 163 (1960).
51. D. A. MacInnes, *J. Amer. Chem. Soc.* **41**, 1086 (1919).
52. R. M. Garrels, *in* "Glass Electrodes for Hydrogen and other Cations, Principles and Practice" (G. Eisenman, ed.), Chapter 13. Dekker, New York, 1967.

53. R. H. Stokes and R. A. Robinson, *J. Amer. Chem. Soc.* **70**, 1870 (1948).
54. R. G. Bates, B. R. Stapes, and R. A. Robinson, *Anal. Chem.* **42**, 867 (1970).
55. R. A. Robinson and R. H. Stokes, "Electrolyte Solutions," Appendix 8-10. Butterworths, London and Washington, D.C., 1959.
56. R. A. Robinson, W. C. Duer, and R. G. Bates, *Anal. Chem.* **43**, 1862 (1971).
57. R. A. Robinson and R. G. Bates, *Anal. Chem.* **45**, 1666 (1973).
58. R. A. Robinson and R. G. Bates, *Anal. Chem.* **45**, 1668 (1973).
59. R. A. Robinson and R. H. Stokes, "Electrolyte Solutions," p. 441 (eq. 15-9). Butterworths, London and Washington, D.C., 1959.
60. J. V. Leyendekkers, *Anal. Chem.* **43**, 1835 (1971).
61. J. N. Butler and R. Huston, *Anal. Chem.* **42**, 1308 (1970).
62. J. Bagg and G. A. Rechnitz, *Anal. Chem.* **45**, 271 (1973).
63. J. Bagg and G. A. Rechnitz, *Anal. Chem.* **45**, 1069 (1973).
64. G. J. Moody and J. D. R. Thomas, *in* "Ion Selective Electrodes" (E. Pungor, ed.), p. 97. Akademiai Kaido, Budapest, 1973.
65. E. Pungor and K. Toth, *Anal. Chem. Acta* **47**, 291 (1969).
66. K. Srinivasan and G. A. Rechnitz, *Anal. Chem.* **41**, 1203 (1969).
67. G. Eisenman, D. O. Rudin, and J. V. Cosby, *Science* **126**, 831 (1957).
68. G. A. Rechnitz, *Chem. Eng. News* **45**(25), 146 (1967).
69. G. A. Rechnitz, M. R. Kresz, and S. B. Zamochnik, *Anal. Chem.* **38**, 973 (1966).
70. T. S. Light and J. L. Swartz, *Anal. Lett.* **1**, 825 (1968).
71. G. A. Rechnitz and M. R. Kresz, *Anal. Chem.* **38**, 1786 (1966).
72. N. A. Lange (ed.), "Hand Book of Chemistry," 10th Ed., p. 1088. McGraw Hill, New York, 1961.
73. J. N. Butler, *in* "Ion Selective Electrodes" (R. A. Durst, ed.), Chapter 5. Nat. Bur. Std. Spec. Publ. 314, Washington, D.C., 1969.
74. R. P. Buck, *Anal. Chem.* **46**, 28R (1974).
75. A. Sher, R. Solomon, K. Lee, and M. W. Muller, *Phys. Rev.* **144**, 593 (1966).
76. R. W. Ure, *J. Chem. Phys.* **26**, 1363 (1957).
77. H. M. Stahr and D. O. Clardy, *Anal. Lett.* **6**, 211 (1973).
78. J. Vesely, *J. Electroanal. Chem. Interfacial Electrochem.* **41**, 134 (1973).
79. M. J. D. Brand and G. A. Rechnitz, *Anal. Chem.* **40**, 509 (1968).
80. J. Vesely, *Chem. Listy* **65**, 86 (1971).
81. J. J. Lingane, *Anal. Chem.* **39**, 881 (1967).
82. R. A. Durst and J. K. Taylor, *Anal. Chem.* **39**, 1483 (1967).
83. R. Bock and S. Strecker, *Z. Anal. Chem.* **235**, 322 (1968).
84. R. E. Mesmer, *Anal. Chem.* **40**, 463 (1968).
85. N. Parthasarathy, J. Buffle, and D. Monnier, *Anal. Chim. Acta* **68**, 185 (1974).
86. J. Buffle, N. Parthasarathy, and W. Haerdi, *Anal. Chim. Acta* **68**, 253 (1974).
87. J. Vesely and K. Stulik, *Anal. Chim. Acta* **73**, 157 (1974).
88. J. J. Lingane, *Anal. Chem.* **40**, 935 (1968).
89. G. Neumann, *Ark. Kemi* **32**, 229 (1970).
90. W. E. Bazzelle, *Anal. Chim. Acta* **54**, 29 (1971).
91. E. W. Baumann, *Anal. Chim. Acta* **42**, 127 (1968).
92. A. F. Berndt and R. Stearns, *Anal. Chim. Acta* **74**, 446 (1975).
93. R. A. Durst, *in* "Ion Selective Electrodes" (R. A. Durst, ed.), Chapter 11. Nat. Bur. Std. Spec. Publ. 314, Washington, D.C., 1969.
94. L. G. Bruton, *Anal. Chem.* **43**, 579 (1971).
95. G. Graw, *Analyst* **77**, 661 (1952).
96. F. J. Rossotti and H. Rossotti, *J. Chem. Ed.* **42**, 475 (1965).

97. A. Liberti and M. Mascini, *Anal. Chem.* **41**, 676 (1969).
98. W. Selig, *Mikrochim. Acta* 87 (1973).
99. M. S. Front and J. W. Ross, Jr., *Anal. Chem.* **40**, 1169 (1968).
100. R. A. Durst, *Mikrochim. Acta* **3**, 611 (1969).
101. R. A. Durst and J. W. Ross, Jr., *Anal. Chem.* **40**, 1343 (1968).
102. E. W. Baumann, *Anal. Chim. Acta* **54**, 189 (1971).
103. T. B. Warner, *Anal. Chem.* **41**, 527 (1969).
104. H. Zentner, *Chem. Ind.* 480 (1973).
105. K. Srinivasan and G. A. Rechnitz, *Anal. Chem.* **40**, 509 (1968).
106. K. Srinivasan and G. A. Rechnitz, *Anal. Chem.* **40**, 1818 (1968).
107. B. A. Raby and W. E. Sunderland, *Anal. Chem.* **40**, 939 (1968).
108. R. A. Durst, *Anal. Chem.* **41**, 2089 (1969).
109. J. C. Van Loon, *Anal. Lett.* **1**, 393 (1968).
110. D. Weiss, *Chem. Listy* **63**, 1152 (1969).
111. C. R. Edmond, *Anal. Chem.* **41**, 1327 (1969).
112. J. L. Guth and R. Wey, *Bull. Soc. Fr. Mineral. Cristallogr.* **92**, 105 (1969).
113. R. T. Oliver and A. G. Clayton, *Anal. Chim. Acta* **51**, 409 (1970).
114. B. L. Ingram, *Anal. Chem.* **42**, 1825 (1970).
115. J. A. Blay and J. H. Ryland, *Anal. Lett.* **4**, 653 (1971).
116. M. A. Peters and D. M. Ladd, *Talanta* **18**, 665 (1971).
117. P. Rinaldo and P. Montesi, *Chim. Ind.* **53**, 26 (1971).
118. D. R. Simpson, *Amer. Mineral.* **54**, 1711 (1969).
119. L. Evans, R. D. Hoyle, and J. B. Macaskill, *N.Z.J. Sci.* **13**, 143 (1970).
120. J. Trusl, *Chem. Listy* **64**, 322 (1970); *J. Ass. Offic. Agr. Chem.* **53**, 267 (1970).
121. E. J. Duff and J. L. Stuart, *Anal. Chim. Acta* **52**, 155 (1970); **57**, 233 (1971); *Talanta* **19**, 76 (1972).
122. R. Ponget, *Chim. Anal.* **53**, 479 (1971).
123. R. L. Clements, G. A. Sergeant, and P. J. Webb, *Analyst* **96**, 51 (1971).
124. T. S. Light and R. F. Mannion, *Anal. Chem.* **41**, 107 (1969).
125. H. J. Francis, Jr., J. H. Deonarine, and D. D. Persing, *Microchem. J.* **14**, 580 (1969).
126. J. Pavel, R. Kuebler, and H. Magnor, *Microchem. J.* **15**, 192 (1970).
127. M. E. Aberlin and C. A. Bunton, *J. Org. Chem.* **35**, 1825 (1970).
128. D. A. Shearer and G. F. Morris, *Microchem. J.* **15**, 199 (1970).
129. W. Selig, *Z. Anal. Chem.* **249**, 30 (1970).
130. J. L. Stuart, *Analyst* **95**, 1032 (1970).
131. M. B. Terry and F. Kasler, *Mikrochim. Acta* 569 (1971).
132. T. B. Warner, *Science* **165**, 178 (1969).
133. T. Anfalt and D. Jagner, *Anal. Chim. Acta* **53**, 13 (1971).
134. R. H. Babcock and K. A. Johnson, *J. Amer. Water Works Ass.* **60**, 953 (1968).
135. N. T. Crosby, A. L. Dennis, and J. G. Stevens, *Analyst* **93**, 643 (1968).
136. N. T. Crosby, *J. Appl. Chem.* **19**, 100 (1969).
137. T. S. Light, R. F. Mannion, and K. S. Fletcher, *Talanta* **16**, 1441 (1969).
138. D. E. Collis and A. A. Diggens, *Water Treat. Exam.* **18**, 192 (1969).
139. J. E. Harwood, *Water Res.* **3**, 273 (1969).
140. T. S. Light, *in* "Ion Selective Electrodes" (R. A. Durst, ed.), Chapter 10. Nat. Bur. Std. Spec. Publ. 314, Washington, D.C., 1969.
141. S. J. Patterson, N. G. Bunton, and N. T. Crosby, *Water Treat. Exam.* **18**, 182 (1969).
142. T. B. Warner, *Water Res.* **5**, 459 (1971).
143. P. K. Ke and L. W. Reiger, *Anal. Chim. Acta* **53**, 23 (1971).
144. E. Belack, *J. Amer. Water Works Ass.* **64**, 62 (1972).

145. F. Brudevold, E. Moreno, and Y. Bakhos, *Arch. Oral. Biol.* **17**, 1155 (1972).
146. L. A. Elfers and C. E. Decker, *Anal. Chem.* **40**, 1658 (1968).
147. R. S. Yunghans and T. B. McMuller, *Fluoride* **3**, 143 (1970).
148. M. Buck and G. Rensman, *Fluoride* **4**, 5 (1971).
149. A. Liberti and M. Mascini, *Fluoride* **4**, 49 (1971).
150. R. C. Harris and H. H. Williams, *Appl. Meteorol.* **8**, 299 (1969).
151. E. F. Croomes and R. C. McNutt, *Analyst* **93**, 729 (1968).
152. M. S. Frant, *Plating* **54**, 702 (1967); *Galvanotech.* **63**, 745 (1972).
153. D. E. Jordan, *J. Ass. Offic. Agr. Chem.* **53**, 447 (1970).
154. H. Boesch and H. Weingerl, *Z. Anal. Chem.* **262**, 104 (1972).
155. H. H. Moeken, H. Eschrich, and G. Willeborts, *Anal. Chim. Acta* **45**, 233 (1969).
156. B. A. Raby and W. E. Sunderland, *Anal. Chem.* **39**, 1304 (1967).
157. V. Westerlund-Helmerson, *Anal. Chem.* **43**, 1120 (1971).
158. F. C. Chang, H. T. Tsai, and S. C. Wu, *Anal. Chim. Acta* **71**, 477 (1974).
159. T. A. Palmer, *Talanta* **19**, 1141 (1972).
160. J. Thomas, Jr., and H. J. Gluskoter, *Anal. Chem.* **46**, 1321 (1974).
161. C. W. Louw and J. F. Richards, *Analyst* **97**, 334 (1972).
162. K. E. Macleod and H. L. Crist, *Anal. Chem.* **45**, 1272 (1973).
163. T. B. Warner and D. J. Bressan, *Anal. Chim. Acta* **63**, 165 (1973).
164. N. Shiraishi, Y. Murata, G. Nakagawa, and K. Kodama, *Anal. Lett.* **6**, 893 (1973).
165. C. E. Robertson and J. D. Hem, *Geochim. Cosmochim. Acta* **32**, 1343 (1968).
166. J. N. Wilson and C. Z. Marczewski, *Anal. Chem.* **45**, 2409 (1973).
167. T. Eriksson, *Anal. Chim. Acta* **65**, 417 (1973).
168. J. R. Entwistle, C. J. Weedon, and H. J. Hayes, *Chem. Ind.* 433 (1973).
169. P. Venkateswarlu, *Anal. Chem.* **46**, 878 (1974).
170. D. R. Taves, *Talanta* **15**, 1015 (1968).
171. H. C. McCann, *Arch. Oral Biol.* **13**, 475 (1968).
172. F. W. Barnes and J. Runcie, *J. Clin. Pathol.* **21**, 668 (1968).
173. C. W. Weber and B. L. Reid, *J. Nutr.* **97**, 90 (1968).
174. P. J. Ke, L. W. Regier, and H. E. Power, *Anal. Chem.* **41**, 1081 (1969).
175. W. I. Rogers and J. Wilson, *Anal. Biochem.* **32**, 31 (1969).
176. B. F. Erlanger and R. A. Sack, *Anal. Biochem.* **33**, 318 (1970).
177. J. D. Neefus, J. Cholak, and B. E. Saltzman, *Amer. Ind. Hyg. Ass. J.* **31**, 96 (1970).
178. G. J. Kakabadse, B. Manohin, J. M. Bather, E. C. Weller, and P. Woodbridge, *Nature (London)* **229**, 626 (1971).
179. L. Singer and W. D. Armstrong, *Anal. Chem.* **40**, 613 (1968).
180. G. Bang, T. Kristoffersen, and K. Meyer, *Acta Pathol. Microbiol. Scand.* **A78**, 49 (1970).
181. P. Grøn, K. Yao, and M. Spinelli, *J. Dent. Res. Suppl. to No.* 5 **48**, 709 (1969).
182. P. Hotz, H. R. Muhlemann, and A. Schait, *Helv. Odontol. Acta* **14**, 26 (1970).
183. D. R. Taves, *Nature (London)* **217**, 1050 (1968).
184. L. Singer and W. D. Armstrong, *Arch. Oral Biol.* **14**, 1343 (1969).
185. B. W. Fry and D. R. Taves, *J. Lab. Clin. Med.* **75**, 1020 (1970).
186. R. Aasenden, F. Brudevold, and B. Richardson, *Arch. Oral Biol.* **13**, 625 (1968).
187. P. Grøn, H. G. McCann, and F. Brudervold, *Arch. Oral Biol.* **13**, 203 (1968).
188. L. Singer, W. D. Armstrong, and J. J. Vogel, *J. Lab. Clin. Med.* **74**, 354 (1969).
189. M. W. Sun, *Amer. Ind. Hyg. Ass. J.* **2**, 133 (1969).
190. A. A. Cernik, J. A. Cooke, and R. J. Hall, *Nature (London)* **227**, 1260 (1970).
191. J. Tusl, *Clin. Chim. Acta* **27**, 216 (1970); *Anal. Chem.* **44**, 1693 (1972).
192. N. Shane and D. Miele, *J. Pharm. Sci.* **57**, 1260 (1968).
193. M. J. Larsen, M. Kold, and F. R. Von der Fehr, *Caries Res.* **6**, 193 (1972).

194. B. C. Jones, J. E. Heveran, and B. Z. Senkowski, *J. Pharm. Sci.* **58**, 607 (1969).
195. W. P. Ferren and N. A. Shane, *J. Food Sci.* **34**, 317 (1969).
196. L. Torma and B. E. Ginther, *J. Ass. Offic. Agr. Chem.* **51**, 1181 (1968).
197. R. L. Baker, *Anal. Chem.* **44**, 1326 (1972).
198. J. S. Jacobson and L. I. Heller, *Environ. Lett.* **1**, 43 (1971).
199. D. A. Levaggi, W. O. Yung, and M. Feldstein, *J. Air Pollut. Ass.* **21**, 277 (1971).
200. J. J. Zipper, B. Fleet, and S. P. Perone, *Anal. Chem.* **46**, 2111 (1974).
201. R. A. Durst, *Anal. Chem.* **40**, 931 (1968).
202. O. Klockow, H. Ludwig, and M. A. Girando, *Anal. Chem.* **42**, 1682 (1970).
203. C. Harzdorf, *Z. Anal. Chem.* **245**, 67 (1969).
204. I. Sekerka and J. F. Lechner, *Talanta* **20**, 1167 (1973).
205. D. J. Curran and K. S. Fletcher, *Anal. Chem.* **41**, 267 (1969).
206. T. Eriksson and G. Johansson, *Anal. Chim. Acta* **52**, 465 (1970).
207. T. Eriksson, *Anal. Chim. Acta* **58**, 437 (1972).
208. T. Anfalt, D. Dyrssen, and D. Jagner, *Anal. Chim. Acta* **43**, 487 (1968).
209. W. Selig, *Mikrochim. Acta* 229 (1970).
210. B. Jaselskis and M. K. Bandemer, *Anal. Chem.* **41**, 855 (1969).
211. E. W. Baumann, *Anal. Chem.* **42**, 110 (1970).
212. S. E. Manahan, *Anal. Chem.* **42**, 128 (1970).
213. H. R. Bronstein and D. L. Manning, *J. Electrochem. Soc.* **119**, 125 (1972).
214. H. Heckel and P. F. Marsh, *Anal. Chem.* **44**, 2347 (1972).
215. K. M. Stelting and S. E. Manahan, *Anal. Chem.* **46**, 592 (1974).
216. K. M. Stelting and S. E. Manahan, *Anal. Chem.* **46**, 2118 (1974).
217. P. A. Evans, G. J. Moody, and J. D. R. Thomas, *Lab. Prac.* **20**, 644 (1971).
218. J. B. Orenberg and M. D. Morris, *Anal. Chem.* **39**, 1776 (1967).
219. E. W. Baumann, *Anal. Chem.* **40**, 1731 (1968).
220. W. Selig, *Mikrochim. Acta* 349 (1973).
221. T. Anfalt and D. Jagner, *Anal. Chim. Acta* **47**, 483 (1969).
222. F. J. C. Rossotti and H. Rossotti, "The Determination of Stability Constants," Chapter 5. McGraw-Hill, New York, 1961.
223. N. E. Vanderborgh, *Talanta* **15**, 1009 (1968).
224. R. E. Mesmer and C. F. Baes, Jr., *Inorg. Chem.* **8**, 618 (1969).
225. E. W. Baumann, *J. Inorg. Nucl. Chem.* **31**, 3155 (1969).
226. A. Aziz and S. J. Lyle, *Anal. Chim. Acta* **47**, 49 (1969).
227. K. Kleboth, *Monatsh. Chem.* **101**, 767 (1970).
228. P. R. Patel, M. C. Moreno, and J. M. Patel, *J. Res. Nat. Bur. Std.* **A75**, 205 (1971).
229. P. Klotz, A. Mukherji, S. Feldberg, and L. Newman, *Inorg. Chem.* **10**, 740 (1971).
230. K. Srinivasan and G. A. Rechnitz, *Anal. Chem.* **40**, 1818 (1968).
231. K. Srinivasan and G. A. Rechnitz, *Anal. Chem.* **40**, 1955 (1968).
232. S. L. Grassino and D. N. Hume, *J. Inorg. Nucl. Chem.* **33**, 421 (1971).
233. Y. Moriguchi and I. Hosokawa, *Nippon Kagaku Zasshi* **92**, 56 (1971).
234. B. Noren, *Acta Chem. Scand.* **23**, 931 (1969).
235. F. M. Hall and S. J. Slater, *Aust. J. Chem.* **21**, 2663 (1968).
236. A. M. Bond and G. Hefter, *J. Inorg. Nucl. Chem.* **33**, 429 (1971).
237. B. Elgquist, *J. Inorg. Nucl. Chem.* **32**, 937 (1970).
238. H. Gamsjager, P. Schindler, and G. B. Kleinert, *Chimia* **23**, 229 (1969).
239. S. P. Tanner, J. B. Walker, and G. R. Choppin, *J. Inorg. Nucl. Chem.* **30**, 2067 (1968).
240. A. M. Bond and J. A. O'Donnell, *J. Electroanal. Chem., Interfacial Electrochem.* **26**, 137 (1970).
241. A. M. Bond and G. Hefter, *Inorg. Chem.* **9**, 1021 (1970).

242. G. Hefter, *J. Electroanal. Chem., Interfacial Electrochem.* **39**, 345 (1972).
243. J. W. Swaddle and W. E. Jones, *Can. J. Chem.* **48**, 1054 (1970).
244. K. D. Asmus and J. H. Fendler, *J. Phys. Chem.* **72**, 4285 (1968); **73**, 1583 (1969).
245. T. Anfalt and D. Jagner, *Anal. Chim. Acta* **70**, 365 (1974).
246. I. M. Kolthoff and H. L. Sanders, *J. Amer. Chem. Soc.* **59**, 416 (1937).
247. P. L. Markovic and J. O. Osburn, *Amer. Inst. Chem. Eng. J.* **19**, 504 (1973).
248. K. Toth and E. Pungor, *Anal. Chim. Acta* **51**, 221 (1970).
249. K. Toth, J. Gavaller, and E. Pungor, *Anal. Chim. Acta* **57**, 131 (1971).
250. E. Pungor and K. Toth, *Pure Appl. Chem.* **34**, 105 (1973).
251. K. Toth and E. Pungor, *Anal. Chim. Acta* **64**, 417 (1973).
252. R. Rangarajan and G. A. Rechnitz, *Anal. Chem.* **47**, 324 (1975).
253. E. Pungor and K. Toth, *Pure Appl. Chem.* **31**, 521 (1972).
254. E. Pungor and K. Toth, *Pure Appl. Chem.* **36**, 441 (1973).
255. P. L. Bailey and E. Pungor, *Anal. Chim. Acta* **64**, 423 (1973); *in* "Ion Selective Electrodes" (E. Pungor, ed.), p. 167. Akademiai Kaido, Budapest, 1973.
256. E. Pungor, K. Toth, and J. Havas, *Acta Chim. Acad. Sci. Hung.* **48**, 17 (1966); *Mikrochim. Acta* 689 (1966).
257. J. Havas, E. Papp, and E. Pungor, *Acta Chim. Acad. Sci. Hung.* **58**, 9 (1968).
258. E. Pungor and E. Papp, *Acta Chim. Acad. Sci. Hung.* **66**, 19 (1970).
259. W. E. Morf, G. Kahr, and W. Simon, *Anal. Chem.* **46**, 1538 (1974).
260. Z. Puhony, K. Toth, and E. Pungor, *Magyr. Kem. Folyoirat* **76**, 206 (1970); *Acta Chim. Acad. Sci. Hung.* **68**, 177 (1971).
261. K. Toth, *in* "Ion Selective Electrodes" (E. Pungor, ed.), p. 145. Akadimiai Kaido, Budapest, 1973.
262. M. H. Sorrentino and G. A. Rechnitz, *Anal. Chem.* **46**, 943 (1974).
263. B. Fleet and H. Von Storp, *Anal. Chem.* **43**, 1575 (1971).
264. A. Marton and E. Pungor, *Magyr. Kem. Folyoirat* **77**, 390 (1971); *Anal. Chim. Acta* **54**, 209 (1971).
265. N. A. Kazarjan and E. Pungor, *Anal. Chim. Acta* **51**, 213 (1970); *Magyr. Kem. Folyoirat* **77**, 105 (1971); *Anal. Chim. Acta* **60**, 193 (1972).
266. W. H. Ficklin and W. C. Gofschall, *Anal. Lett.* **6**, 217 (1973).
267. H. Mallissa and G. Jellinek, *Z. Anal. Chem.* **245**, 70 (1969).
268. N. Bottazzini and V. Crespi, *Chimi. Ind.* **52**, 866 (1970).
269. D. A. Katz and A. K. Mukherji, *Microchem. J.* **13**, 604 (1968).
270. D. Kuttel, O. Szabadka, B. Csakvary, K. Meszaros, J. Havas, and E. Pungor, *Magyr. Kem. Folyoirat* **75**, 181 (1969).
271. H. F. Wastgestian, *Z. Anal. Chem.* **246**, 237 (1969).
272. D. Weiss, *Chem. Listy* **65**, 305 (1971).
273. R. C. Harris and H. H. Williams, *Appl. Meteorol.* **8**, 299 (1969).
274. R. P. Buck and V. R. Shepard, Jr., *Anal. Chem.* **46**, 2097 (1974).
275. W. Krijgsman, J. F. Mansveld, and B. Griepink, *Clin. Chim. Acta* **29**, 575 (1970).
276. E. Papp and E. Pungor, *Z. Anal. Chem.* **246**, 26 (1970).
277. H. Dahms, R. Rock, and D. Seligson, *Clin. Chem.* **14**, 859 (1969).
278. A. Shatkay, *Talanta* **17**, 381 (1970).
279. B. W. Hipp and G. W. Langdale, *Commun. Soil Sci. Plant Anal.* **2**, 237 (1971).
280. A. R. Selmer-Olsen and A. Øien, *Analyst* **98**, 412 (1973).
281. T. W. Florence, *J. Electroanal. Chem., Interfacial Electrochem.* **31**, 77 (1971).
282. P. J. Muldoon and B. J. Liska, *J. Dairy Sci.* **54**, 117 (1971).
283. L. Hansen, M. Buechele, J. Koroschec, and W. J. Warwick, *Amer. J. Clin. Pathol.* **49**, 834 (1968).

284. L. Kopito and H. Schwachmann, *Pediatria* **43**, 794 (1969).
285. V. H. Holsinger, L. P. Posati, and M. J. Pallansch, *J. Dairy Sci.* **50**, 1189 (1967).
286. R. L. LaCroix, D. R. Keeney, and L. M. Walsh, *Soil Sci. Plant Anal.* **1**, 1 (1970).
287. E. Papp and E. Pungor, *Z. Anal. Chem.* **250**, 31 (1970).
288. Y. M. Dessouky, K. Toth, and E. Pungor, *Analyst* **95**, 1027 (1970).
289. J. C. Van Loon, *Analyst* **93**, 788 (1968).
290. W. Potmann and E. A. M. F. Dahmen, *Mikrochim. Acta* 303 (1972).
291. L. Vajda and J. Kovacs, *Hung. Sci. Instrum.* **20**, 31 (1971).
292. T. G. Lee, *Anal. Chem.* **41**, 391 (1969).
293. M. I. Brittan, N. W. Hanf, and R. R. Libenberg, *Anal. Chem.* **42**, 1306 (1970).
294. J. C. Van Loon, *Anal. Chim. Acta* **54**, 23 (1971).
295. W. Selig, *Mikrochim. Acta* 46 (1971).
296. A. M. Knevel and P. F. Kehr, *Anal. Chem.* **44**, 1863 (1972).
297. I. Simonyi and I. Kalman, *in* "Ion Selective Electrodes" (E. Pungor, ed.), p. 253. Akadimia Kaidi, Budapest, 1973.
298. R. A. Carter, *Proc. Ass. Clin. Biochem.* **5**, 67 (1968).
299. H. J. Degenhart, G. Abetn, B. Bevaart, and J. Baks, *Clin. Chim. Acta* **38**, 217 (1972).
300. D. L. Turner, *J. Food Sci.* **37**, 791 (1972).
301. J. R. Pearson and K. J. Humphreys, *J. Pharm. Pharmacol.* **22**, 126 S (1970).
302. W. Selig, *Makrochim. Acta* 453 (1973).
303. B. Paletta, *Mikrochim. Acta* 1210 (1969).
304. B. Paletta and K. Panzenbeck, *Clin. Chim. Acta* **26**, 11 (1969).
305. J. Havas, M. Huber, I. Szabo, and E. Pungor, *Hung. Sci. Instrum.* **9**, 1923 (1967).
306. H. Arino and H. H. Kramer, *Nucl. Appl.* **4**, 356 (1968).
307. D. Weiss, *Z. Anal. Chem.* **262**, 28 (1972).
308. K. Burger and B. Pinter, *Hung. Sci. Instrum.* **8**, 11 (1966).
309. J. H. Woodson and H. A. Liebhafsky, *Anal. Chem.* **41**, 1894 (1969); *Nature (London)* **224**, 690 (1969).
310. J. D. Burger and H. A. Liebhafsky, *Anal. Chem.* **45**, 600 (1973).
311. W. L. Hoover, J. R. Melton, and P. A. Howard, *J. Ass. Office Anal. Chem.* **54**, 760 (1971).
312. T. Braun, C. Ruiz de Pardo, and E. Salazar, *Radiochem. Radianal. Lett.* **3**, 397 (1970).
313. R. F. Overman, *Anal. Chem.* **43**, 616 (1971).
314. A. Altinata and B. Perkin, *Anal. Lett.* **6**, 667 (1973).
315. S. Ikeda, *Anal. Lett.* **7**, 343 (1974).
316. R. Wawro and G. A. Rechnitz, *Anal. Chem.* **46**, 806 (1974).

Chapter 6

ELECTRODES SELECTIVE TO OTHER ANIONS

There are a number of anion-selective electrodes that are derived from both solid and liquid ion exchangers. The solid electrodes can be homogeneous or heterogeneous, although there are very few homogeneous solid electrodes and the only one known to possess exceptional qualities is the silver sulfide electrode.

Silver sulfide has two modifications,[1] α-Ag_2S and β-Ag_2S. The α form is cubic and stable above 176°C, and is an electronic conductor. The β form is stable at lower temperatures and is monoclinic. The silver ions of the β form are the major charge carriers at ordinary room temperature,[2–4] particularly if the membrane is in contact with an electrolyte on both sides.[5] This good conductivity and low solubility product of the β form ($S_p = 1.5 \times 10^{-51}$)[6] make the Ag_2S electrode one of the most reliable sensors. It has excellent resistance to oxidation and reduction, and furthermore can be easily fabricated into a dense polycrystalline membrane by conventional pressing techniques.[7] The electrode has also been prepared by the incorporation of the silver sulfide precipitate into a silicone rubber matrix, as done by Pungor,[8] and by thermomolding a mixture of silver sulfide and thermoplastic polymer such as polythene.[9]

The silver sulfide precipitate incorporated into a ceramic membrane has been used as an electrode to estimate concentrations of sulfide in the range 1–10^{-7} M.[10] A compact membrane made by pressing polycrystalline Ag_2S is used in electrodes supplied by the firms of Beckman,[11] Coleman,[12] Corning,[13] Foxboro,[14] Orion Research,[15] Philips,[16] and Monokrystaly, the Research Institute of Single Crystals, Turnov, Czechoslavakia (Crytur).

A. PREPARATION OF MEMBRANE ELECTRODES

In 1958 Fischer and Babcock,[17] following Tendeloo and Krips,[18] used paraffin as a matrix to hold the $BaSO_4$ precipitate and formed a membrane that was selective neither to anions nor to cations. Similarly, Pungor and co-workers used paraffin[19] and later silicone rubber to incorporate the required precipitate or the ion exchange resin granules into the rubber matrix.[20–22] The $BaSO_4$ membrane electrode was prepared in this way, as well as the $MnPO_4$ and $BiPO_4$ membrane electrodes.[23, 24] These membranes, which were used as electrodes selective to SO_4^{2-} and PO_4^{3-} ions, suffered from interferences from other common anions.[25] Prototypes of these membranes supplied by Pungor were used by Rechnitz *et al.*[26] who found the sulfate electrode response to changes in SO_4 concentration in the range 10^{-1}–10^{-6} *M* very encouraging (24–29 mV/pSO_4). However, the electrode is not commercially available. The phosphate electrode showed poor stability.

A membrane prepared from a powdered mixture of Ag_2S (32 mole %), PbS (31 mole %), $PbSO_4$ (32 mole %), and Cu_2S (5 mole %) by pressing to form a disk, has been found to be selective to sulfate ions.[27, 28]

Inorganic phosphate salts (in excess KH_2PO_4, $AlPO_4$, $CrPO_4$, or $FePO_4$) incorporated into silicone rubber have been tried as a membrane electrode responsive to phosphate ions.[29] Unfortunately, this electrode lacked selectivity as it responded to all anions. However, an improved one based on a silver coordination reaction has been developed.[30] The reactions involved are

$$Ag^+ + 2\,CS(NH_2)_2 \rightarrow Ag[CS(NH_2)_2]_2^+$$

$$2Ag[CS(NH_2)_2]_2^+ + HPO_4^{2-} = \{Ag[CS(NH_2)_2]_2\}_2HPO_4$$

Unfortunately, this complex compound of silver is hygroscopic. To obviate this, thiourea is polymerized with glutaraldehyde to give polythiourea glutaraldehyde (ptg) which behaves in the same manner as thiourea. The silver complex of ptg is soluble in water but forms an insoluble salt with dibasic phosphate. This salt, ground with Ag_2S, can be pressed to form a membrane that, as an electrode, responds linearly to the HPO_4^{2-} ion in the concentration range 10^{-4}–10^{-1} *M* and shows good selectivity over SO_4^{2-}, NO_3^-, ClO_4^-, and CH_3COO^- ions.

A silver thiocyanate electrode selective to both thiocyanate and silver ions has been prepared[31] by thermomolding a mixture of silver thiocyanate and a thermoplastic polymer such as polythene in a press. This method has been used to prepare a number of electrodes whose properties have been reviewed by Liberti.[32]

An electrode selective to NO_3^- ion has been prepared by polymerizing a mixture of phenol, formaldehyde, ammonia, and nickel nitrate directly to a film.[33] It has been found to be responsive to NO_3^- ion and unresponsive to SO_4^{2-}, other multivalent anions, and most cations. However, it is slightly responsive to H^+ ions. A universal ion-selective electrode based on graphite powder has been described.[34] Graphite paste was prepared from a liquid ion exchanger containing the ion of interest (Aliquat 336 in NO_3^- form) and clean commercial graphite powder. This paste, packed into a tubing, formed the electrode. In a similar way Orion 92-07-02 liquid ion exchanger mixed with wax-treated carbon powder has been packed into a holder to form a membrane selective to NO_3^- ions.[35]

A coated wire electrode has also been developed.[36,37] A platinum wire (diameter ~ 1 mm) whose tip was melted to form a spherical button was soldered to a length of RG-58 coaxial cable. The Pt wire was dipped into Aliquat in nitrate form—plastic mixture (polymethyl methacrylate solution in methyl acetate, 4 ml plus 1 ml Aliquat in nitrate form) several times to coat it uniformly. The solution of Aliquat in nitrate form was made by dissolving Aliquat 336 (15 ml) in decanol (2 ml) with 10 ml of 1.0 *M* KNO_3 solution. Aliquat can be converted to other anionic forms. It is therefore possible to make a number of anion-selective electrodes, some of which (perchlorate, thiocyanate, oxalate, acetate, benzoate, sulfate, halides, salicylate, and anions of phenylalanine and leucine) have already been prepared and tested.[36]

A solid state perchlorate-sensitive membrane electrode based on a radical ion salt derived from *N*-ethylbenzothiazole-2, 2′-azine has been prepared.[38] It has high selectivity for perchlorate over many common anions. Other solid state perchlorate-selective electrodes are based on the perchlorate salts of *p*-diamines such as *N*,*N*,*N*′,*N*′,-tetramethyl-*p*-phenylenediamine, *N*,*N*,*N*′,*N*′,-tetra-*n*-butyl-*p*-phenylenediamine, *o*-toluidine, *o*-dianisidine, benzidine, and *N*, *N*, *N*′, *N*′, -tetramethyl benzidine. These salts are used to prepare the electrodes (see Fig. 1) in the same way the Selectrodes are prepared.[39]

Commercially available Orion perchlorate liquid ion exchanger mixed with polyvinyl chloride formed into a disk or coated on a Pt wire acts as an electrode selective to perchlorate ions.[40] A coated platinum wire electrode responsive to anionic detergents has been similarly prepared.[41] The tip of the fine Pt wire is fused to make a ball 1.5 mm in diameter; it is dipped several times in the coating mixture which is a 3 : 1 mixture of PVC dissolved in cyclohexanone and a decanol solution of the ion association complex. This complex is prepared by shaking a mixture [50% (v/v)] of a 0.1 *M* aqueous solution of a sodium salt of a detergent anion and a solution of methyltricapryl ammonium chloride.

By incorporating an alkyl benzene sulfonate–ferroin complex into a polyvinyl chloride matrix, a membrane electrode selective to sulfonate ion

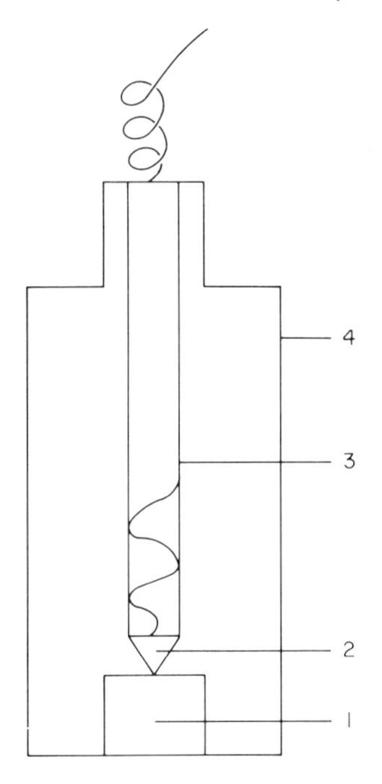

Fig. 1. Construction of a solid state perchlorate electrode: 1, Teflon-carbon rod coated with the perchlorate salt; 2, silver billet; 3, steel contact rod; and 4, polyvinyl chloride body.

in the concentration range 10^{-6}–10^{-2} M in the presence of sulfate, phosphate, nitrate, and chloride ions (10^{-4}–10^{-2} M) has been prepared.[42] Similarly, silicone rubber has been used as a membrane matrix to incorporate powdered hexadecyltrimethyl ammonium dodecylsulfate.[43] This electrode responded to cationic detergents but not to anionic detergents. It exhibited a strong memory effect on going from high to low concentrations.

B. CYANIDE-SELECTIVE MEMBRANE ELECTRODES

Any halide membrane electrode can in theory be converted into a cyanide electrode, but in practice it is found that the AgI-based cyanide electrode is suitable because of its high selectivity for the determination of the cyanide ion. Before measurement, the cyanide electrode must be pretreated by soaking it in 10^{-2} M NaOH solution overnight[44] and then washed with distilled water before use.

The interferences the electrode experiences with other anions and its response to the cyanide ion have already been pointed out. Although Eq. (55) of Chapter 5 has been shown to predict the potential of the electrode, Koryta[45] has derived an alternate expression involving the diffusion coefficients of the cyanide and the iodide ions. Although the electrode senses the free cyanide ion, the stable complexes it forms with metal ions such as Ni(II), Hg(II), Cu(II), Zn(II), and Ag^+ at pH $\gg$ 11 do not interfere with the electrode response.[46] On the other hand, if the pH is less than 11, when some of the cyanide is protonated, the total cyanide $(CN)_t$ cannot be measured with the electrode. For example, in the case of HCN whose pK_a is 9.21 at 25°C, the concentration of the CN^- measured with the help of the cyanide-selective electrode as a function of pH followed the relation[44]

$$-\log(CN^-) = pK_a - \log(CN)_t + \log[(H^+) + K_a] \quad (1)$$

where (H^+) is the hydrogen ion concentration. This equation follows directly from the definition of the dissociation constant. Thus

$$HCN \rightleftharpoons CN^- + H^+ \quad (2)$$

If the total cyanide in solution is $(CN)_t$, then

$$(CN)_t = (CN^-) + (HCN) \quad (3)$$

The dissociation constant K_a is given by

$$K_a = \frac{(H^+)(CN^-)}{(CN)_t - (CN^-)} \quad (4)$$

which on rearrangement yields

$$(CN^-) = \frac{K_a (CN)_t}{(H^+) + K_a} \quad (5)$$

Taking logarithms gives Eq. (1).

It is generally believed that the electrode reaction involved is a replacement reaction at the electrode surface,[47] namely

$$AgI + 2CN^- \rightleftharpoons Ag(CN)_2^- + I^- \quad (6)$$

Consequently, the potential is given by

$$E = E^\circ - (RT/F)\ln a_{I^-} \quad (7)$$

Since the surface concentration of cyanide is low, the liberated I^- determines the value of E.[48] It is possible that in addition to reaction (6), the reaction

$$AgI + CN^- \rightleftharpoons AgCN + I^- \quad (8)$$

also plays a role in the mechanism by which the electrode senses the cyanide ion.[49]

The selectivity constant K_{I-CN} determined for the Ag_2S–AgI membrane electrode (see Table 12 of Chapter 5) by Toth and Pungor is 1.0, whereas Bound *et al.*[50] determined a value of 1.3–1.5. The theoretical value according to the stoichiometry of reaction (6) should be 2. The low value obtained could be due to the competing reaction (8). Consequently, the existence of a diffusion layer in the Ag_2S–AgI membrane electrode has been postulated[50] since the use of an AgI membrane only which had no Ag_2S acting as a barrier to diffusion gave a value of 1.73 for K_{I-CN}.

The influence of pH on the response of the cyanide ion-selective membrane electrode has been further discussed by Mascini.[51] The potential of the electrode at 25°C is given by Eq. (7) or by

$$E = E^\circ - 0.059 \log(\mathrm{CN}^-) \times \tfrac{1}{2} \tag{9}$$

according to reaction (6). Substitution of Eq. (5) into Eq. (9) gives

$$E = E^\circ - 0.059 \log\left[\frac{K_a(\mathrm{CN})_t}{(\mathrm{H}^+) + K_a}\right] \times \frac{1}{2} \tag{10}$$

Comparing Eq. (10) with Eq. (7) yields

$$2(\mathrm{I}^-) = (\mathrm{CN}^-) = \frac{K_a(\mathrm{CN})_t}{(\mathrm{H}^+) + K_a} \tag{11}$$

Mascini examined Eq. (11) and found the experimental data to deviate from its predictions. Consequently, to explain the experimental results, the following reaction was suggested:

$$\mathrm{AgI} + 2\mathrm{HCN} \rightleftharpoons \mathrm{Ag(CN)_2^-} + 2\mathrm{H}^+ + \mathrm{I}^- \tag{12}$$

The equilibrium constant ($\overline{K}$) of this reaction is given by

$$\overline{K} = \frac{[\mathrm{Ag(CN)_2^-}](\mathrm{H}^+)^2(\mathrm{I}^-)}{(\mathrm{HCN})^2} \tag{13}$$

Expressing Eq. (13) in terms of the dissociation constant of HCN, $K_a(= 5 \times 10^{-10})$, the equilibrium constant of $\mathrm{Ag(CN)_2^-}(= 10^{-21})$ and the solubility product of $\mathrm{AgI}(= 8.3 \times 10^{-17})$, one obtains

$$\overline{K} = S_{p(\mathrm{AgI})} K_a^2 / K_{\mathrm{Ag(CN)_2}} = 2.1 \times 10^{-14} \tag{14}$$

Assuming that $[\mathrm{Ag(CN)_2^-}] = (\mathrm{I}^-)$ and substituting the value of $\overline{K}$ derived in Eq. (14) into Eq. (13) gives

$$\frac{(\mathrm{I}^-)(\mathrm{H}^+)}{(\mathrm{HCN})} = 1.4 \times 10^{-7} \tag{15}$$

Mass balance gives the total cyanide, $(\mathrm{CN})_t$. Thus

$$(\mathrm{CN})_t = (\mathrm{HCN}) + (\mathrm{CN}^-) + 2[\mathrm{Ag(CN)_2^-}] \tag{16}$$

that is,

$$(\mathrm{CN})_t = (\mathrm{HCN}) + (\mathrm{CN}^-) + 2(\mathrm{I}^-) \tag{17}$$

Expressing (CN^-) in terms of K_a, Eq. (17) becomes

$$(\mathrm{CN})_t = 2(\mathrm{I}^-) + (\mathrm{HCN})\left[1 + \frac{K_a}{(\mathrm{H}^+)}\right] \tag{18}$$

Substituting for (HCN) from Eq. (15) gives, on rearrangement,

$$(I^-) = \frac{1.4 \times 10^{-7}(CN)_t}{2.8 \times 10^{-7} + (H^+) + K_a} \tag{19}$$

Equation (19) has been found to predict the experimental values very well for solutions whose pH $\leqslant$ 9. At pH $\geqslant$ 9, both Eqs. (11) and (19) were found to fit the experimental data.

This treatment has been extended to include the influence of metal cyanide complexes on the response of the cyanide-sensitive halide membrane electrodes.[52] If $(CN)_t$ and $(M)_t$ (M is the metal ion) are the total concentrations at the surface of the cyanide-sensitive iodide membrane electrode, then

$$\begin{aligned}(CN)_t &= 2[Ag(CN)_2^-] + (CN^-) + (HCN) + \sum_i i[M(CN)_i] \\ &= 2[Ag(CN)_2^-] + (CN^-) + K_f(H^+)(CN^-) + \sum_i i\beta_i(M)(CN^-)^i\end{aligned} \tag{20}$$

$$(M)_t = (M) + \sum_i M(CN)_i = M\left[1 + \sum_i \beta_i(CN^-)^i\right] \tag{21}$$

where K_f is the formation constant of HCN and the β_i's are the overall formation constants of $M(CN)_i$ complexes. Other side reactions, if any, are neglected.

The equilibrium constant of reaction (6) is large ($\bar{K} = 10^4$; $\bar{K} = 10^8$ and 10^{10} for AgBr and AgI reactions, respectively) and one can write

$$(I) = \sqrt{\bar{K}}\,(CN^-) \tag{22}$$

so that Eq. (20) and (21) become

$$\begin{aligned}(CN)_t &= 2(I^-) + (CN^-)\left[1 + K_f(H^+) + \sum_i i\beta_i(M)(CN^-)^{i-1}\right] \\ &= (I^-)\left[2 + \bar{K}^{-1/2} + \bar{K}^{-1/2}K_f(H^+) + \sum_i i\beta_i\bar{K}^{-1/2}(M)(I^-)^{i-1}\right]\end{aligned} \tag{23}$$

$$(M)_t = (M)\left[1 + \sum_i \beta_i\bar{K}^{-1/2}(I^-)^i\right] \tag{24}$$

From Eq. (23), it is seen that, as observed in the case of the iodide-selective electrode in alkaline solutions,[46] the maximum iodide concentration (I^-) is $(CN)_t/2$. If equilibrium (6) is not considered, then Eqs. (23) and (24) become

$$(CN)_t = (CN^-)\left[1 + K_f(H^+) + \sum_i i\beta_i(M)(CN^-)^{i-1}\right] \qquad (23a)$$

$$(M)_t = (M)\left[1 + \sum_i \beta_i (CN^-)^i\right] \qquad (24a)$$

With the help of Eqs. (23) and (24), a family of theoretical curves of $\log(I^-)$ vs. $\log(M)_t$ at constant $(CN)_t$ and pH can be constructed by substituting fixed values of (I^-) in the range $(CN)_t/2$ to 10^{-6} M and solving Eq. (23) for (M) and Eq. (24) for $(M)_t$. Similarly, theoretical curves of $\log(CN^-)$ vs. $\log(M)_t$ can be constructed by using Eqs. (23a) and (24a). In these calculations the literature values for β_i (formation constants for metal–cyanide complexes) are used.

Experimentally, Mascini and Napoli,[52] using polythene-molded silver halide electrodes,[9, 53] potentiometrically titrated solutions of cyanide with solutions of Cd(II), Zn(II), or Ni(II) at a pH (regulated between 6 and 8) that was measured with a glass electrode. Since the halide ion concentration is given by Eq. (7), a plot of $\log(X^-)$ vs. $\log(M)_t$ can be obtained and compared with the theoretical curves (see Fig. 2) in whose construction the following values of the β_i's were used:

Cd(II):	$\log \beta_1 = 5.5$,	$\log \beta_2 = 10.6$,	$\log \beta_3 = 15.3$	$\log \beta_4 = 18.9$
Zn(II):		$\log \beta_2 = 11.0$,	$\log \beta_3 = 16.0$	$\log \beta_4 = 19.6$
Ni(II):				$\log \beta_4 = 30$

The results of Fig. 2 show that in the case of the silver iodide electrode, and the Cd(II)–CN^- system, good agreement was found between the experimental values and the theoretical results calculated from Eqs. (23) and (24), whereas the agreement with the theoretical curves calculated from Eqs. (23a) and (24a) was unsatisfactory. Use of the silver bromide electrode with the same system [Cd(II)–CN^-] gave results (see Fig. 2b) that deviated from the theoretical curves calculated from both sets of equations (23)–(24) and (23a)–(24a). This was attributed to slow kinetics at the electrode surface of the reaction between AgBr and the complex species in the solution. In the case of the Zn(II)–CN^- system, the theoretical curves calculated from Eqs. (23) and (24) agreed with the experimental results obtained with the AgBr electrode over the entire range of metal concentrations (Fig. 2c). But the AgI electrode gave values that agreed with the theoretical curves only at high metal ion concentrations. Again

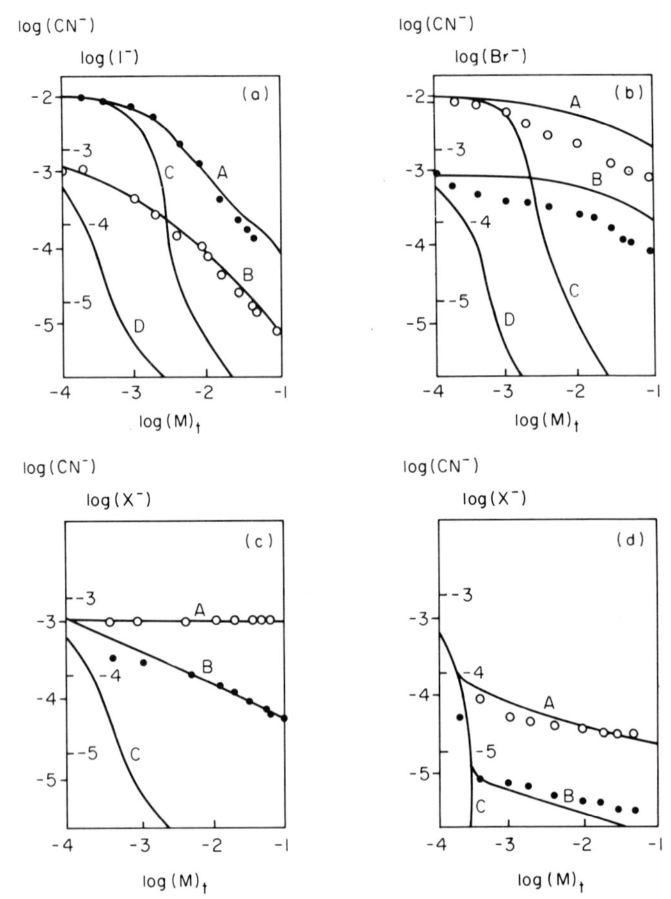

Fig. 2. Theoretical curves and experimental points for the responses of the silver halide electrodes in cyanide solutions in the presence of metal ions. (a) The response of the silver iodide electrode to cyanide ions in the presence of cadmium(II) ions for two total cyanide concentrations: curve C = 10^{-2} *M* and curve D = 10^{-3} *M*. Curves A and B correspond to iodide concentrations of 10^{-2} and 10^{-3} *M*, respectively. (b) The response of the silver bromide electrode to cyanide ions in the presence of cadmium(II) ions for two total cyanide concentrations: curve C = 10^{-2} *M* and curve D = 10^{-3} *M*. Curves A and B correspond to bromide concentrations of 10^{-2} and 10^{-3} *M*, respectively. (c) The responses of the silver iodide and bromide electrodes to cyanide ions in the presence of zinc(II) ions for a total cyanide concentration of 10^{-3} *M*. Curve A represents the response of the bromide electrode, curve B represents that of the iodide electrode, and curve C represents the theoretical response: plot of $\log(X^-)$ vs. $\log(M)_t$ at a total cyanide concentration of 10^{-3} *M*. (d) The responses of the silver iodide and bromide electrodes to cyanide ions in the presence of nickel(II) ions for a total cyanide concentration of 10^{-3} *M*. Curve A represents the response of the bromide electrode, curve B represents that of the iodide electrode, and curve C represents the theoretical response: plot of $\log(X^-)$ vs. $\log(M)_t$ at a total cyanide concentration of 10^{-3} *M*.

the AgBr electrode gave good agreement with the theoretical curves with the Ni(II)–CN^- system (Fig. 2d).

The AgCl membrane electrode used in the metal–cyanide systems just described gave erratic results that were attributed to the high dissolution rate[54] at the phase boundary resulting in the absence of a complexing agent at the electrode surface. On the basis of the results given in Fig. 2, the following conclusions have been drawn:

(1) Halide membrane electrodes can be used in direct potentiometry to sense the total cyanide or free cyanide. This depends on the stability of the complexes and on the metal ion concentration.

(2) When weak complexes are formed (see Figs. 2c and 2d) AgBr is a better sensor than AgI, and both metal complexes and free cyanide react with AgBr or AgI to give Br^- or I^- ions which determine the potential.

(3) When strong complexes are formed (Figs. 2a and 2b), AgI is outstanding in sensing the free cyanide in a wider range of pCN. But with excess metal ions present, the electrode response is not indicative of the free cyanide in solution.

As already indicated, the cyanide-selective membrane electrode, besides being used in the direct estimation of cyanide ion activity, can be used as an indicator electrode in titrations. It has been used as a continuous monitor of cyanide ions[49] in the concentration range 5×10^{-5}–10^{-2} *M*. In this procedure the Gran plot standard addition technique has been used.[55] Alternatively, a silver sulfide electrode responsive to Ag^+ ions has been used indirectly to estimate low levels of cyanide ions.[47] The commercially available Orion cyanide electrode has been used to determine the amount of cyanide present in the hydrolysate obtained after emulsion hydrolysis of cyanogenic glycosides in Sudan grasses.[56] The electrode has been used to determine thiocyanate present in water and other samples.[57] Water samples containing CNS and other ionic impurities such as Fe^{2+}, Fe^{3+}, Co^{2+}, Co^{3+}, and Cd^{2+} were passed through a cation exchange column to remove these ionic impurities; the CNS was then converted to CN by treating 7 ml of the water sample with 20% H_3PO_4. Bromine water was added, the mixture was stirred, and excess bromine was removed by phenol. BrCN produced in the reaction

$$SCN^- + 4Br_2 + 4H_2O \rightarrow CNBr + SO_4^{2-} + 7Br^- + 8H^+$$

was reduced by adding a SO_2-saturated solution:

$$CNBr + SO_2 + 2H_2O \rightarrow HCN + Br^- + SO_4^{2-} + 3H^+$$

Addition of 4 *M* NaOH ensured complete hydrolysis of HCN to CN which was determined with the help of the cyanide-sensitive electrode.

The cyanide electrode has been used to estimate the cyanide in industrial waste water,[44] silver plating baths,[58] and various plants and fruit brandies.[59]

C. SULFIDE-SELECTIVE MEMBRANE ELECTRODES

These electrodes respond only to sulfide, silver, and mercuric ions and to a certain extent to cyanide ions also.

Calibration curves obtained by Ag_2S membrane electrodes are Nernstian and the range extends from saturated solutions to silver and sulfide levels of the order of 10^{-8} *M*.[4] The lower limit of detection is only limited by the experimental difficulties in preparing extremely dilute solutions of ions without ionic adsorption on and desorption from the surfaces of the container vessels and the electrodes.

When excess levels of Ag or sulfide exist in solutions in which there is high complexation, very low levels of the free ions can be estimated. In Table 1 is presented the potential response of the electrode to various solutions as obtained by Durst.[60] A plot of this data gives a straight line with a slope of 59.2 mV/pAg. The characteristics of this electrode as described by Swartz[61, 62] are given in Table 2.

It was demonstrated that the Ag_2S electrode and not the $Ag–Ag_2S$ membrane electrode gave the correct end point in the titration of chloride

TABLE 1

Response of the Ag_2S Membrane Electrode in Various Solutions

Solution composition	Response of the electrode E (mV)	Concentration of Ag as pAg (calculated)
10^{-1} *M* $AgNO_3$	+ 550	1.1
10^{-3} *M* $AgNO_3$	+ 438	3
10^{-4} *M* $AgNO_3$	+ 385	4
10^{-5} *M* $AgNO_3$	+ 323	5
10^{-6} *M* $AgNO_3$	+ 260	6
10^{-7} *M* $AgNO_3$	+ 225	7
10^{-8} *M* $AgNO_3$	+ 213	8
Saturated AgI	+ 150	8.2
Saturated AgI + 10^{-6} *M* KI	+ 21	10.3
Saturated AgI + 10^{-4} *M* KI	− 91	12.3
Saturated AgCl + 1 *M* $Na_2S_2O_3$	− 256	14.2
Saturated AgCl + 0.1 *M* KI	− 298	15.5
0.1 *M* Na_2S + 1 *M* NaOH	− 872	24.9

TABLE 2

Some Properties of the Ag_2S Membrane Electrode

Concentration range sensed:	1–10^{-5} M total sulfide
	1–10^{-20} M free sulfide
	1–10^{-5} M total silver
	1–10^{-23} M free silver
Slope	29 mV/pS^{2-} and 59 mV/pAg^+
pH range	0–14
Transient time	5 msec
Resistance	50–100 kΩ
Interferences	(S^{2-})—none
	(Ag^+), Hg^{2+} ($K_{Ag, Hg} = 0.08$)
Temperature coefficient	+ 0.05 mV/°C in 0.1 M Na_2S (in 1 M NaOH)
	− 0.4 mV/°C in 0.1 M $AgNO_3$

solutions with $AgNO_3$ in highly acidic and oxidizing solutions (e.g., in a solution of 0.01 M Cl^- in 0.1 M Fe^{3+} and 1 M HNO_3). Schmidt and Pungor[63] determined the selectivities of the sulfide membrane electrode to various anions in solutions of sulfide and to various cations in solutions of sliver nitrate by titrating solutions of Na_2S and an interfering anion with a $AgNO_3$ solution and of solutions of $AgNO_3$ and an interfering cation with a Na_2S solution, respectively. The selectivities were determined experimentally and then calculated from the literature values of the solubilities of the compounds involved.[64, 65]

The selectivities of the sulfide electrode were evaluated using the equation

$$K_{S,X} = \frac{S^{1/2}_{(Ag_2S)}}{S^{1/n}_{Ag_nX_m}} = \frac{a^{1/2}_{S^{2-}}}{a^{m/n}_{X}} \tag{25}$$

where S_{ij} is the solubility product of the compound, n and m are the stoichiometric constants, and the a_i's are the activities of the concerned ions evaluated experimentally. Both the experimental and calculated values of $K_{S,X}$ given in Table 3 as p$K_{S,X}$ ($-\log K_{S,X}$) show reasonable agreement. Similarly, the p$K_{Ag,Y}$ values evaluated as

$$K_{Ag,Y} = \frac{S_{Ag_2S}}{S^{1/m}_{Y_nS_m}} = \frac{a^2_{Ag^+}}{a^{n/m}_{Y}} \tag{26}$$

are given in Table 4. The selectivity of the Ag_2S membrane electrode was also evaluated for some anions which formed complexes during titration

TABLE 3

Measured and Calculated Selectivity Constants for the Ag_2S Membrane Electrode

Interfering ion (X)	$pK_{S, X}$ Measured	$pK_{S, X}$ Calculated
I^-	7.8	7.85–8.60
Br^-	11.6	11.51–12.09
Cl^-	14.2	14.10–14.50
OH^-	16.2	16.00–16.73
SCN^-	12.1	11.86–12.13
SO_4^{2-}	21.2	22
PO_4^{3-}	16.1	14.75–18.45

TABLE 4

Measured and Calculated Selectivity Constants for the Ag_2S Membrane Electrode

Interfering ion (Y)	$pK_{Ag, Y}$ Measured	$pK_{Ag, Y}$ Calculated
Tl^+	24.2	24.48–29.25
Cu^{2+}	13.7	10.91–13.3
Pb^{2+}	21.8	20.93–21.8
Cd^{2+}	22.4	20.46–22.3
Ni^{2+}	23.3	21.70–28.90
Zn^{2+}	27.9	22.27–26.8
Fe^{2+}	31.5	26.82–31.2
Mn^{2+}	36.2	33.24–38.80
La^{3+}	40.3	44.17

with silver nitrate. These values of $pK_{S, Z}$, evaluated as

$$K_{S, Z} = \frac{S_{Ag_2S}^{1/2}}{S_{AgZ}^{1/n}} = \frac{a_{S^{2-}}^{1/2}}{a_Z^{m/n}} a_{S, Z}^{1/n} \tag{27}$$

are given in Table 5. In Eq. (27), S_{AgZ} is the instability constant of the complex whose activity is $a_{S, Z}$. In addition a value of 48.4 for the solubility product of silver sulfide (pS_p) has been determined. A value of 50.83 at zero ionic strength has been determined by Hseu and Rechnitz[6] who took into consideration the hydrolysis of Na_2S solution:

$$S^{2-} + H_2O \rightleftharpoons HS^- + OH^- \tag{28}$$

This is a significant reaction as $K_{h_1} = K_w/K_2 = 2.773$ at an ionic strength

TABLE 5

Measured and Calculated Selectivity Constants for the Ag_2S Membrane Electrode with Complexing Anions

Complexing anion	Complex in solution	$pK_{S,z}$ Measured	$pK_{S,z}$ Calculated
CN^-	$Ag(CN)_2^{2-}$	3.20	3.3–5.98
$S_2O_3^{2-}$	$Ag(S_2O_3)^{3-}$	10.70	10.47–11.42

of $I = 0.1\ M$. K_{h_1}, K_w, and K_2 are the hydrolysis constant, water dissociation constant, and dissociation constant of HS^-, respectively. The further hydrolysis reaction

$$HS^- + H_2O \rightleftharpoons H_2S + OH^- \tag{29}$$

is not that significant since $K_{h_2} = K_w/K_1 = 0.628 \times 10^{-7}$ at 25°C.

The Ag_2S electrode response to Ag^+ ions is given by[6]

$$E = 0.5576 + 0.059 \log(Ag^+) \tag{30}$$

In terms of (S^{2-}), which is given by

$$(Ag^+) = \sqrt{[S_p/(S^{2-})]}$$

Eq. (30) becomes

$$E = \text{const} - 0.0296 \log(S^{2-}) \tag{31}$$

where the constant is equal to $0.5576 + 0.0296 \log S_p$.

In order to evaluate the constant in Eq. (31) a sulfide membrane electrode was used in the cell

Sulfide electrode	Na_2S solution (10^{-1}–$3 \times 10^{-5}\ M$) with 0.01 N NaOH; pH = 11.4–11.8; I = 0.1–0.5 M $NaNO_3$	SCE

and the cell emf E was measured. The concentration of S^{2-} was calculated from the dissociation constant K_2 of the reaction

$$HS^- \rightleftharpoons H^+ + S^{2-} \tag{32}$$

K_2 is given by

$$K_2 = \frac{(S^{2-})(H^+)}{(HS^-)} \tag{33}$$

Since the total sulfide concentration $(S)_t$, which is known, is given by

$$(S)_t = (S^{2-}) + (HS^-) \tag{34}$$

Eq. (33), on rearrangement, gives

$$(S^{2-}) = \frac{(S)_t}{1 + ((H^+)/K_2)} \tag{35}$$

Since the pH is known from measurement, (S^{2-}) can be known provided K_2 is known. Since the value of K_2 is determined by the ionic strength I of the medium used, its value at any given I can be calculated from the Debye–Hückel relation[66] which relates the thermodynamic ionization constant K_2' to K_2, namely

$$pK_2 = pK_2' - \frac{2\sqrt{I}}{1 + \sqrt{I}}$$

The values of K_2 calculated in this way, for values of I equal to 0.1, 0.3, and 0.5 M used in the measurements, are 3.635×10^{-15}, 6.134×10^{-15}, and 8.10×10^{-15}, respectively. Hseu and Rechnitz[6] used a value of 1.2×10^{-15} at 25°C for K_2'. Use of this value has been questioned by Pungor and Toth[65] since they measured a value of $pK_2 = 12.65$ at 25°C and $I = 0.1$ M. The Hseu and Rechnitz value at 0.1 M corresponds to 14.44. From the values of (S^{2-}) thus computed, plots were made of E vs. $-\log(S^{2-})$. These plots at the three ionic strengths used gave straight lines according to Eq. (31) with slopes of 30.0 mV/pS^{2-} at $I = 0.1$ M and 29.7 mV/pS^{2-} at $I = 0.3$ and 0.5 M. The intercepts on the potential axis of these lines were -0.911, -0.883, and -0.864. A plot of these intercepts versus $I^{1/2}$ extrapolated to $I^{1/2} = 0$ gave a straight line intercept of -0.946, which is equal to the constant $(0.5576 + 0.0296 \log S_p)$ in Eq. (31). Thus a value of 1.48×10^{-51} is derived for the solubility product of silver sulfide at 25°C.

Equation (35) substituted into Eq. (31) gives

$$E = \text{const} - 0.0296 \log\left[\frac{(S)_t}{1 + ((H^+)/K_2)}\right] \tag{36}$$

When I and $(S)_t$ are held constant for the condition $[(H^+)/K_2] \gg 1$, Eq. (36) becomes

$$E = \text{const} - 0.0296\ \text{pH} \tag{37}$$

The Ag_2S electrode in sulfide solutions obeyed Eq. (37) with a slope of 29.5 mV/pH. At a given pH, the electrode followed the sulfide concentration. Thus the electrode can be used to determine either the concentration of free sulfide ion or the total sulfide concentration provided the pH of the sample solution is measured at the same time. This procedure was used by Hseu and Rechnitz to study the formation of the thiostannate complex

$$SnS_2(s) + S^{2-} \rightleftharpoons SnS_3^{2-}$$

Free (S^{2-}) and (H^+) were measured and since K_2 was known, $(S)_t$ was determined with the help of Eq. (35). The concentration of SnS_3^{2-} was equal to the sum of the known concentration of Na_2S used in the experiment and $(S)_t$. Thus a value of 2.062×10^5 was determined for the formation constant $K_f[= (SnS_3^{2-})/(S^{2-})]$ at 25°C and $I = 0.1$ *M*.

In view of the Ag_2S membrane electrode response to both silver and sulfide ions, it has been used in the estimations of silver in the following analytical operations:

(1) The determination of traces of silver[67] and silver adsorbed on glass.[68]

(2) The determination of CN^- and Cl^- by titration with silver.[69]

(3) The consecutive determinations of alkali bromide and thiocyanate in their mixtures.[70] The mixture was treated with copper acetate and *L*-ascorbic acid in the presence of 6 *N* HNO_3 to form cuprous bromide and cuprous thiocyanate. The former is soluble in an excess of NH_4OH, whereas the latter is not. Bromide and thiocyanate can be estimated by titrating with a $AgNO_3$ solution using the Ag_2S electrode.

(4) The indirect determination of Cl^- in cleaning solutions for power plant boilers.[71] The addition of excess Ag^+ ions results in the precipitation of AgCl; the excess Ag^+ can be sensed by the electrode.

(5) The potentiometric determination of the Ag present in ZnS, CdS, ZnSe, and CdSe.[72]

(6) In the study of the complexation of Ag^+ ions with acetonitrile.[73] The values for β_1 and β_2 for the complex species $AgCH_3CN^+$ and $Ag(CH_3CN)_2^+$ shown in Table 6 were derived from potentiometric titrations at an ionic strength of 0.1 *M* $NaClO_4$ using the various electrode systems.

The Ag_2S membrane electrode has been used

(a) to estimate sulfide, sulfite, and polysulfides[74] present in pulping liquor,[62]

(b) to determine sulfides in general[75, 76] and in nanogram amounts,[77] in waste waters,[60, 78, 79] in natural seawater,[80] and in an automatic semicontinuous titration procedure in which a density gradient of mercuric nitrate is employed,[81]

(c) in the determination of standard potentials of electrodes made from mixed metal sulfides of the Ag_2S–MS type,[82] and

(d) in the potentiometric titration of sulfide ions,[83, 84] of thiourea in 1 or 0.1 *M* NaOH with standard $AgNO_3$,[85, 86] of phenyl thiourea and

TABLE 6

Silver–Acetonitrile Complexes[a]

Electrode system	β_1	β_2	Comments
Ag_2S–4 M calomel	2.6	6.0	Corrected for liquid junction arising from addition of acetonitrile
Ag wire–4 M calomel	2.6	4.4	No correction for liquid junction
Ag_2S–4 M calomel	2.6	5.2	No correction for liquid junction
Ag_2S–perchlorate	2.3	2.4	No correction for liquid junction
Ag_2S–fluoride	2.8	1.6	

[a]Measured at 25°C in 0.1 M $NaClO_4$.

N,N-diphenyl thiourea in alkaline, neutral, and acidic media with $AgNO_3$,[87] of *p*-urazine in 1 M NaOH with standard $AgNO_3$,[88] of thiols[89, 90] and thiol–H_2S mixtures by themselves and when they are present in petroleum products,[90] and of thioacetamide in the concentration range $10^{-1} - 10^{-3}$ M with $AgNO_3$ in acid and alkaline media.[86, 91]

Alexander and Rechnitz have used the silver sulfide electrode in automated protein determinations,[92] in monitoring and analyzing individual proteins and protein mixtures in serum,[93] and finally in monitoring proteins involved in antibody and antigen reactions.[94] The electrode has been used to study the properties of an enzyme, *o*-acetylserine sulfhydrylase, obtained from germinating rapeseed.[95] The kinetics of enzyme-catalyzed reactions have been followed by using the electrode.

D. MEMBRANE ELECTRODES SELECTIVE TO OTHER ANIONS

Heterogeneous solid membrane electrodes responsive to anions which have not reached a stage of importance because of their lack of selectivity and/or other properties, have been described in the beginning of this chapter. Some additional properties are outlined in the following paragraphs.

The coated platinum wire electrode[37] selective to NO_3^- has been used to test the effects of foreign ions on the response of the electrode. Potential measurements were made in solutions of 5×10^{-3} M NO_3^- ion containing 0.04, 0.09, and 0.12 M concentrations of the interfering ion of interest. First the potential was measured in the NO_3^- solution, the interfering ion was then added, and the potential was remeasured. This change in potential is given by

$$\Delta E = \text{slope} \times \log \left[1 + K_{ij}^{\text{pot}} \frac{a_i^{1/z}}{a_{NO_3}} \right] \quad (38)$$

where the a's are the activities of the ions concerned. Some of the values measured in this way with both the wire electrode and the commercially available Orion electrode are shown in Table 7. Chlorate ion interferes most strongly with the response of the NO_3 electrode, but the sulfate ion does not interfere at all.

TABLE 7

Selectivity Constants log K of NO_3-Selective Membrane Electrodes

	log K	
Interfering ion	Wire electrode	Orion liquid membrane electrode
Cl^-	− 1.4	− 2.2
NO_2^-	− 0.8	− 1.2
ClO_3^-	0.26	0.3
SO_4^{2-}	− 3.1	− 3.2

In Table 8 are given the selectivity constants determined for the carbon paste (treated with Orion liquid 92-07-02) NO_3-selective electrode along with the values obtained for the other commerical liquid NO_3-selective electrodes.[35]

The wire electrode has been used to estimate NO_x present in ambient air.[37] The estimation consisted of collecting air in a gas wash bottle containing 2% H_2O_2. The solution was treated with MnO_2 to destroy excess

TABLE 8

Selectivity Constants K_{ij}^{pot} of NO_3-Selective Membrane Electrodes

Interfering ion	Beckman No. 39618	Corning No. 476134	Orion No. 92-07-02	Carbon paste
$H_2PO_4^-$			3.0×10^{-4}	3.0×10^{-4}
SO_4^{2-}	1.0×10^{-5}	1.0×10^{-3}	6.0×10^{-4}	7.0×10^{-5}
Cl^-	1.0×10^{-2}	4.0×10^{-3}	6.0×10^{-3}	3.0×10^{-3}
HPO_4^{2-}			8.0×10^{-5}	6.0×10^{-5}
Br^-	2.8×10^{-1}	1.0×10^{-1}	9.0×10^{-1}	4.0×10^{-2}
I^-	5.6	25	20	4
ClO_4^-	10^2	$> 10^3$	10^3	14

H_2O_2, and the NO_3 was estimated potentiometrically. The method can be used in the presence of a 40-fold excess of SO_2 and SO_3.

Other coated Pt wire electrodes selective to perchlorate ion[40] and anionic detergents[41] have been evaluated for interferences from other ions. The perchlorate electrode had a Nernstian response in the concentration range 10^{-1}–10^{-4} M. The hydroxide interference was $K_{ij}^{pot} = 1.3 \times 10^{-3}$ for the disk electrode and 1.2×10^{-3} for the Orion commercial electrodes in 0.1 M NaOH solution. The other selectivity constants for the perchlorate electrode due to the interfering ions I^-(10^{-1} M), Br^- (10^{-2} M), and NO_3^- (10^{-2} M) were 5×10^{-3}, 10^{-6}, and 2.9×10^{-5}, respectively, whereas the commercial electrode had selectivity constants of 1.2×10^{-2}, 5.6×10^{-4}, and 1.5×10^{-3}, respectively. Laurylbenzene sulfonate-responsive coated Pt wire electrodes[41] had the following values for the selectivity constant K_{ij}^{pot}: j = Cl^-, 0.12; SO_4^{2-}, 0.006; NO_3^-, 0.93; ClO_4^-, 0.81; CH_3COO^-, 0.59; lauryl sulfate, 1.36; lauryl sulfonate, 0.81; and p-toluene sulfonate, 0.75.

The selectivity of the SO_4-selective electrode[28] to various ions in different concentrations is given in Table 9.

TABLE 9

Selectivity Constants K_{ij}^{pot} of the SO_4-Selective Electrode in Mixed Solutions

Interfering ion concentration (M)		Concentration of SO_4 in solution	K_{ij}^{pot}
ClO_4^-	9.8×10^{-2}	1.96×10^{-3}	4.6×10^{-3}
NO_3^-	1.0×10^{-1}	1.0×10^{-3}	5.2×10^{-3}
Cl^-	1.18×10^{-1}	1.47×10^{-2}	1.2×10^{-2}
Br^-	1.43×10^{-1}	1.43×10^{-2}	0.2

The thermoplastic molded silver thiocyanate membrane electrode was found to be selective to Ag^+ ions[31] following the equation

$$E = E^\circ + 0.059 \log(Ag^+) \tag{39}$$

If no silver ions are present but only CNS^- ions, the resulting Ag^+ ion activity will depend on the activity of the CNS^- ion and Eq. (39) becomes

$$E = E^\circ + 0.059 \log S_{p(AgCNS)} - 0.059 \log(CNS^-) \tag{40}$$

A value of 1.585×10^{-12} has been derived for the solubility product S_p of AgCNS. The interferences from the halide ions (K_{CNS-X}^{pot}) in a solution of CNS^- of concentration 10^{-5} M were 2.8×10^{-3} with 0.1 M KCl, 1.0 with 10^{-3} M Br^-, and 1.7×10^2 with 10^{-5} M I^-. The electrode gave a

Nernstian response to CNS^- ions in the concentration range $0.1–10^{-4}$ M in aqueous–nonaqueous media such as water–acetone, methanol, and acetonitrile. In these media, titrations of KCNS with nitrates of Cu(II), Cd(II), Co(II), Ni(II), and Fe(III) have been followed by using the electrode. Similarly, it has been used as an end point indicator in titrations with silver and mercuric nitrates. Furthermore, the electrode has been used to follow the kinetics of alkaline hydrolysis of a thiocyanate complex, $Co(NH_3)_5SCN^{2+}$. The reaction is

$$Co(NH_3)_5SCN^{2+} + OH^- \rightleftharpoons Co(NH_3)_5OH^{2+} + SCN^-$$

The thiocyanate concentration was obtained as a function of time by following the change in emf with time of the electrochemical cell in which the thiocyanate-selective electrode was used. As a first-order reaction, the hydrolysis followed the equation $\ln(a/(a-x)) = kt$, where a, the initial concentration of the complex, was determined by the estimation of cobalt and $a - x$ was the concentration of SCN^- obtained at different times. The plot of $\ln(a/(a-x))$ versus time t gave a straight line whose slope yielded a value of 4.3×10^{-4} sec^{-1} for k, the kinetic rate constant.

Coated wire electrodes selective to Cl^- and SCN^- ions have also been prepared and their behavior evaluated.[96] The electrodes were prepared as already indicated by dipping a Pt wire in a solution containing quaternary ammonium chloride or thiocyanate salt and a solvent additive such as decanol or nitrobenzene. The thiocyanate wire electrode prepared from octadecyldimethylbenzyl ammonium thiocyanate exhibited the selectivities shown in Table 10.

TABLE 10

Selectivity Constants K_{ij}^{pot} for the Coated Wire Electrode Selective to SCN^-(i) Ion

Interfering ion (j)	Solvent additive [10% (v/w)]	
	Decanol	Nitrobenzene
ClO_4^-	1.1	1.1
I^-	1.2	0.5
ClO_3^-	1.3×10^{-1}	2.2×10^{-1}
NO_3^-	1.7×10^{-1}	1.5×10^{-1}
Br^-	1.0×10^{-1}	1.1×10^{-1}
Cl^-	5.0×10^{-2}	4.0×10^{-2}
SO_4^{2-}	8.0×10^{-3}	7.0×10^{-3}

Silicone rubber membranes or another matrix containing $BaSO_4$[17, 97] or other precipitates such as metal phosphates,[20, 22, 24] have been used as potential-indicating electrodes in titrations. $BaSO_4$ incorporated into parchment paper[98] and $BaCrO_4$ embedded in silicone rubber[99] have been used in some studies to follow the current–voltage characteristics of $BaSO_4$ membranes and the selectivity characteristics of chromate membranes to various interfering ions such as Cl^-, NO_3^-, HPO_4^{2-}, and $Cr_2O_7^{2-}$.

Some of the nonselective membrane electrodes such as powdered polythene mixed with spectral graphite and thermosealed to a polythene tube[100] or silicone rubber-based graphite, or even precipitate-based silicone rubber ion-selective electrodes,[101] have been used as electrodes in voltametry.

REFERENCES

1. P. Rahlfs, *Z. Phys. Chem. B* **31**, 157 (1936).
2. M. H. Hebb, *J. Chem. Phys.* **20**, 185 (1952).
3. C. Wagner, *J. Chem. Phys.* **21**, 1819 (1953).
4. J. W. Ross, Jr., *in* "Ion Selective Electrodes" (R. A. Durst, ed.), Chapter 2. Nat. Bur. Std. Spec. Publ. 314, Washington, D.C., 1969.
5. K. S. Fletcher, III and R. F. Mannion, *Anal. Chem.* **42**, 285 (1970).
6. T. M. Hseu and G. A. Rechnitz, *Anal. Chem.* **40**, 1054, 1661 (1968).
7. Instruction Manual, Sulfide Ion Electrode, *Model 94-16 (1967),* Orion Research Inc., Cambridge, Massachusetts.
8. E. Pungor, *Anal. Chem.* **39**, 28A (1967).
9. M. Mascini and A. Liberti, *Anal. Chim. Acta* **51**, 231 (1970).
10. I. C. Popescu, C. Liteanu, and L. Savici, *Rev. Roum. Chim.* **18**, 1451 (1973).
11. Select Ion Electrodes, Bulletin 7145-A. Beckman Instruments Inc., Fullerton, California.
12. Sulfide, Silver Ion-Selective Electrode 3-805 (1968). Coleman Instruments, Oakbrook, Illinois.
13. pH and Ion-Selective Electrodes. Corning-EEL Scientific Instruments, Corning Glass Works, Corning, New York.
14. Sulfide Ion-Monitoring Electrode, GS 1-3F1D (1970). Foxboro Co., Foxboro, Massachusetts.
15. Analytical Methods Guide (1971). Orion Research Inc., Cambridge, Massachusetts.
16. Ion Selective Solid State Electrode for Sulfide and Silver, Type IS 550 S/Ag. Philips Electronic Instruments Inc., Mount Vernon, New York.
17. R. B. Fischer and R. F. Babcock, *Anal. Chem.* **30**, 1732 (1958).
18. H. J. C. Tendeloo and A. Krips, *Rec. Trav. Chim.* **76**, 703, 946 (1957); **77**, 406, 678 (1958).
19. E. Pungor and E. Hollos-Rokosinyi, *Acta Chim. Acad. Sci. Hung.* **27**, 63 (1961).
20. E. Pungor, J. Havas, and K. Toth, *Acta Chim. Acad. Sci. Hung.* **41**, 239 (1964).
21. E. Pungor and J. Havas, *Acta Chim. Acad. Sci. Hung.* **50**, 77 (1966).
22. E. Pungor, J. Havas, and K. Toth, *Z. Chem.* **5**, 9 (1965); *Instrum. Contr. Syst.* **38**, 105 (1965).
23. E. Pungor and K. Toth, *Mikrochim. Acta* 656 (1964).
24. E. Pungor, K. Toth, and J. Havas, *Mikrochim. Acta* 689 (1966).
25. G. A. Rechnitz, *Chem. Eng. News* **45**, (25) 146 (1967).
26. G. A. Rechnitz, Z. F. Lin, and S. B. Zamochnick, *Anal. Lett.* **1**, 29 (1967).

27. G. A. Rechnitz, G. H. Fricke, and M. S. Mohan, *Anal. Chem.* **44**, 1098 (1972).
28. M. S. Mohan and G. A. Rechnitz, *Anal. Chem.* **45**, 1323 (1973).
29. G. G. Guilbault and P. J. Brignac, Jr., *Anal. Chem.* **41**, 1136 (1969).
30. F. R. Shu and G. G. Guilbault, *Anal. Lett.* **5**, 559 (1972).
31. M. Mascini, *Anal. Chim. Acta* **62**, 29 (1972).
32. A. Liberti, *in* "Ion Selective Electrodes" (E. Pungor, ed.), p. 37. Akademiai Kiado, Budapest, 1973.
33. T. N. Bobbelstein and H. Diehl, *Talanta* **16**, 1341 (1969).
34. J. P. Sapio, J. F. Colaruotolo, and J. M. Bobbitt, *Anal. Chim. Acta* **67**, 240 (1973); **71**, 222 (1974).
35. G. A. Qureshi and J. Lindquist, *Anal. Chim. Acta* **67**, 243 (1973).
36. H. James, G. Carmack, and H. Freiser, *Anal. Chem.* **44**, 856 (1972).
37. B. M. Kneebone and H. Freiser, *Anal. Chem.* **45**, 449 (1973).
38. M. Sharp, *Anal. Chim. Acta* **62**, 385 (1972).
39. M. Sharp, *Anal. Chim. Acta* **61**, 99 (1972).
40. T. J. Rohm and G. G. Guilbault, *Anal. Chem.* **46**, 590 (1974).
41. T. Fujinaga, S. Okazaki, and H. Freiser, *Anal. Chem.* **46**, 1842 (1974).
42. T. Tanaka, K. Hiiro, and A. Kawahara, *Anal. Lett.* **7**, 173 (1974).
43. A. G. Fogg, A. S. Pathan, and D. T. Burns, *Anal. Chim. Acta* **69**, 238 (1974).
44. E. Pungor and K. Toth, *Analyst* **95**, 625 (1970).
45. J. Koryta, *Anal. Chim. Acta* **61**, 329 (1972).
46. K. Toth and E. Pungor, *Anal. Chim. Acta* **51**, 221 (1970).
47. B. Fleet and H. Von Storp, *Anal. Lett.* **4**, 425 (1971).
48. D. H. Evans, *Anal. Chem.* **44**, 875 (1972).
49. B. Fleet and H. Von Storp, *Anal. Chem.* **43**, 1575 (1971).
50. G. P. Bound, B. Fleet, H. Von Storp, and D. H. Evans, *Anal. Chem.* **45**, 788 (1973).
51. M. Mascini, *Anal. Chem.* **45**, 614 (1973).
52. M. Mascini and A. Napoli, *Anal. Chem.* **46**, 447 (1974).
53. M. Mascini and A. Liberti, *Anal. Chim. Acta* **47**, 339 (1969).
54. W. Jaenicke, *Z. Elektrochem.* **55**, 648 (1951).
55. B. Fleet and A. Y. W. Ho, *Talanta* **20**, 793 (1973); *in* "Ion Selective Electrodes" (E. Pungor, ed.), p. 17. Akademiai Kiado, Budapest, 1973.
56. W. J. Blaedel, D. B. Easty, L. Anderson, and T. R. Farrell, *Anal. Chem.* **43**, 890 (1971).
57. G. Nota, *Anal. Chem.* **47**, 763 (1975).
58. L. N. Lapatnik, *Anal. Chim. Acta* **72**, 430 (1974).
59. B. Gyorgy, L. Andre, L. Stehl, and E. Pungor, *Anal. Chim. Acta* **46**, 318 (1969).
60. R. A. Durst, *in* "Ion Selective Electrodes" (R. A. Durst, ed.), p. 403. Nat. Bur. Std. Spec. Publ. 314, Washington, D.C., 1969.
61. T. S. Light and J. L. Swartz, *Anal. Lett.* **1**, 825 (1968).
62. J. L. Swartz and T. S. Light, *Tappi* **53**, 90 (1970).
63. E. Schmidt and E. Pungor, *Anal. Lett.* **4**, 641 (1971).
64. E. Pungor and K. Toth, *Anal. Chim. Acta* **47**, 291 (1969).
65. E. Pungor and K. Toth, *Pure Appl. Chem.* **34**, 105 (1973).
66. H. A. Laitinen, "Chemical Analysis," p. 20. McGraw-Hill, New York, 1960.
67. D. C. Muller, P. W. West, and R. H. Muller, *Anal. Chem.* **41**, 2038 (1969).
68. R. A. Durst and B. T. Duhart, *Anal. Chem.* **42**, 1002 (1970).
69. F. J. Conrad, *Talanta* **18**, 952 (1971).
70. J. E. Burroughs and A. I. Attia, *Anal. Chem.* **40**, 2052 (1968).
71. J. G. Frost, *Anal. Chim. Acta* **48**, 321 (1969).
72. B. G. Iofis, N. I. Savvin, A. V. Vishnyakov, and A. V. Gordievskii, *Zavod. Lab.* **39**, 267 (1973).

73. K. M. Stelting and S. E. Manahan, *Anal. Chem.* **46**, 592 (1974).
74. E. Papp and J. Havas, *Magy. Kem. Foly.* **76**, 307 (1970).
75. R. Bock and H. J. Puff, *Z. Anal. Chem.* **240**, 381 (1968).
76. A. Mirna, *Z. Anal. Chem.* **254**, 114 (1971).
77. J. Slanina, E. Buysman, J. Agterdenbos, and B. Griepink, *Mikrochim. Acta* 657 (1971).
78. T. S. Light, *in* "Ion Selective Electrodes" (R. A. Durst, ed.), p. 349. Nat. Bur. Std. Spec. Publ. 314, Washington, D.C., 1969.
79. E. W. Baumann, *Anal. Chem.* **46**, 1345 (1974).
80. E. Mor, V. Scotto, G. Marcenaro, and G. Alabiso, *Anal. Chim. Acta* **75**, 159 (1975).
81. B. Fleet and A. Y. W. Ho, *Anal. Chem.* **46**, 9 (1974).
82. M. Koebel, *Anal. Chem.* **46**, 1559 (1974).
83. R. Naumann and C. Weber, *Z. Anal. Chem.* **253**, 111 (1971).
84. L. C. Green and B. S. Harrap, *J. Soc. Leather Trade's Chem.* **55**, 131 (1971).
85. M. K. Papay, K. Toth, and E. Pungor, *Anal. Chim. Acta* **56**, 291 (1971).
86. M. K. Papay, K. Toth, and E. Pungor, *in* "Ion Selective Electrodes" (E. Pungor, ed.), p. 225. Akademiai Kiado, Budapest, 1973.
87. M. K. Papay, V. P. Izvekov, K. Toth, and E. Pungor, *Anal. Chim. Acta* **69**, 173 (1974).
88. V. P. Izvekov, M. K. Papay, K. Toth, and E. Pungor, *Analyst* **97**, 634 (1972).
89. L. C. Green and B. S. Harrap, *Anal. Biochem.* **42**, 377 (1971).
90. F. Peter and R. Rosset, *Anal. Chim. Acta* **64**, 397 (1973).
91. M. K. Papay, K. Toth, E. Izvekov, and E. Pungor, *Anal. Chim. Acta* **64**, 409 (1973).
92. P. W. Alexander and G. A. Rechnitz, *Anal. Chem.* **46**, 860 (1974).
93. P. W. Alexander and G. A. Rechnitz, *Anal. Chem.* **46**, 250 (1974).
94. P. W. Alexander and G. A. Rechnitz, *Anal. Chem.* **46**, 1253 (1974).
95. T. T. Ngo and P. D. Shargool, *Anal. Biochem.* **54**, 247 (1973).
96. T. Stworzewicz, J. Czapkiewicz, and M. Leszko, *in* "Ion Selective Electrodes" (E. Pungor, ed.), p. 259. Akademiai Kiado, Budapest, 1973.
97. E. B. Buchanan, Jr., and J. L. Seago, *Anal. Chem.* **40**, 517 (1968).
98. C. Liteanu and I. C. Popescu, *Rev. Roum. Chim.* **18**, 319 (1973).
99. E. J. Hakoila, U. O. Lukkari, and H. K. Lukkari, *Suom. Kemistilehti B* **46**, 170 (1973).
100. M. Mascini, F. Pallozzi, and A. Liberti, *Anal. Chim. Acta* **64**, 126 (1973).
101. F. Feher, G. Nagy, K. Toth, and E. Pungor, *Analyst* **99**, 699 (1974).

Chapter 7

ELECTRODES SELECTIVE TO CATIONS

Ion exchange membrane systems were the first to be tried as membrane electrodes.[1] The response of a cation (anion) exchange membrane to particular cation (anion) is the general rule, but the response of an ion exchange electrode to a particular ion in the presence of other ions of like charge is an exceptional property that has been noted in only a few cases, an example of which is the glass electrode selective to hydrogen ions[2, 3] (see Chapter 9). Therefore, efforts have been made to discover membrane systems that respond specifically to particular ions in the presence of other ions. The work undertaken to fabricate solid membrane electrodes selective to cations is described in this chapter.

A. PREPARATION OF ELECTRODES

In general two methods are used to prepare solid membrane electrodes selective to cations. Method one consists of using single crystal or compacted disks as the membrane electrode. An example is the silver sulfide crystal in which the silver ions are the mobile species. By itself it can be used to detect either Ag^+ ions or the sulfide ions. Similarly, the Ag_2S precipitate can be compressed into a pellet and used as the electrode. Ag_2S has a very low solubility and so can be used to act as an inert matrix to hold other metallic sulfides. By compression, a membrane selective to the second metal ion can be formed.[4] If the sample solution contains no silver ions initially, the activity of the silver ion (a_{Ag^+}) at the membrane–solution interface is given by

$$a^2_{Ag^+} a_{S^{2-}} = S_{p(Ag_2S)} \tag{1}$$

$$a_{M^{2+}} a_{S^{2-}} = S_{p(MS)} \tag{2}$$

where the S_p's are the solubility products.

Eliminating $a_{S^{2-}}$ from Eqs. (1) and (2) gives

$$a_{Ag^+} = \left[\frac{S_{p(Ag_2S)}}{S_{p(MS)}} a_{M^{2+}} \right]^{1/2} \tag{3}$$

For a silver ion-selective membrane electrode, the potential is given by the Nernst equation

$$E = \text{const} + \frac{RT}{F} \ln a_{Ag} \tag{4}$$

Substituting Eq. (3) into Eq. (4) gives

$$E = \text{const} + \frac{RT}{2F} \ln a_{M^{2+}} \tag{5}$$

Provided the solubility of MS is greater than that of Ag_2S, and at the same time small enough relative to the ions in the sample solution, the electrode would follow Eq. (5). The equilibria governing Eqs. (1) and (2) must be established soon enough for Eq. (5) to work. Such systems responsive to Cu(II), Cd(II), and Pb(II) have been prepared.[4] Even microelectrodes responding to these ions and others such as Ag^+, S^{2-}, Br^-, Cl^-, and I^- have been prepared.[5]

An electrolytic method using a silver rod as the anode in a dilute solution of sodium sulfide until its surface was coated with Ag_2S has been used.[6] The Ag_2S electrode has been made sensitive to a specific metal ion by immersing it in a dilute solution of sodium sulfide and then adding a solution of the concerned metal nitrate until the free metal ion concentration was greater than the free S^{2-} concentration. The electrode was rinsed and stored in a saturated solution of the metal sulfide. This method has been used in attempting to prepare electrodes sensitive to Cu, Cd, Pb, Hg(II), Ni, Co and Zn ions.[6]

In the second method, the active material is dispersed in an inert binder or matrix and formed into a membrane.[7] In some cases, the active material was smeared on the surface of graphite as in the preparation of halide Selectrodes (see Chapter 5). The sulfides of Ag and Cu, Hg, Cd, Pb, etc., may be deposited on the surface of graphite rod and dried at 200°C. Finally the surface is made hydrophobic by treatment with an organic solvent such as carbon tetrachloride, benzene, or mesitylene.[8] Membrane electrodes called Selectrodes have been found to be very responsive to the cations concerned.[9] Some of the membrane Selectrodes have been used in measurements of the activity of H^+, Cl^-, and Cu^{2+} ions.[10]

In some cases platinum wire has been coated with electroactive materials and used as electrodes selective to some ions. For example, it has been found that a mixture of stearic acid and methyl-tri-*n*-octyl ammonium stearate coated on a platinum wire and immersed in aqueous phosphate solutions acted as an electrode whose potential varied, according to a simple thermodynamic theory, as a linear function of pH from pH 2 to 12 with a slope of 59 mV/pH.[11]

A number of membrane electrodes respond in general to a number of cations. Electrodes based on 7,7,8,8-tetracyanoquinodimethane (I) (tcnq)

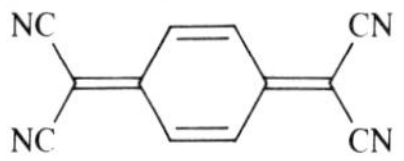

(I) tcnq

radical salts have been prepared.[12] The salts, for example Ag(tcnq), $Cu(tcnq)_2$, tetraethyl ammonium (tcnq), tetraethyl ammonium $(tcnq)_2$, and tetraphenyl arsonium $(tcnq)_2$, were ground and made into pellets to be used as electrodes. Weidenthaler and Pelinka[13] have shown that the response of these electrodes toward changes in the activity of certain metal ions M^{n+} in solution resulted from an equilibrium between solid $M^{n+}(tcnq^-)$ and its ions at the electrode–electrolyte interface. The electrode potential is given by an equation similar to Eq. (5). Thus,

$$E = E^\circ - \frac{RT}{nF} \ln S_p + \frac{RT}{nF} \ln a_{M^{n+}} \tag{6}$$

where S_p is the solubility product of the salt.

Ag(tcnq) was tested in various solutions for responses to Ag^+, H^+, and Na^+ ions. For Ag^+ ions a Nernstian response with a theoretical slope was obtained whereas the slope for H^+ and Na^+ ions was less. Cl^-, Br^-, I^-, S^{2-}, and CN^- ions interfered by film formation. Similarly, $Cu(tcnq)_2$ was tested for its response to H^+ and Ni^{2+} ions. In this case Ag^+ and Hg^+ ions interfered. In the case of $(C_2H_5)_4N(tcnq)$, $(C_2H_5)_4N(tcnq)_2$ electrodes, H^+ ion interference was noted. Tetraphenyl arsonium $(tcnq)_2$ gave a Nernstian response to the arsonium ion.

Some solid state selective electrodes based on 11,11,12,12-tetracyanonaphtho-2, 6-quinodimethane (II, tnad), 9-dicyanomethylene-2,4,7-trinitrofluorene (III, dtf), and 2,4,5,7-tetranitrofluorene-$\Delta^{9,\alpha}$- malononitrile (IV, tfm) have been prepared and tested for responses to Pb^{2+}, Cu^{2+}, and tetraphenyl arsonium ions.[14] These electrodes gave better responses than those based on tcnq. The calibrated graphs were excellent for Pb^{2+}, Cu^{2+}, and tetraphenyl arsonium ions. Polyvinyl chloride (PVC) membrane plasticized with tricresyl phosphate has been found to respond to Ag^+, Tl^+, Hg^{2+}, Na^+, and K^+ ions.[15] Some metal chalcogenides treated with silver sulfide and compacted into disks acted as electrodes responsive to the corresponding metal cations such as Pb^{2+}, Cr^{3+}, Ni^{2+},

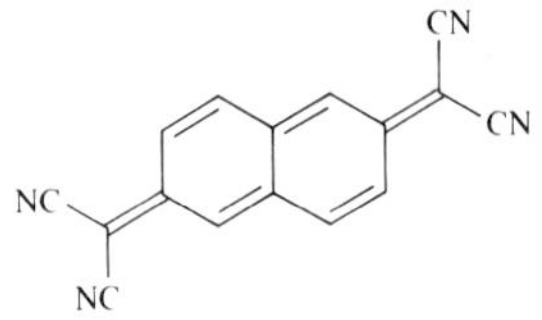

(II) tnad

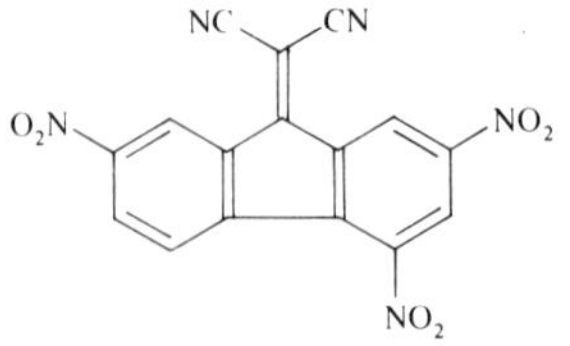

(III) dtf

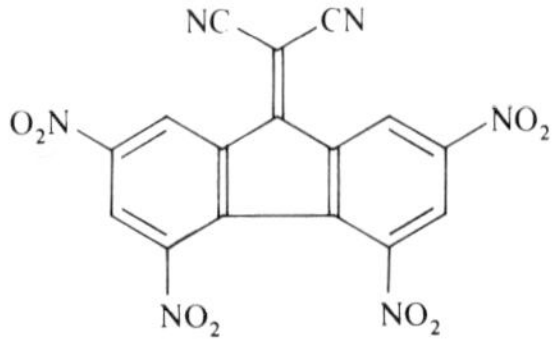

(IV) tfm

Co^{2+}, Cd^{2+}, Zn^{2+}, Cu^{2+}, and Mn^{2+}.[16] Compacted disk electrodes were less sensitive than the sintered disks. The response to the Ag^+ ion was 59.5, mV/pAg, to Pb^{2+}, Ni^{2+}, Cd^{2+}, Zn^{2+}, and Cu^{2+} it was 29.5 mV/pM; and to Cr^{3+} it was 20 mV/pM; Co(II) and Mn(II) had a non-Nernstian response. Silicone rubber membrane electrodes containing hexadecyl trimethyl ammonium dodecyl sulfate that were responsive to cationic detergents have been prepared.[17] They showed strong memory effects when solutions were changed.

B. AMMONIUM-SELECTIVE MEMBRANE ELECTRODES

There are two types of electrodes, one formed from solid membranes and the other from liquid membranes (see Chapter 8). The most familiar solid membrane electrode is the pH type of glass membrane electrode selective to NH_4^+ ions.[18] The alkali metal cation electrode behavior in liquid ammonia at $-38°C$ has been studied.[19] It responds to protonated solvent (NH_4^+) and so can be used to measure the activity of the NH_4^+ ion provided correction for alkaline error is applied. Beckman has put out a solid organic membrane electrode that is selective to NH_4^+ ions (Beckman Catalog No. 39626). The sensing organic material (composition or name of compound is not revealed) is contained at the end of a glass inert tube.[20] It measures NH_4^+ ion activity in the presence of other cations, and has a constant electrode response in the pH range 2–8.5. It is used to assay the urea content of blood serum by treatment with urease (see Chapter 11). In contrast to these electrodes, which sense only the NH_4^+ ion, an arrangement with the glass pH electrode has been worked out to sense ammonia.[21]

The sensing surface of a flat-ended glass pH electrode is pressed tightly against a hydrophobic polymer membrane which acts as a seal trapping a film of NH_4Cl solution. The arrangement is as follows:

Ag–AgCl	NH_4Cl solution	0.1 *M* NH_4Cl film	Glass electrode
Reference electrode	Bulk	Membrane	

The hydrophobic membrane is permeable only to NH_3 and not to NH_4^+ or any other ions. At equilibrium the partial pressures of NH_3 on either side of the membrane are equal. The emf of this cell is given by

$$E = E^\circ_{\text{glass}} + \frac{RT}{F} \ln \alpha_{H^+} - E^\circ_{\text{Ag-AgCl}} + \frac{RT}{F} \ln \alpha_{Cl^-} \tag{7}$$

$$E = E' + \frac{RT}{F} \ln a_{H^+} \tag{8}$$

The dissociation reaction

$$NH_4^+ \rightleftharpoons NH_3 + H^+ \tag{9}$$

has a constant

$$K = \frac{a_{H^+} a_{NH_3}}{a_{NH_4^+}} = 10^{-9.25} \tag{10}$$

For the changes generally observed in the values of a_{NH_3} it is shown[21] that $a_{NH_4^+}$ remains practically constant. Thus substitution of Eq. (10) into Eq. (8) at constant pH gives

$$E = \text{const} - \frac{RT}{F} \ln a_{NH_3} \tag{11}$$

Diffusion of NH_3 from the alkaline test solution through the membrane into the trapped solution alters the NH_3 concentration, causing a pH change that is monitored by the pH glass electrode. This type of electrode has been marketed by the Orion Company. The hydrophobic membrane could be of collodion, leucine–methionine, or silicone–polycarbonate.[22]

The hydrophobic membrane has been eliminated in a recent modification called the air-gap electrode (see Fig. 5, Chapter 10). Instead, a thin film of the required electrolyte solution is deposited on the glass membrane, and only a gap of air separates this film of electrolyte solution from the sample solution, which is kept well stirred. A compact unit containing the reference electrode and the glass membrane electrode has been described.[23] It has been used in the determination of the ammonia content of waste waters,[24] hydrogen sulfite content in wine,[25] and total inorganic and total organic carbon contents in water.[26]

The Orion ammonia probe and the Beckman NH_4^+-selective membrane electrode have been used in the estimation of NH_4^+ ions present in airborne particulates after filtering the air, and then collecting and extracting the particles. The electrodes yielded almost the same results.[27] The electrodes have been used to detect the NH_4^+ ion produced in the urea–urease reaction,[22] which has been utilized in an automated process to determine serum urea.[28]

The Orion electrode (gas permeable) has been used to determine ammonia in aquaria, seawater,[29] waste water,[30] natural and waste waters on an automated continuous flow system with on-line mini computer and printer,[31] condensed steam and boiler feed water,[18, 32] and tobacco and tobacco smoke.[33]

C. CADMIUM-SELECTIVE MEMBRANE ELECTRODES

As already indicated the solid electrode selective to Cd^{2+} ions can be obtained by mixing CdS and Ag_2S and subjecting the mixture to pressure and/or temperature to form a membrane.[4, 34] Compression of sulfides of Cd, Ag, and Cu(I) to form pellets or, even better, putting these sulfides on a ceramic plate and sintering them, has given electrodes responsive to Cd ions.[16, 35] Even incorporating mixtures of Cd and Ag_2S into silicone rubber gave membrane electrodes selective to Cd ions.[36] Separately prepared precipitates of Ag_2S and CdS used to form a membrane gave poor electrodes, whereas use of coprecipitated Ag_2S and CdS which was sintered gave good electrodes whose response was Nernstian to Cd^{2+} in the concentration range 10^{-1}–10^5 *M*. Ag, Cu(II), Hg(II), Fe(III), S^{2-}, and I^- interfered with the response of the electrode.

Hot pressing a mixture of cadmium and silver sulfides with polythene in a molding press has been used by Mascini and Liberti,[37] who investigated the effects of various procedures used in the preparation of the metal sulfides on the final responses of the electrodes fabricated from them. To prepare Cd electrodes, sulfides were obtained from eight procedures, as follows:

(1) CdS precipitated by adding Na_2S to neutral solutions of $Cd(NO_3)_2$;
(2) CdS precipitated from an acidic solution by bubbling H_2S;
(3) precipitate obtained in (1) subject to heat treatment at 600°C for 6 hr in a stream of H_2S;
(4) precipitate obtained in (2) subject to heat treatment as in (3);
(5) coprecipitation of Ag_2S–CdS by adding Na_2S to silver and cadmium nitrate solutions [0.1 *M* $AgNO_3$, 0.05 *M* $Cd(NO_3)_2$];
(6) coprecipitation of Ag_2S–CdS from an acidic solution by bubbling H_2S;

(7) Ag_2S–CdS obtained in (5) subject to heat treatment as in (3);
(8) Ag_2S–CdS obtained in (6) subject to heat treatment.

The electrode response to Cd ions in the presence of 1 *M* $NaNO_3$ was Nernstian only in case 8. Heat treatment seemed to rearrange the crystal structure in the precipitate which when formed into a membrane gave an ideal response of 29.1 mV/decade of concentration at an ionic strength of 1.0 *M*.

The electrode could be used in mixed solvents, acetone–water, and dioxane–water. Its use in the alkaline pH range is limited by the formation of cadmium hydroxides. Ag^+, Hg^{2+}, and Cu^{2+} interfered with the electrode response by reacting with the membrane electrode material.

The Orion 94-48 Cd electrode[4, 38] has been evaluated by Brand *et al.*[39] The selectivity constants determined for the Cd electrodes using both cations and anions are given in Table 1 along with similar values determined by Mascini and Liberti.[37] The responses of the electrode in water–dimethyl sulfoxide (DMSO) solutions of $Cd(NO_3)_2$ have also been examined.[39] It showed a near Nernstian response even in 100% DMSO.

TABLE 1

Selectivity Constants for the Cd Electrodes

Interfering cation M^{z+}	$K_{Cd-M^{z+}}$ [a]	$K_{Cd-M^{z+}}$ [b]	Interfering anion X^{z-}	$K_{Cd-X^{z-}}$ [a]
Cd^{2+}	1	1		
H^+	2.41	5×10^{-4}	S^{2-}	3.77×10^{-22}
Na^+	3.21×10^{-8}		CN^-	5.37×10^{-16}
K^+	6.69×10^{-8}		OH^-	1.49×10^{-6}
Mg^{2+}	1.63×10^{-4}		I^-	6.06×10^{-6}
Ca^{2+}	2.24×10^{-4}		CO_3^{2-}	1.72×10^{-5}
Zn^{2+}	4.14×10^{-4}	10^{-4}	CrO_4^{2-}	5.07×10^{-5}
Co^{2+}	2.03×10^{-2}	5×10^{-5}	SO_3^{2-}	6.98×10^{-5}
Ni^{2+}	3.24×10^{-2}	5×10^{-6}	IO_3^-	26.3
Al^{3+}	1.34×10^{-1}		$Cr_2O_7^{2-}$	1×10^{11}
Mn^{2+}	2.68		F^-Cl^-	$\approx 10^{-2}$
Pb^{2+}	6.08	5×10^{-1}	Br^-	$\approx 10^{-2}$
Tl^+	122		SO_4^{2-}	$\approx 10^{-2}$
Fe^{2+}	196		ClO_4^-	$\approx 10^{-2}$
Fe^{3+} (in perchloric acid)		3×10^{-2}		

[a]Values from Brand *et al.*[39]
[b]Values from Mascini and Liberti.[37]

The response of the Selectrode, solid state Cd(II) electrode, has also been measured.[40] Cadmium buffers have been used to calibrate the membrane electrode which displayed a Nernstian response with a sensitivity close to the theoretical limit imposed only by the solubility product of cadmium sulfide; that is, its response was linear up to pCd 9 at pH 6.7 and up to pCd 11 at pH 9.

These electrodes have been used in nonaqueous media[37, 39] and to detect end points in EDTA[39, 40] and other titrations involving nitrilotriacetic acid and 8-hydroxy quinoline.[37]

D. CALCIUM-SELECTIVE MEMBRANE ELECTRODES

The importance of Ca ions in physiological fluids inspired many investigators to search for a calcium sensor. Tendeloo[41] in 1936 used natural calcium fluorite to sense Ca ions. Although the electrode gave a linear response it was not Nernstian. Later Tendeloo and Krips[42] used calcium salts of low solubility (calcium oxalate and other calcium salts) held in a paraffin matrix containing a nonionic detergent on a gauze for estimation of Ca^{2+} ions in solutions. These showed poor specificity as did a calcium stearate membrane electrode, which, however, showed a stronger response to Ca^{2+} than did the oxalate electrode.[43] This poor selectivity was also noted by other workers[44, 45] who used calcium oxalate in paraffin with nonionic detergent to form the membranes. However, a membrane with good specificity toward Ca ions was prepared by Gregor and Schonhorn[46] who used the Langmuir–Schaefer technique[47] to form the multilayer calcium stearate membrane between the edges of a precisely cracked glass plate. Because of this complicated method of preparation, this multilayer electrode system has found little use. Shatkay and co-workers[45, 48] used polymeric membranes to incorporate electroactive substances into the membrane matrix. It was found that both membranes, one formed by the impregnation of an inert polyvinyl chloride (PVC) matrix with tributyl phosphate and the other formed by the impregnation of PVC matrix with tributyl phosphate (TBP) plus thenoyl trifluoroacetone (TTA) (chelating agent for Ca^{2+}), exhibited high selectivity for the Ca^{2+} ion in the presence of Na, Mg, and Ba ions. The selectivity constants for the PVC–TBP (1 : 3) membrane were

$$K_{\text{Ca-Mg}} = 0.115 \qquad \text{and} \qquad K_{\text{Ca-Na}} = 0.4$$

whereas the value of $K_{\text{Ca-Na}}$ for the PVC–TBP–TTA (1 : 3 : 1) membrane was 5.6×10^{-3}. These membranes were formed by mixing a 10% (w/w) solution of PVC in cyclohexanone with tributyl phosphate in a PVC/TBP ratio of 1 : 3 (w/w). The mixture was poured into a petri dish to form the membrane. When a 25% solution of TTA (the chelating agent for Ca) in tributyl phosphate was used to form the membrane, the final composition of the membrane after evaporation was PVC : TBP : TTA = 1 : 3 : 1.

A similar electrode selective to Ca^{2+} ions with little interference from Na^+ and NH_4^+ ions in molal concentrations 100–200-fold higher than the Ca^{2+} concentration has been described.[49] However, Mg^{2+} interfered with the electrode response at a molal concentration 20-fold higher than that of the Ca^{2+} ion and H^+ ions had little effect on its response to Ca^{2+} ions.

About the time Shatkay's work was published (1966–1967), a liquid ion exchange membrane system[50, 51] (see Chapter 8) was discovered; it is marketed by Orion Research and Corning Glass Works.

A solid ion exchanger of simple construction which responds rapidly and selectively to Ca^{2+} ions has been prepared.[52] The Ca salt of dioctylphosphoric acid or didecylphosphoric acid was dissolved in three to five parts by weight of an alcohol–ether solution of collodion. A film of this collodion formed over the end of a glass tube which, in contact with a 0.01 *M* $CaCl_2$ solution with Ag–AgCl wire in it, served as the membrane electrode. It showed a favorable response to Ca^{2+} ions over Mg^{2+}, Ba^{2+}, Na^+, and K^+ ions.

Another membrane electrode selective to Ca^{2+} ions has been prepared by Moody *et al.*[53] who used a PVC matrix to hold an ion exchanger selective to Ca^{2+} ions. Didecylphosphoric acid dissolved in dioctylphenyl phosphonate was converted to the Ca form by treatment with $CaCl_2$ solution. This exchanger was mixed with polyvinyl chloride, dissolved in tetrahydrofuran, and poured into a glass ring on a glass plate to form a membrane. In a similar way Orion ion exchanger 92-20-02 has been incorporated into the polyvinyl matrix to form the membrane electrode. In another study[54] the optimum composition of the polyvinyl chloride matrix membranes selective to Ca^+ ions was investigated. The PVC matrix [28.8%(w/w)] containing dioctylphenyl phosphonate and monocalcium dihydrogen tetra(didecyl phosphate) in 10 : 1 proportion gave membranes that had good electrode characteristics. Membranes containing only didecyclphosphoric acid gave sluggish electrodes with short linear response ranges. The monocalcium di(decyl phosphate) membrane sensor produced better electrode characteristics than didecylphosphoric acid. No practically useful electrodes were observed with cellulose acetate, ethyl cellulose, collodion, or pyroxylin when these were used as polymer matrix materials in place of PVC. The possibility of using a platinum wire electrode coated with a mixture of 5% PVC dissolved in cyclohexanone and 0.1 *M* calcium didecyl phosphate in dioctyl phosphonate has been explored and found to be quite promising.[55]

The Ca^{2+}-selective PVC matrix membrane of Moody *et al.*[53] has been used without an internal reference solution by keeping the membrane in intimate contact with the end of a graphite rod most of which was covered with a hydrophobic material.[56] The graphite rod was forced through a Tygon tubing to bulge the membrane. The electrode gave a value of 29.58 mV for the Nernstian factor. A simple and sturdy Selectrode (see Fig. 1), without either a reservoir of organic ion exchanger or a bulky aqueous inner reference system, has been described.[57] The electrochemical cell was composed of a calcium Selectrode with a solid state calomel-based inner reference system. That is,

$$\mathrm{TG}\,|\underbrace{\mathrm{Hg{-}Hg_2Cl_2{-}KCl(s)CaSO_4\cdot 2H_2O(s)}}_{\text{Paste}}|\mathrm{Membrane}|\mathrm{Ca^{2+}\,sample}|\mathrm{KCl(s)Hg_2Cl_2{-}Hg} \quad (12)$$

Calomel reference paste applied to the Teflon graphite conductor (TG) was tightly pressed against the membrane prepared according to Moody *et al.*[53] This contained the Ca salt of di-*n*-octylphenylphosphoric acid. The electrode characteristics of the PVC-based Ca-selective membrane are given in Table 2 along with those of three commercial electrodes.

The selectivity constants evaluated by different methods using the Ca-selective electrodes described above are listed in Tables 3 and 4 along with those available for the commercial electrodes. The values of K_{ij} determined for the same i and j ions vary from method to method and electrode to electrode. Thus the values given are only approximate quantities which serve as a general guide to electrode behavior in the various solutions at the concentrations used in the measurements.

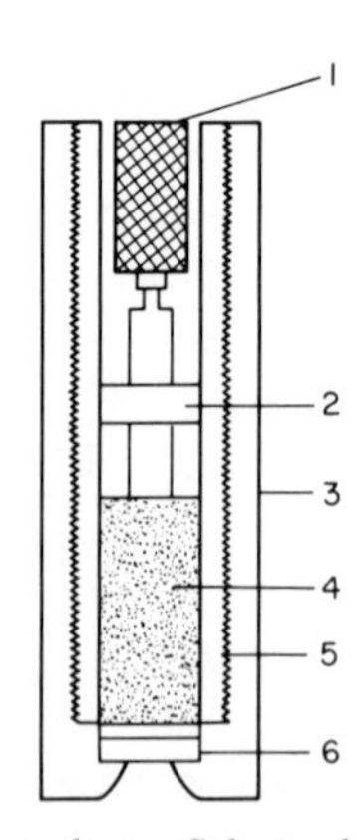

Fig. 1. Construction of a solid membrane Selectrode (after Ruzicka *et al.*[57]). 1, Screened cable; 2, metallic contact; 3, outer tube of Teflon; 4, Teflon–graphite cylinder; 5, Selectrode body; 6, PVC membrane reinforced with a nylon net. Solid state calomel-based inner reference system [Hg_2Cl_2, KCl(s), $CaSO_4$, $2H_2O$(s)] exists between the surface of 4 and 6. The calomel paste applied to the surface of 4 is tightly pressed against PVC membrane 6.

TABLE 2

Ca-Selective Membrane Electrode Characteristics

Electrode	Concentration range (M)	pH range	mV/decade response	Resistance (25°C, MΩ)	Dynamic response (sec)	Temperature range (°C)	Lifetime
Orion 92-20 (liquid membrane)	10^{-5}–1	5.5–11	26.5–29.58	< 25	< 6–30	0–50	30 days
Corning No. 476041 (liquid membrane)	10^{-5}–1	5–10	~ 30	500	< 10	10–60	15 days
Beckman No. 39608 (solid)	5×10^{-4}–1	5–11	29.3–30	< 500	< 10	− 5 to 50	70 days
PVC (Moody *et al.*[53])	5×10^{-5}–10^{-1}	5–9.5	~ 30	~ 25	~ 6	0–48	> 18 months

TABLE 3

Selectivity Constant K_{Ca-M} Determined by Two Methods Using Different Ca-Selective Membrane Electrodes[a]

Interfering ion[b]	Orion 92-20 liquid membrane		PVC electrode containing commercial ion exchanger (Orion 92-20-02) incorporated		PVC	
	Method 1	Method 2	Method 1	Method 2	Method 1	Method 2
Mg^{2+}	0.01–0.29	0.007–0.04	0.0044–0.14	0.0035–0.01	0.0051–0.13	0.0045–0.0097
Ba^{2+}	0.0059–0.18	0.0036–0.027	0.0029–00088	0.002–0.003	0.003–0.090	0.0023–0.0074
Zn^{2+}	0.011–0.65	0.007–0.32	0.0094–0.88	0.0069–0.62	0.0098–0.9	0.0056–0.5
Na^{+}	0.0058–0.42	0.005–0.006	0.0054–0.29	0.006–0.0068	0.0052–0.27	0.0054–0.0061
K^{+}	0.002–0.37	0.0013–0.0016	0.0027–0.41	0.0028–0.0069	0.0026–0.39	0.0027–0.0055
H^{+}	330–590	590	—		29–27	26

[a] Method 1: Eq. (31), Method 2: Eq. (32), of Chapter 5. Values from Moody *et al.*[53]
[b] The activity range of M^{2+} is 1.9×10^{-2}–9.7×10^{-6} and of M^{+} is 10^{-1}–10^{-2}.

TABLE 4

Selectivity Constants K_{Ca-M} Measured in Ca^{2+}–M^{2+} Solution Using Different Ca-Selective Membrane Electrodes[a]

Interfering ion			Orion 92-20	Orion 92-20	Beckman No. 39608	Beckman No. 39608	PVC	PVC membrane Selectrode	PVC graphite	Coated Pt wire
Mg^{2+}			1.4×10^{-2}	4×10^{-2}	1.2×10^{-1}	3.4×10^{-1}	0.22–0.036	2.5×10^{-4}	6×10^{-4}	1.4×10^{-2}
Sr^{2+}			1.7×10^{-2}	7×10^{-2}	9.3×10^{-2}	—	—	1.7×10^{-2}	—	2.1×10^{-2}
Ba^{2+}			1×10^{-2}	4×10^{-2}	7.9×10^{-2}	9×10^{-1}	0.013–0.005	2.5×10^{-4}	1.8×10^{-3}	3.6×10^{-3}
Cu^{2+}			2.7×10^{-1}					1.6×10^{-4}		1.5×10^{-1}
Ni^{2+}			8×10^{-2}					—	3×10^{-3}	3.9×10^{-3}
Pb^{2+}			6.3×10^{-1}					—	2.9	1.86
Fe^{2+}			8×10^{-1}					—		
Zn^{2+}			3.2				0.065–0.045	6×10^{-2}	2.7×10^{-1}	32.3
Cd^{2+}	3×10^{-2}	—	—	3×10^{-2}				3×10^{-4}		
Li^{+}	—	—	—	—				5.8×10^{-5}		
K^{+}			10^{-4}	10^{-3}		3.4×10^{-2}	2.2×10^{-5}	2×10^{-6}		
Na^{+}			10^{-4}	10^{-3}	1.5×10^{-2}	2.9×10^{-2}	6.7×10^{-5}	6.3×10^{-6}		
NH_4^{+}			10^{-4}		—	—				
TEA^{+}			$< 10^{-3}$		4.6×10^{-1}					
H^{+}	10^{8}	2×10^{5}[b]	10^{5}		72	1.5×10^{4}	40–25	1.6×10^{4}		
Ref.	57	58	51[c]	59	60	52	53	57	56	55

[a] The concentration of M in solution is kept constant and that of Ca^{2+} is varied [Eq. (37) of Chapter 5].
[b] Used in a microelectrode.
[c] Includes manufacturer's data.

The Ca-selective Selectrode[57] has been thoroughly evaluated for its response using Ca buffers, and has been compared with the commercial Orion liquid ion exchange membrane electrode. Measurements of the emf of cell (12) using Ca buffers have been carried out over a period of weeks to check the stability and the response of the Selectrode and of the commercial Orion electrode. The cell emf may be expressed by the equation

$$E = E^\circ + \frac{RT}{2F} \ln a_{Ca^{2+}} \tag{13}$$

The linear part of the calibration curve has been taken to compute values for the slope and E°. The values so derived over a period of weeks are given in Table 5. The calibration of the Orion electrode

$$[\mathrm{pH} > 6, \quad \text{ionic strength} = 0.1, \quad \mathrm{pNa} = 1]$$

had a Nernstian response of 27 mV/pCa unit over three weeks of operation. During this period the electrode was stored in a pCa 2, pNa 1 solution. After three weeks the electrode slowly deteriorated and at the end of seven weeks a slope of 23.6 mV/pCa was observed. The Selectrode definitely has better features than the Orion electrode.

Many of the electrodes described above have been used in the determination of Ca^{2+} ion activities and as indicator electrodes in EDTA titrations.[57, 60] Rechnitz and Hseu[60] have used the Beckman electrode in the study of the formation constants of calcium malate ($\log K_f = 2.66$), calcium citrate ($\log K_f = 3.67$), and Ca-DCTA (calcium *trans*-1, 2-diaminocyclohexane *N,N,N′,N′*-tetraacetic acid, $\log K_f = 12.65$). The PVC electrode of Moody *et al.*[53] has been used to estimate the amount of Ca

TABLE 5

Stability of Potential and Response of Ca Selective Electrodes

	Age of electrode (weeks)							Range
	0	1	3	5	7	3^a	7^a	(pCa)
Orion liquid membrane								
Slope	26.5	27.5	26.5	23.6	23.6	27.0 ± 0.58	25.54 ± 1.66	2–3.9
E°	84.5	86.8	81.1	69.8	75.8	83.7 ± 2.49	78.81 ± 5.76	
Selectrode (di-*n*-octyl phenyl phosphoric acid)								
Slope	30.66	31.27	30.87	30.75	31.37	30.96 ± 0.25	30.89 ± 0.32	2–5.57
E°	59.78	65.22	60.40	58.77	59.15	62.05 ± 2.56	60.56 ± 2.75	

[a]All values calculated by regression analysis and stated in millivolts: mean ± standard deviation.

present in water. Shatkay has used his electrode and the commercial electrode in thermodynamic studies concerned with the determination of Ca^{2+} ion activities in mixed solutions[61] and in pure aqueous solutions of $CaCl_2$.[62]

The evaluation of ion activities and selectivities is not without ambiguity. Assumptions and/or conventions (see Chapter 5) about the activity coefficient of Ca^{2+} in standardizing solutions which contain a number of other ions have to be made. The results depend on the assumptions made in the evaluation.[45,61] The finer points involved in the evaluation are brought out in the correspondence between Shatkay[63] and others.[64,65]

E. CESIUM-SELECTIVE MEMBRANE ELECTRODES

Cesium-12-molybdophosphate precipitate was mixed with GE silicone, partly dried, and pressed at 1600 psi into a circular disk about 3 mm thick. This membrane was fixed to one end of a glass tube to form the membrane electrode.[66] The electrode response gave a slope of 24.5 mV/pCs at 25°C. Despite this low value it could still be used to determine Cs concentrations as low as 10^{-5} *M*. The electrode took about 2 min to reach steady potential. At a Cs concentration of 10^{-2} *M*, the selectivity constants determined by varying the concentration of the interfering ion from 10^{-5} to 10^{-1} *M* were as shown in the accompanying tabulation.

M	$K_{Cs\text{-}M}$	M	$K_{Cs\text{-}M}$
Li^+	1.2×10^{-1}	Rb^+	4.3×10^2
Na^+	7.2×10^{-2}	NH_4^+	6.3
K^+	1.8×10^{-1}	Tl^+	5.6×10^2

The electrode has been used in precipitation reactions involving Cs^+ and 12-molybdophosphoric acid.[67] The titration gave no sharp end point, but when the Gran plot (see Chapter 5) of the data was carried out, the end point could be easily detected. Precipitation took place in three stages:

$$Cs + H_3PMo_{12}O_{40} \rightleftharpoons CsH_2PMo_{12}O_{40} + H^+$$

$$Cs + CsH_2PMo_{12}O_{40} \rightleftharpoons Cs_2HPMo_{12}O_{40} + H^+$$

$$Cs + Cs_2HPMo_{12}O_{40} \rightleftharpoons Cs_3PMo_{12}O_{40} + H^+$$

The electrode could be used to follow the titration of the Cs^+ ion with sodium tetraphenyl borate solution. Thus, by employing the standard addition technique (see Chapter 5) the concentration of Cs in solutions could be determined.

F. COPPER-SELECTIVE MEMBRANE ELECTRODES

A PVC membrane containing cuprous iodide precipitate when preconditioned in 0.2 M KI solution responded to activities of Cu^{2+}, Cu^{+}, and I^{-} ions.[68] Values of the slopes of the plots of the potential versus log(concentration) were 32 mV for Cu^{2+} and 40 mV for Cu^{+} in the concentration range 10^{-2}–$10^{-5}M$. It is suggested that the electrode would be useful for use in ion exchange chromatography. A cuprous sulfide membrane electrode selective to copper ions has been prepared from lower oxides of copper.[69] Ferric and not ferrous ion interfered with the response of the electrode. Amines that do not form stable complexes with Cu never interfered with the response of the electrode. The electrode has been used in studies involving oxidation–reduction systems.[70]

Cuprous sulfide precipitate hot pressed (300°C for 24 hr) under a pressure of 7 ton/cm^2 gave a pellet which, embedded in a glass tube with araldite, formed the membrane electrode. There was no internal solution and the leading wire was directly connected to the pellet.[71] This electrode showed a Nernstian behavior for Cu(I) ion concentrations of 10^{-5} M in pure solutions, to 3×10^{-7} M in the presence of complexing ligands. The electrode has been used in the measurement of stepwise formation constants of copper(I) complexes with halides in acetonitrile. Copper(I) perchlorate solution in a 0.1 M perchlorate supporting electrolyte was titrated with tetraethyl ammonium halides. Overall complex formation constants β_1 and β_2 were calculated by combining formation constant expressions (L is ligand)

$$Cu^+ + L^- \rightarrow CuL; \qquad \beta_1 = \frac{(CuL)}{(Cu^+)(L^-)} \tag{14}$$

$$Cu^+ + 2L^- \rightarrow CuL_2^-; \qquad \beta_2 = \frac{(CuL_2^-)}{(Cu^+)(L^-)^2} \tag{15}$$

with the mass balance equations

$$(Cu)_t = (Cu^+) + (CuL) + (CuL_2^-) \tag{16}$$

$$(L)_t = (L^-) + (CuL) + 2(CuL_2^-) + K_f(Cu^+)(L^-) \tag{17}$$

where K_f is the association constant of the cation of the supporting electrolyte and the ligand. The concentration of CuL was found to be small when the ratio of total ligand to total Cu was greater than 2. Therefore Eq. (16) becomes

$$(Cu)_t - (Cu^+) = (CuL_2^-) \tag{18}$$

Substitution of Eq. (18) into Eq. (17) on rearrangement becomes

$$(L^-) = \frac{(L)_t - 2[(Cu)_t - (Cu^+)]}{1 + K_f(Cu^+)} \tag{19}$$

Substitution of Eqs. (18) and (19) into Eq. (15) gives

$$\beta_2 = \frac{(Cu)_t - (Cu^+)}{(Cu^+)}\left[\frac{(L)_t - 2[(Cu)_t - (Cu^+)]}{1 + K_f(Cu^+)}\right]^{-2} \tag{20}$$

Since the values for all the parameters in Eq. (20) are known, β_2 can be evaluated. To evaluate β_1, the ratio of total ligand to total copper should be less than 2. Equations (15) and (16) give the relation

$$(CuL_2^-) = \beta_2(Cu^+)(L^-)^2 = (Cu)_t - (Cu^+) - (CuL) \tag{21}$$

Substitution for (CuL) from Eq. (14) into Eq. (21) on rearrangement gives

$$\beta_1 = \frac{(Cu)_t - (Cu^+) - \beta_2(Cu^+)(L^-)^2}{(Cu^+)(L^-)} \tag{22}$$

In terms of total and free ligand concentrations, Eq. (17), substituting for (CuL) and (CuL_2^-) from Eqs. (14) and (15) and rearranging, becomes

$$\beta_1 = \frac{(L)_t - (L^-)[1 + K_f(Cu^+) + 2\beta_2(Cu^+)(L^-)]}{(Cu^+)(L^-)} \tag{23}$$

Equating and rearranging Eq. (22) and (23) gives the relation

$$\beta_2(Cu^+)(L^-)^2 + (L^-)[1 + K_f(Cu^+)] + (Cu)_t - (Cu^+) - (L)_t = 0 \tag{24}$$

Since all the quantities are known, the quadratic can be solved for the free ligand concentration (L^-) which substituted into Eq. (22) or (23) gives the value for β_1. The values thus calculated are given in Table 6, along with the formation constant of the complex $Cu(S = C(NH_2)_2)_2^+$. The usefulness of the electrode in the study of the interactions of Cu(I) ions with organic ligands in acetonitrile has been illustrated for thiourea.

TABLE 6

Cu(I) Complexes in Acetonitrile

Titrant	K_f for titrant salt ($mole^{-1}$)	Supporting electrolyte	log β_1	log β_2
Et_4NCl	35	0.1 *M* Et_4NClO_4	4.8	10.6
Et_4NBr	10	0.1 *M* Et_4NClO_4	3.8	7.7
Et_4NI	5	0.1 *M* Et_4NClO_4	3.2	6.4
NaI	0	0.1 *M* $NaClO_4$	3.1	6.2
Thiourea	—	0.1 *M* Et_4NClO_4	—	6.3

Hirata and Date[72] prepared a copper(I) sulfide membrane electrode by incorporating copper powder and copper(I) sulfide into a silicone rubber membrane. The response of both this electrode and the copper(II) sulfide electrode to Cu^{2+} ions was investigated as functions of pH, interfering ions, temperature, and membrane matrix. The copper(II) sulfide electrode was found to be less satisfactory. An entirely different method has been used to prepare the so-called ceramic membranes.[73] A fine powder of copper(I) sulfide was formed by heating a mixture of copper powder and sulfur (the molar ratio 2 : 1) at 400–800°C for 2 hr in an atmosphere of H_2S and compressed at 150 kg/cm^2 to form a tablet of 15 mm diameter and 2–3 mm thickness. It was heated again at 500–900°C for 1–5 hr in an atmosphere of H_2S or nitrogen. After the tablet cooled to room temperature, its surfaces were polished and washed with an ultrasonic wave cleaner. The leading wire was attached to the membrane, thereby eliminating the internal solution. The different factors such as distribution of Cu(I) particle size, sulfur powder, pressure, temperature, nature of the atmosphere during annealing, and the technique of polishing affecting the final quality of the membrane electrode have been investigated. When x in $Cu_{2-x}S$, which describes the composition of copper sulfide ceramic, was increased, the response time of the electrode became shorter and it became difficult to make nonporous ceramic. When x was decreased, the response time became larger and the slope of the potential versus log $a_{Cu^{2+}}$ became lower, thus indicating the formation of pores. The optimum composition of the membrane ceramic was found to be $Cu_{1.79}S$.

The response time of the electrodes so prepared was short even at low concentrations, and the potentials were steady and reproducible. The Nernst equation was obeyed in the concentration range 10^{-1}–10^{-6} M. Cu(II), Ag^+, NH_4^+, Fe^{3+}, and I^- interfered with the response of the electrode. The other characteristics of the electrode are pH range 1–7 and temperature range 0–80°C.

A compact membrane made of a single crystal with the empirical composition $Cu_{1.8}Se$ is available commercially (Monokrystaly, The Research Institute of Single Crystals, Turnov, Czechaslavakia, "Crytur"). Its response ($K_{Cu\text{-}Pb}$) to Cu^{2+} ions (0.01 M) in the presence of Pb^{2+} ions (0.01 M) has been found to be 1.1×10^{-3} at pH 3 and 6.6×10^{-4} at pH 5.[74] The electrode has been used in the complexometric determination of aluminum, based on backtitration with copper(II) solution.[75]

The solid state membrane electrode selective to Cu(II) ion, like the Cd-selective electrode, can be prepared from dispersion of copper sulfide in the Ag_2S matrix.[4] The electrode surface has been examined[76] microscopically when it was subject to the action of oxidizing agents which produced pits at various locations in the material. Consequently, a mixed electrode potential was observed. The slope of the plot of the potential versus log(concentration), the stability, and the speed of response of the electrode were low when the electrode surface had pits. However, diamond polishing the surface improved the electrode significantly.

Similarly, it was found that the normal shiny surface of Cu(II) electrode became tarnished when exposed to Cl^- ions.[77] However, polishing the surface easily removed the dullness and restored the electrode characteristics. When nonchloride reference electrodes were used, no such tarnish appeared except in the presence of added chloride. However, premixed chloride had little effect, since it was probably bound as copper–chloro complexes. Ross[4] has shown that for the mixed cupric–silver sulfide membrane in the presence of Cu^{2+} and Cl^- ions, the reaction

$$Ag_2S(s) + Cu^{2+} + 2Cl^- \rightleftharpoons 2AgCl(s) + CuS(s)$$

may take place. If the Ag_2S on the membrane surface is stable, the condition that

$$a_{Cu}(a_{Cl})^2 < \frac{S_{p(Ag_2S)}}{S^2_{p(AgCl)}S_{p(CuS)}}$$

must be satisfied. Otherwise, if the activity product $a_{Cu}a^2_{Cl}$ exceeds the given value, Ag_2S at the electrode surface will be converted to AgCl. This will be followed by an abrupt change in electrode function.

The electrode behavior in solutions containing submicromolar levels (10^{-6}–10^{-9} *M*) of Cu^{2+} ions has been described.[78] At low concentrations, there is a surface concentration of Cu^{2+} ions that is in equilibrium with the bulk concentration. It is the surface Cu^{2+} ion concentration that determines the potential. At low levels of Cu^{2+} ion (10^{-5} *M*) EDTA has been found to interfere with the electrode function. On the other hand, H^+, Ca^{2+}, Zn^{2+}, Al^{3+}, and Fe^{3+} ions interfered only slightly at the 10^{-7} *M* level of the Cu^{2+} ion.

The electrode has been used in the study of the formation constants of copper(II) complexes of ethane-1-hydroxy–1,1-diphosphonic acid (EDPA, V).[79]

```
HO     OH  OH
  \    |   /
O=P—C—P=O
  /    |   \
HO    CH3  OH
```

V

Formation constants of CuY^{2-}, $CuHY^-$, and CuH_2Y have been determined in an aqueous solution whose ionic strength was maintained at 0.1 *M* with tetraethyl ammonium perchlorate at 25°C.

Titration of EDPA using the glass electrode was carried out with tetraethyl ammonium hydroxide to determine the acid dissociation constants, which were found to be pK_2 (2.31), pK_3 (6.99), and pK_4 (10.93). The pK_1 value was too large to be determined. The formation constants determined for the Cu–EDPA (1 : 10) complexes were log $K_{CuH_2Y}^{H_2Y}$ = 4.8, log K_{CuHY}^{HY} = 7.47, and log K_{CuY}^{Y} = 11.84, where

$$K_{CuH_2Y}^{H_2Y} = \frac{(CuH_2Y)}{(Cu^{2+})(H_2Y^{2-})}$$

where the EDPA molecule is represented by Y. On the basis of these results, it has been suggested that the metal chelate CuY^{2-} probably has the more stable ring structure (VI):

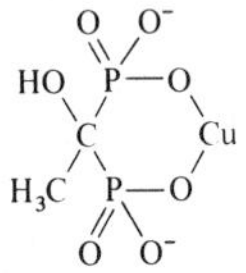

O O⁻ / HO P—O / C Cu / H₃C P—O / O O⁻

(VI) CuY^{2-} structure

The Orion 94-29 cupric ion-selective electrode has been used in the potentiometric measurement of Cu in seawater.[80] The Cu(II) electrode has been used in the analysis of seawater and river water to determine their Ca and Mg contents. Very early in the development of ion-selective membrane electrodes, the Ca liquid ion exchange membrane electrode was used for these estimations. Some of the components of the water interfered with the Ca electrode. Thus a more precise end point location in the titrations of Ca by EDPA in the presence of those species which interfered with the Ca electrode has been accomplished with the use of the solid state Cu(II)-selective membrane electrode.[81]

The electrode has been used for the end point detection in chelometric titrations of metal ions.[82] In these titrations a wide variety of chelating agents employing ordinary titration equipment have been used without much concern about chloride or redox interference. Similarly, reactions such as

$$M^{n+} + Cu\text{–}EDTA^{2-} \rightleftharpoons M\text{–}EDTA^{n-4} + Cu^{2+}$$

have been followed.[83] The position of equilibrium depends on the relative values of the stability constants of the two metal complexes. Thus

$$K = \frac{(M\text{–}EDTA)(Cu)}{(M)(Cu\text{–}EDTA)} = \frac{K_m}{K_{Cu}}$$

where K_m and K_{Cu} are the formation constants of M–EDTA and Cu–EDTA. The method consists in using the electrode to follow the titration of the sample containing the metal ion in the presence of Cu–EDTA with the standard EDTA solution. When $K_m > K_{Cu}$ (e.g., Th, Zr, Fe ions) a drop of Cu–EDTA is used to detect the end point. When $K_m < K_{Cu}$ (e.g., Sm, La, Ca), 1 ml of Cu–EDTA is used. Other complexometric titrations of Cu^{2+} ions with several ligands in nonaqueous media (methanol, acetone, and acetonitrile) have been performed using the electrode.[84] The advantages of using the solid membrane electrode in place of the liquid membrane electrode have been illustrated. Nitrilotriacetic acid has been quantitatively determined via complexometric titration with Cu^{2+} solution.[85] Values of conditional stability of Cu(II)–Erichrome Red B complex in aqueous solutions of pH 3–6 have been determined.[86] Copper(II) sulfide with $Cu_{2-x}S$, where $0 > x > 1$, embedded in silicone rubber, exhibited a Nernstian response in the concentration range 10^{-1}–10^{-6} M.[87, 88] Its impedance was 1.7 kΩ and it worked[88] in the pH range 2–6.4. Measurements with copper(II) sulfate, nitrate, and chloride solutions indicated that the anions had no effect on the electrode response to Cu^{2+} ions.[87, 88] The divalent ions Pb^{2+}, Cd^{2+}, Zn^{2+}, Co^{2+}, Ni^{2+}, and Mn^{2+} exhibited no interference with the Cu^{2+} ion. However, other metal cations (e.g., Ag^+, Hg^{2+}, Bi^{3+}) which formed less soluble precipitates than copper sulfide, interfered with the electrode response by reacting with its surface.[89] The electrode has been used for the direct potentiometric determination of Cu^{2+}, complexometric titrations of Cu^{2+} and Th^{4+} ions,[87] precipitation titrations of Cu^{2+} and S^{2-},[87] and other potentiometric titrations involving sulfides, thioacetamide, and EDTA solutions.[88]

Copper(II) and silver sulfides thermomolded with polyethylene and heat sealed to polythene tubing with 10^{-3} M $CuCl_2$ internal reference solution containing Ag–AgCl served as a membrane electrode selective to Cu^{2+} ions.[90] Membranes of different sulfide composition were prepared and only those membranes prepared from simultaneously precipitated copper and silver sulfides followed the equation

$$E = E^\circ + 0.0296 \log a_{Cu^{2+}}$$

When the electrode was used with a SCE reference electrode a value of 230 ± 5 mV was derived for E° at $\log a_{Cu^{2+}} = 0$. The electrode response was not affected by Zn^{2+}, Co^{2+}, Ca^{2+}, Mg^{2+}, Na^+, and K^+ ions ($K_{Cu\text{-}M} < 10^{-4}$), whereas Hg(I, II) and Ag^+ ions interfered. The selectivity constants of the electrode in the presence of halide ions are shown in the accompanying table. The electrode served as a good indicator electrode for Cu(II) in potentiometric EDTA titrations in water or in mixed solvents.

X^-	K_{Cu^{2+}, X^-}
Cl^-	$\geqslant 10^{-1}$
Br^-	$\geqslant 10^{-4}$
I^-	$\geqslant 10^{-6}$

The Selectrode prepared by depositing the electroactive material [Cu(II)S + Ag_2S] on the surface of a graphite rod and polishing it served as an electrode selective to Cu^+ ions.[91] Other Selectrodes prepared by using other active materials such as Cu^{2+} and CuS, were also evaluated. Their responses to Cu^{2+} ions along with those of other commerical electrodes are shown in Table 7. The Selectrode has been used in titrations involving

TABLE 7

Sensitivity and Standard Potential of Cu(II)-Selective Membrane Electrodes[a]

		pCu		$E°$ (vs. SCE)
Electrode	Material	At pH 4.7	At pH 8.9	mV/pCu = 0
Selectrode	Cu	4–1	4–1	+ 96
Selectrode	CuS	8–1	12–1	+ 351
Selectrode	CuS–Ag_2S	10–1	13–1	+ 369
Selectrode	CuS–HgS	8–1	11–1	+ 345
Selectrode	CuS–CdS	4–1	6–1	+ 352
Orion	CuS–Ag_2S	6–1	11–1	+ 269
Beckman	CuS–Ag_2S	2–1	2–1	+ 295
Radiometer[b]	CuSe–Cu_2Se	6–1	10–1	+ 195

[a]Data from Hansen *et al.*[91]

[b]Preliminary Instruction Manual for Cupric Selenide Electrode F 1112 Cu Radiometer A/S Copenhagen, 1971.

copper(II) and EDTA, nitrilotriacetic acid, and ethylenediaminedi(*o*-hydroxyphenylacetic acid).[91] The Selectrode calibrated in a series of Cu^{2+} buffers at various pH levels has been used in the determination of the stability constants of Cu^{2+} complexes of glycine and EGTA at an ionic strength of 0.1 *M*.[92] Methods for the calculation of stability constants of chelate complexes from pH and pM values are given for the Cu(II)–EGTA system. The values derived are $\log K_{CuL}^{Cu,L} = 16.80$ and $\log K_{CuHL}^{H,CuL} = 5.3$. The Cu(II) Selectrode has been used in continuous in situ measurements of copper(II) activity in soil by burying it in a soil–solution system.[92] The total Cu(II) concentration has been assayed by a standard addition technique.

G. IRON-SELECTIVE MEMBRANE ELECTRODES

Chalcogenide glass (60% Se, 28% Ge, and 12% Sb) doped with Fe, Co, or Ni acts as an electrochemical sensor for ferric ions.[93] A Nerstian response has been noted in the concentration range 10^{-5}–10^{-1} *M*. It has been found to have good selectivity to Fe^{3+} in the presence of Fe^{2+}. The electrode response behavior supported a redox potential mechanism rather than an ion exchange mechanism. This was based on the variation of membrane potential with the concentration of Fe^{3+} ion. For a redox mechanism, the potential E is given by

$$E = E_R + \frac{2.303RT}{nF} \log \frac{(M^{n+})}{(M)} \tag{25}$$

where n is the number of electrons involved in the reduction of M^{n+} to M, whose standard potential is represented by E_R, (M^{n+}) and (M) are the concentrations, and the slope is 60 mV for $n = 1$. On the other hand, for an ion exchange mechanism the potential E is given by

$$E = E^\circ + \frac{2.303RT}{zF} \log \frac{(M^{z+})_{\text{solution}}}{(M^{z+})_{\text{membrane}}} \tag{26}$$

where the slope is 20 mV for $z = 3$.

For Fe^{3+} solutions the slope was found to be 60 mV. Another membrane electrode of chalcogenide glass of composition $Fe_nSe_{60}Ge_{28}Sb_{12}$ (n between 1.3 and 2) properly prepared and activated responded to changes in Fe^{3+} ion concentration in perchlorate, chloride, and nitrate solutions with an average Nernstian slope of 57.6 mV/decade over the concentration range 10^{-2}–10^{-5} *M* Fe^{3+}.[94] A useful response was found down to at least 10^{-6} *M* Fe^{3+}. To develop the active sensor for monitoring Fe^{3+} ion concentration, the electrode surface was etched in a caustic solution and exposed to a high concentration ($\leqslant 10^{-3}$ *M*) of Fe^{3+}. Thus the process of activation involved oxidation of the surface and further chemical reaction with the Fe^{3+} ion.

The electrode responds selectively to the Fe^{3+} ion in the presence of Fe^{3+} ion–sulfate complexes, some of which are $FeSO_4^+$, $Fe(SO_4)_2^-$, and HSO_4^-, in addition to other hydroxides of iron. It has been shown[95] that it is possible to monitor the SO_4^{2-} ion in aqueous solution (concentration $< 10^{-3}$ *M*) by merely adding Fe^{3+} in the concentration range 5×10^{-4}–1×10^{-3} *M* and adjusting the pH to 2.1. Alternatively, the Fe^{3+}–sulfate complexes can be broken and precipitated as $BaSO_4$ by adding Ba^{2+} ions incrementally.[96] The liberated Fe^{3+} ion is sensed by the electrode, and the appearance of a constant potential signals the end point. In this way the electrode has been used to estimate sulfate ion in natural waters.

A coated platinum wire responding to Fe^{3+} ion in the concentration range 10^{-1}–10^{-4} *M* has been described.[97] The electroactive membrane coated onto Pt wire is composed of the tetrachloroferrate(III) salt of the quaternary ammonium compound tricaprylmethyl ammonium chloride, Aliquat 336S, and polyvinyl chloride dissolved in tetrahydrofuran or cyclohexanone. The electrode was found to be very selective to tetrachloroferrate(III) and had small interference from Cu^{2+}, Al^{3+}, NO_3^-, F^-, and SO_4^{2-}. Of the remaining cations, Sn^{2+}, Hg^{2+}, Fe^{2+}, and Zn^{2+}, Sn^{2+} had the highest interference. This electrode has been used in the analysis of iron ores.[97]

H. LEAD-SELECTIVE MEMBRANE ELECTRODES

This, like the Cd^{2+} and Cu^{2+} electrodes, can be prepared from sulfides of Pb and Ag by pressing them together into a pellet.[4] Flow-through electrodes selective to Cu, Cd, and Pb have also been prepared.[98] The PbS–Ag_2S membrane electrode has been found to have interference from Hg^{2+}, Ag^+, and Cu^{2+} ions.[4] The Orion electrode has been used in the direct titration of sulfate in the presence of 50% *p*-dioxane.[99] For the estimation of sulfate, $PbSO_4$ must be absent and phosphates (if any) in the sample must be removed prior to titration of sulfate with lead perchlorate solution. Cl^- and NO_3^- interfere seriously if they are present in a 100-fold excess. Sulfur in organic compounds in 60% *p*-dioxane[100] and semimicroquantities of oxalate in 40% *p*-dioxane[101] have been determined using the lead-selective electrode. Also microquantities of orthophosphate have been determined by direct potentiometric titration[102] with lead perchlorate. The solution was buffered to pH 8.25–8.75 and nitrate and SO_4^{2-} ions interfered little with the electrode response. The presence of Cl^- and F^- interfered slightly and caused higher values for phosphorus to be determined. As a diagnostic tool, the Pb-selective electrode can be used to detect lead poisoning in children,[103] and has also been used in the determination of SO_2 in flue gases[104] and of the association constant of $PbSO_4$ ($K_f = 531$ liter/mole)[105] and of sulfate in the range 20–3000 ppm in mineral water and seawater.[106] The latter determination involved separation of Cl^- and HSO_3^- from the sample by passing it first through a cation exchange resin in Ag form and secondly through a cation exchange resin in acid form. The recovered solution was titrated with standard lead nitrate solution.

The solid Pb-selective membrane electrode has been evaluated for performance in aqueous and nonaqueous media.[107] A differential measuring technique was used to eliminate the aqueous–nonaqueous liquid junction. Lead in blood, saliva, and human urine has been estimated.

Silicone membranes impregnated with lead sulfide have been used as electrodes selective to lead ion.[108] This was prepared in the same way as the copper sulfide-impregnated copper-selective electrode with a direct solid contact to the electrode eliminating the internal reference solution. It has a response time of less than 2 min, and can be used in the pH range 2.8–7.0 and the temperature range 10–70°C. Ordinary ions showed little interference with the response of the electrode in the concentration range 10^{-2}–10^{-5} *M*; the slope was 29 mV/pM. A mixture of lead and silver sulfides, thermomolded with a polymer to form a membrane, has been found to respond selectively to lead ions.[109] Compression of sulfides of Pb, Ag, and Cu(I) to form pellets, or even better, compressing the sulfides at a pressure of 10 ton/cm^2 and sintering them at 350–500°C for 3 hr, has yielded electrodes that exhibit a Nernstian response to lead ions in the concentration range 10^{-1}–10^{-6} *M* of Pb^{2+} ions.[110] However, Ag^+, Cu^{2+}, Hg^{2+}, Fe^{3+}, S^{2-}, and I^- ions interfered seriously with the electrode response to lead ions.[111] Even a mixture of lead selenide or lead telluride with silver sulfide compacted into pellets or compacted and sintered at 100–600°C gave membranes selective to Pb^{2+} ions.[112] Sintered electrodes were found to be superior in response to compacted pellets. Among the common ions, Ag, Cu(II), Hg(II), Fe(III), S^{2-}, and Cl^- interfered with the electrode response. Similarly, the response of single crystal and precipitated lead-sensitive chalcogenide electrodes (PbS, PbSe, and PbTe) in lead ion buffer solutions was studied.[113] The sensitivity to Pb^{2+} ions decreased in the order PbS $\sim$ PbSe $>$ PbTe, while the sensitivity to pH followed the order PbS $<$ PbSe $\ll$ PbTe. The presence of Ag_2S in the electrode decreased the sensitivity of the electrode. Although the chalcogenide electrodes are not advantageous to use in direct potentiometry, they can be used to advantage in precipitation and complexometric titrations of lead.[113]

A Selectrode selective to lead ions has also been prepared, like the Selectrode selective to Cu(II), by using PbS–Ag_2S as the electroactive material deposited on the surface of a graphite rod.[114] It exhibited a Nernstian response in the pPb range 2–11. Consequently, it is suitable for measuring Pb(II) ion activity and so can be used for indicating the equivalence point in potentiometric titrations of EDTA and NYA. Titrations of anions such as SO_4^{2-}, CrO_4^{2-}, WO_4^{2-}, $C_2O_4^{2-}$, $Fe(CN)_6^{4-}$, and $P_2O_7^{4-}$ can also be performed using the electrode.

I. POTASSIUM-SELECTIVE MEMBRANE ELECTRODES

The most familiar potassium-selective membrane electrode is the glass electrode that is discussed in Chapter 9. In recent years, with the introduction of the valinomycin-based liquid membrane electrode discussed in Chapter 8, a variety of solid membrane electrodes (without a reservoir of ion exchange liquid saturating an inert matrix) have been prepared and their electrochemical behavior examined.

A solid K^+-selective electrode containing biological materials as sensor has been described.[115] It responded to step changes in K^+ ion activity with a Nernstian response. Another K^+-selective electrode based on the selective property of a polymeric material containing a macrocyclic antibiotic (exact compound not named) has been mentioned.[116] The electrode showed an immediate response to K^+ ion. The selectivity constant K_{K-M} was 10^{-4} for Na and 10^{-3} for NH_4^+.

There has been increased activity in the last two years to develop solid membranes based on chemical compounds that are known to complex with and carry the K ions across the polymer matrix in which the complexing compound is held. Much time and effort seem to have been invested in developing the K^+-selective electrodes, both solid and liquid.

A silicone rubber membrane electrode containing valinomycin (see liquid membranes for structure) with and without softener has been prepared and its selectivity to K^+ ion, stability, and reproductivity have been evaluated.[117] Some of the characteristics of various membranes containing valinomycin are given in Table 8.

TABLE 8

Characteristics of Valinomycin-Based Membrane Electrodes[a]

Electrode's membrane composition	Sensitivity range pK^+	Average slope (mV)	Reproducibility (mV)	Emf drift (Δ mV/time)
PVC + diphenyl ether				
+ dibutyl sebacate				
+ valinomycin	1–5	59	± 20	± 0.4 mV/24 hr
PVC dibutylphthalate	1–4	55	± 2.0	± 0.6 mV/24 hr
+ valinomycin	4–5	40	± 1.0	+ 4.5 mV/24 hr
Silicone rubber (SR)				
+ tetramethoxysilane	1–4	54		
+ valinomycin	4–5	35		
SR + dibutylphthalate				
(17 wt%)	1–4	55	± 0.6	+ 0.8 mV/24 hr
+ valinomycin	4–5	49		
SR + valinomycin	1–5	60		
(5 wt%)	5–6	45	± 0.5	0.1 mV/120 hr
Millipore (Philips type	1–5	59		
IS 560-K)	5–6	45	± 0.5	< ± 1.4 mV/120 hr

[a]The valinomycin-based silicone rubber membrane without a plasticizer has characteristics comparable to the valinomycin-based liquid membrane electrode. (Data from Pick *et al.*[117])

The electrode selectivity for most alkali and alkaline earth ions[118] is comparable (see Table 9) to that of a conventional liquid membrane electrode (Millipore filter saturated with valinomycin in diphenyl ether).[119] A similar silicone rubber membrane electrode but containing potassium zinc ferrocyanide[120] has been evaluated for the determination of alkali metal ions (see Table 9). The response of the electrode to K^+ ions was Nernstian with a slope of 59 mV/pK^+.

A Selectrode that is selective to K^+ ions has been developed. This, like the Ca electrode, containes an inner calomel reference paste applied to a Teflon–graphite surface. The Teflon–graphite conductor is tightly pressed against the membrane, which is formed from polyvinyl chloride and valinomycin dissolved in tetrahydrofuran and poured over a nylon net.[121] In addition, a variety of polymers, such as polyurethane, silicone rubber, polymethyl methacrylate, and different plasticizing solvents (e.g., dioctyladepate, DOA; dioctylphthalate, DOP; diphenyl ether, D; and diethylphthalate, DEP) have been used to incorporate valinomycin. The membranes so prepared were evaluated as Selectrodes and it was found that the best Selectrode characteristics were possessed by the membrane formed from valinomycin–DOP and PVC. The selectivity constants for this Selectrode and also for the commercial liquid membrane electrode (Philips IS560-K) were determined according to Eqs. (32) and (34) of Chapter 5. These values together with some literature values for the liquid membranes are given in Table 10.

A paste was formed from a solution of 5% potassium tetra(*p*-chlorophenyl) borate in a plasticizer and finely divided polyvinyl chloride powder. The paste was put in a mold, leveled, and cured at 200°C to form a membrane which was found selective to K^+ ions.[123] The chemical nature of the plasticizer has been found to have a dominant influence in determining the selectivity of the electrode toward univalent cations. The selectivity of the electrode has been determined by a variation of one of the methods described in Chapter 5.

Two solutions (1 and 2) containing ions i and j are made up in such a way that $a_{i(1)} > a_{i(2)}$ and $a_{j(1)} < a_{j(2)}$. Thus, the potential is given by [see Eq. (121) of Chapter 3]

$$E_1 = E^\circ + 2.303\,\frac{RT}{F}\log\left[a_{i(1)} + K_{ij}^{\text{pot}} a_{j(1)}\right] \tag{27}$$

$$E_2 = E^\circ + 2.303\,\frac{RT}{F}\log\left[a_{i(2)} + K_{ij}^{\text{pot}} a_{j(2)}\right] \tag{28}$$

When $E_1 = E_2$,

$$a_{i(1)} + K_{ij}^{\text{pot}} a_{j(1)} = a_{i(2)} + K_{ij}^{\text{pot}} a_{j(2)} \tag{29}$$

TABLE 9

Selectivity K_{ij}^{pot} of Valinomycin- and Potassium Zinc Ferrocyanide-Based Silicone Rubber and Liquid Membrane Electrodes

Interfering ion	Coated Pt wire containing valinomycin and plasticizer		Silicone rubber membrane containing valinomycin	Millipore filter/diphenyl ether containing valinomycin	Silicone rubber containing potassium zinc ferrocyanide
	DOOP	DDP			
K^+	1	1	1	1	1 (10^{-1} *M* conc.)
H^+	2.7	$<10^{-3}$	1.8×10^{-3}	5×10^{-5}	
Li^+	5×10^{-3}	10^{-3}	6.3×10^{-4}	2×10^{-4}	3×10^{-3}
NH_4^+	1.6×10^{-2}	1.2×10^{-2}	2.3×10^{-2}	10^{-2}	1.8
Na^+	3×10^{-3}	10^{-3}	3.3×10^{-4}	2.5×10^{-4}	2.5×10^{-2}
Rb^+	2.3	2.5	1.9	1.9	3.3
Cs^+	4.2×10^{-1}	4.4×10^{-1}	3.4×10^{-3}	4×10^{-1}	9.5
Be^{2+}	1.2×10^{-2}	$<10^{-3}$	8.5×10^{-4}	2.5×10^{-4}	
Ca^{2+}	2×10^{-3}	$<10^{-3}$	5.4×10^{-4}	—	
Sr^{2+}	$<10^{-3}$	$<10^{-3}$	6.2×10^{-4}	2×10^{-4}	
Mg^{2+}	$<10^{-3}$	$<10^{-3}$	6.2×10^{-4}	2×10^{-4}	
Ba^{2+}					
Ni^{2+}	10^{-2}	$<10^{-3}$	7.2×10^{-4}	6×10^{-5}	
Ref.	124	124	118	119	120

TABLE 10

Selectivity K^{pot}_{K-M} of Valinomycin-Based Membrane Electrodes[a]

	Valinomycin–DOP, PVC Selectrode		Philips IS 560-K liquid			
Interfering ion (M)	Eq. (32)	Eq. (34)	Eq. (32)	Eq. (34)	Liquid membranes	
Li^+	2.4×10^{-4}	4.2×10^{-4}	2.2×10^{-4}	3.9×10^{-4}	2×10^{-4}	—
Na^+	6×10^{-5}	7.8×10^{-5}	9.6×10^{-5}	1.3×10^{-4}	2.5×10^{-4}	7.1×10^{-5}
Rb^+	4.7	4.0	4.7	4.0	1.9	—
Cs^+	4.7×10^{-1}	4.1×10^{-1}	4.7×10^{-1}	4.3×10^{-1}	4.0×10^{-1}	—
NH_4^+	1.3×10^{-2}	2.2×10^{-2}	1.6×10^{-2}	2.8×10^{-1}	1×10^{-2}	2×10^{-2}
Ag^+	4.4×10^{-5}	9.3×10^{-5}	1.7×10^{-4}	3.8×10^{-4}	2×10^{-9} (Highly improbable)	—
Mg^{2+}	4.5×10^{-5}	6.8×10^{-5}	4.1×10^{-5}	5.9×10^{-5}	2×10^{-4}	2×10^{-4}
Ca^{2+}	4.9×10^{-5}	6.9×10^{-5}	4.2×10^{-5}	6.0×10^{-5}	2.5×10^{-4}	2×10^{-4}
Ba^{2+}	1.1×10^{-4}	1.8×10^{-4}	1.0×10^{-4}	1.7×10^{-4}	6.0×10^{-5}	—
Fe^{2+}	1.7×10^{-5}	2.2×10^{-5}	1.4×10^{-5}	1.8×10^{-5}	4.0×10^{-4}	—
Cu^{2+}	3.5×10^{-5}	4.2×10^{-5}	4.1×10^{-5}	5.8×10^{-5}	—	—
Ref.	121	121	121	121	s119	122

[a]Determined by Eqs. (32) and (34) of Chapter 5.

which, on rearranging, gives

$$K_{ij}^{pot} = \frac{a_{i(1)} - a_{i(2)}}{a_{j(2)} - a_{j(1)}} \tag{30}$$

The value of K_{ij}^{pot} thus evaluated for $i = K^+$ and $j = Na^+$ was 1.74×10^{-2}, which agreed with the value for $K_{K\text{-}Na}$ derived using single electrolyte solutions in which $a_i = a_j$, i.e.,

$$\log K_{ij}^{pot} = \frac{E_j - E_i}{2.303RT/F}$$

where

$$E_i = E^\circ + \frac{RT}{F} \ln a_i \qquad \text{and} \qquad E_j = E^\circ + \frac{RT}{F} \ln K_{ij}^{pot} a_j$$

Coated platinum wire electrodes have also been developed. In one method[124] the Pt wire was coated with the following solution and dried in air for two days. The coating mixture contained polyvinyl chloride, a plasticizer [either di(2-ethylhexyl)2-ethylhexyl phosphonate (DOOP) or di-*n*-decylphthalate (DDP)], and valinomycin dissolved in tetrahydrofuran. The uncoated portion of the wire was insulated by coating with paraffin. The selectivity constants of these electrodes are given in Table 9 along with other values derived for K^+-selective electrodes. The plasticizer DDP seems to give better responses for K^+ ions than does DOOP. The Cu^{2+} and Zn^{2+} ions poisoned the electrodes. Using these electrodes the K content of whole blood and of seawater has been analyzed.

In another method a miniature solid state K electrode suitable for serum analysis[125] was developed. A solution containing PVC, dipentylphthalate (DPP), and valinomycin was prepared using redistilled cyclohexanone as the solvent. Silver wire, 18 gauge, was cleaned with isopropanol and chloridized in 0.1 *M* HCl. Polyvinyl alcohol (10% solution in 0.005 *M* KCl at 40–50°C) was dip cast onto the Ag–AgCl wire, which was then dipped into the solution just described. It was found beneficial to have higher PVC and DPP concentrations in this solution.

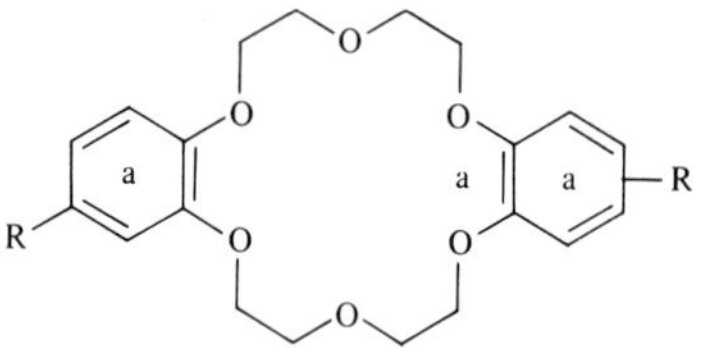

VII. a = cyclohexyl, R = H
VIII. a = benzo, R = H
IX. a = benzo, R = CH_3
X. a = benzo, R = *n*-C_3H_7
XI. a = cyclohexyl, R = CH_3
XII. a = cyclohexyl, R = *n*-C_3H_7

Fig. 2. Structure of 18-crown-6 compounds.

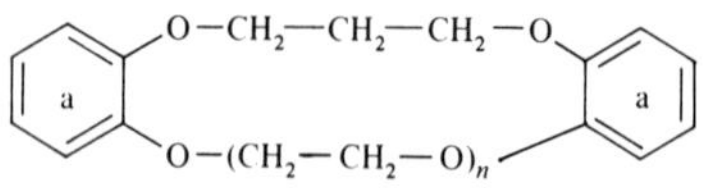

XIII. a = benzo, $n = 2$
XIV. a = benzo, $n = 3$
XV. a = cyclohexyl, $n = 3$
XVI. a = benzo, $n = 4$

Fig. 3. Structure of crown compounds.

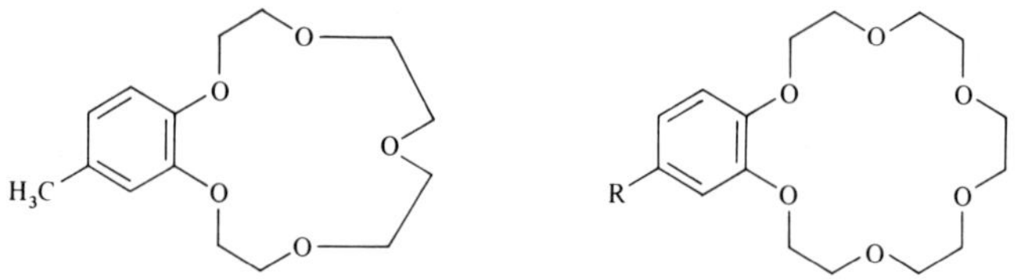

(a) XVII.

(b) XVIII. R = H
XIX. R = NO_2

Fig. 4. Structure of crown compounds.

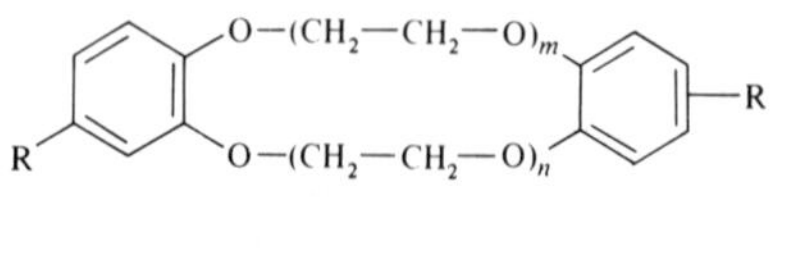

XX. $m = 3, n = 3$, R = CH_3
XXI. $m = 4, n = 4$, R = H
XXII. $m = 4, n = 4$, R = CH_3
XXIII. $m = 4, n = 4$, R = *tert*-C_4H_9
XXIV. $m = 4, n = 4$, R = NO_2
XXV. $m = 5, n = 5$, R = *tert*-C_4H_9
XXVI. $m = 5, n = 5$, R = H

Fig. 5. Structure of crown compounds.

Besides valinomycin, other macrocyclic compounds such as the polyethers discovered by Pedersen,[126] which act as ion-selective carriers have been used in the preparation of ion-selective membrane electrodes. The most important substance of this group is 2,3,11,12-dicyclohexyl-1,4,7,10,13,16-hexaoxacyclo-octadecene (VII) (pages 210, 211) which Pedersen named dicyclohexyl-18-crown-6. The number 18 indicates the number of members in the ring and 6 denotes the number of heterocyclic oxygen atoms. The preparation of this compound is carried out by condensation of bis(*o*-hydroxyphenoxy)ethyl ether with bis(chlorethyl) ether in the presence of NaOH. The dibenzo-18-crown-6 compound formed is converted to the dicyclohexyl-18-crown-6 compound by catalytic hydrogenation. Since the chance discovery of this compound[127] over 60 different cyclic polyethers

TABLE 11

Selectivity Constants of Crown Compounds and Valinomycin in PVC Membranes

	Crown compound	K_{K-Na} [a]	Slope[b] (mV/pK$^+$)
VII.	Dicyclohexyl-18-crown-6	1.1×10^{-2}	58
VIII.	Dibenzo-18-crown-6	7.7×10^{-2}	51
IX.	Dimethyldibenzo-18-crown-6	6.7×10^{-2}	60
X.	Di-*n*-propyldibenzo-18-crown-6	6.3×10^{-2}	59
XI.	Dimethyldicyclohexyl-18-crown-6	1.1×10^{-2}	58
XII.	Di-*n*-propyldicyclohexyl-18-crown-6	1.6×10^{-2}	60
XIII.	Dibenzo-16-crown-5	1.0	
XIV.	Dibenzo-19-crown-6	2.2×10^{-2}	
XV.	Dicyclohexyl-19-crown-6	1.5×10^{-2}	
XVI.	Dibenzo-22-crown-7	3.3×10^{-2}	
XVII.	4-Methylbenzo-15-crown-5	6.7×10^{-2}	
XVIII.	Benzo-18-crown-6	0.53×10^{-2}	
XIX.	4-Nitrobenzo-18-crown-6	0.28×10^{-2}	
XX.	Dimethyldibenzo-24-crown-8	1.0×10^{-1}	59
XXI.	Dibenzo-30-crown-10	0.85×10^{-2}	
XXII.	Dimethyldibenzo-30-crown-10	0.22×10^{-2}	
XXIII.	Di-*tert*-butyldibenzo-30-crown-10	0.24×10^{-2}	
XXIV.	Dinitrodibenzo-30-crown-10	0.95×10^{-2}	
XXV.	Di-*tert*-butyldibenzo-36-crown-12	1.2×10^{-2}	
XXVI.	Dibenzo-36-crown-12	1.6×10^{-2}	
	Valinomycin	0.03×10^{-2} [b]	60

[a]Values from Petranek and Ryba.[138]
[b]Values from Ryba *et al.*[139]

have been prepared.[127–131] The properties of some of these compounds have been reviewed.[126, 128, 132–134]

The compounds form complexes with metal ions by ion–dipole interactions and are soluble in nonpolar solvents.[135] The order of preference of the central ion in the case of 15–18-membered rings is determined by the size of the cavity[127, 129, 131] which is given in the accompanying table. The complexes that these crown compounds form are roughly planar.[136, 137] The K complex with dibenzo-30-crown-6, however, has a wraparound structure, with all oxygen atoms approximately equidistant from the central ion but in different planes.[136]

Polyethers	Size of cavity (Å)
All 14-crown-4	1.2–1.5
All 15-crown-5	1.7–2.2
All 18-crown-6	2.6–3.2
All 21-crown-7	3.4–4.3

In order to study the effect of the structure of these polyethers on the selectivity they exhibit toward K and Na ions. Petranek and Ryba[138] prepared 20 crown compounds (VII–XXVI). Using each of these compounds, membranes were formed by pouring the soltuion [10 mg crown compound + 1 ml dipentylphthalate + 10 ml of 5% (w/v) solution of polyvinyl chloride in cyclohexanone] on a glass plate and evaporating the solvent at room temperature. The values determined for the selectivity constant $K^{\mathrm{pot}}_{\mathrm{K-Na}}$ are given in Table 11 along with the Nerstian slopes obtained for some membranes that were studied by Ryba *et al.*[139] It is seen that the highest selectivities are realized with membranes containing polyethers with six oxygen binding sites in 18-membered rings. An increase in the size of the cavity of the ring (see values given above) does not bring about any increase in selectivity and the crown-containing membranes do not display that high selectivity to the K^+ ion over the Na^+ ion shown by valinomycin-based membranes (see also Tables 9 and 10).

The selectivity constants ($K^{\mathrm{pot}}_{\mathrm{K-M}}$) of membranes containing the dimethyldibenzo-30-crown-10 compound[140] are shown in Table 12.

In another study[141] both valinomycin and the dimethyldibenzo-30-crown-10 compound have been incorporated into a PVC matrix and the membranes so formed have been used as electrodes to study their selectivities to various ions. Both graphical and numerical methods [see Eq. (34), Chapter 5] were used to calculate the values of $K^{\mathrm{pot}}_{\mathrm{K-M}}$, which are also given in Table 12.

The valinomycin-based PVC membrane electrode is definitely more suitable than the crown-based electrode because of its high selectivity. In

TABLE 12

Selectivity K_{K-M} of Dimethyldibenzo-30-Crown-10 and Valinomycin-Based Polyvinyl Chloride Membrane Electrodes

		Crown–PVC membrane				Valinomycin–PVC membrane		
Interfering ion	Concentration (M)	Other	Graphical	Numerical	Concentration (M)	Graphical	Numerical	Concentration (M)
Rb^+		9.2×10^{-1}						
Cs^+		2.5×10^{-1}						
Li^+		5.1×10^{-3}						
Ca^{2+}		3.0×10^{-4}						
Mg^{2+}		1.0×10^{-4}						
Ba^{2+}		$< 1.0 \times 10^{-4}$						
NH_4^+		$> 7.0 \times 10^{-2}$	9.4×10^{-2}	8.2×10^{-2}	10^{-5}	1.5×10^{-2}	1.7×10^{-2}	10^{-6}
			9.4×10^{-2}	7.1×10^{-2}	10^{-4}	1.4×10^{-2}	1.7×10^{-2}	10^{-5}
			6.3×10^{-2}	5.9×10^{-2}	10^{-3}	1.2×10^{-2}	2.1×10^{-2}	10^{-4}
			5.0×10^{-2}	3.6×10^{-2}	10^{-2}	1.1×10^{-2}	2.2×10^{-2}	10^{-3}
Na^+	0.01	3.9×10^{-3}	1.1×10^{-1}	1.4×10^{-1}	10^{-5}	1.1×10^{-4}	1.5×10^{-4}	10^{-6}
	0.1	2.2×10^{-3}				1.6×10^{-4}	1.6×10^{-4}	10^{-5}
						1.1×10^{-4}	3.4×10^{-4}	10^{-4}
Ref.	140	140	141	141	141	141	141	141

yet another study[142] the role of plasticizers used with both valinomycin and cyclic polyethers has been explored. The membranes were prepared from a solution of 0.15–0.2 g of PVC, 5–10 mg neutral carrier, and 0.1–0.4 g plasticizer in 6 ml of tetrahydrofuran. The neutral carriers used were valinomycin (val), dicyclohexyl-18-crown-6, and dibenzo-18-crown-6 cyclic polyethers and the plasticizers used were tributyl phosphate (TBP), dibutylphthalate (DBP), and diphenyl ether (D). Other polymers and dicyclohexyl-18-crown-6 were found to be unsatisfactory. The characteristics of some of the K-selective membrane electrodes are compared with those of other commercial electrodes in Table 13.

J. SILVER-SELECTIVE MEMBRANE ELECTRODES

The silver sulfide membrane electrode is prepared mostly from silver sulfide alone or mixed with other metallic sulfides. There is also the single crystal Ag_2S electrode.

Most of these membrane electrodes are prepared from the sulfide precipitate, sintered, and incorporated or pressed into a ceramic plate or some membrane-forming polymer. These electrodes respond not only to Ag^+ ions but also to sulfide and other ions.[145] Their response to Ag^+ and sulfide ions is Nernstian[146] provided Hg^{2+} and CN^- ions are absent. The electrodes are not influenced by light under ordinary laboratory conditions. The selectivity constants determined for the various Ag_2S membrane electrodes are shown in Table 14.

The silver-selective electrode has been used in the estimation of silver ions (50 ppb)[147] and in the indirect analysis of the Cl^- ion.[148] In the latter procedure, an excess of added Ag^+ ions in equilibrium with a AgCl precipitate is determined potentiometrically. Similarly, the Ag–Ag_2S electrode has been used in the potentiometric titrations of cyanide and chloride consecutively.[149, 150]

When Ag_2S is mixed with PbS or CdS, the electrode responds to Pb or Cd ions[151] and so it is used in potentiometric titrations involving these ions. Ag_2S deposited in ceramic frits has been used as an electrode in redox systems[152] and in acid–base titrations.[153] A silicone rubber-based Ag_2S precipitate membrane electrode has been evaluated potentiometrically for its selectivity by titrations of Na_2S solution with $AgNO_3$ in the presence of I^-, Br^-, Cl^-, OH^-, SCN^-, SO_4^{2-}, CN^-, $S_2O_3^{2-}$, Tl^+, Cu^{2+}, Pb^{2+}, Cd^{2+}, Zn^{2+}, Fe^{2+}, Mn^{2+}, Ni^{2+}, and La^{3+}. The selectivity constants deduced for these ions have been found to agree well with those calculated from the solubility constants.[154]

A polyvinyl chloride membrane containing tricresyl phosphate has been

TABLE 13

Comparison of the Selectivities of PVC–Neutral Carrier Membrane Electrodes with Those of Liquid Membrane Electrodes

Membrane electrode	Ref.	Range of pK^+	Slope (mV/decade)	Interfering ions: K^{pot}_{K-M}						
				Na^+	Rb^+	NH_4^+	Cs^+	Li^+	Ca^{2+}	Mg^{2+}
Valinomycin										
PVC + TBP + D	142	1–2	40	$< 2 \times 10^{-4}$	3	1.2×10^{-2}	2.6×10^{-1}	$\sim 10^{-4}$	$\sim 10^{-4}$	$\sim 10^{-4}$
		2–5	55							
PVC + DBP	142	1–2	40							
		2–5	60	$< 2 \times 10^{-4}$	3	2×10^{-2}	5×10^{-1}	$\sim 10^{-4}$	$\sim 10^{-4}$	$\sim 10^{-4}$
		5–6	30							
PVC + DOA Selectrode	121	1–4	57	7×10^{-5}	4.8	1.3×10^{-2}	6×10^{-1}	3×10^{-4}	5×10^{-5}	5×10^{-5}
PVC + DPP	139	1–5	60	3×10^{-4}						
Liquid membrane	143	—	—	10^{-4}	2.5	2×10^{-2}		$\sim 6 \times 10^{-5}$		
Dibenzo-18-crown-6										
PVC + TBP + D	142	1–2	40							
		2–5	55	7×10^{-2}	10^{-1}	$< 10^{-2}$	10^{-1}			
		5–6	35							
PVC + DBD	142									
		1–4	60	4.5×10^{-1}	10^{-1}	6×10^{-2}	9×10^{-2}			
		4–5	40							
PVC + DPP	139	2.5–5	51	7.7×10^{-2}						
Liquid membrane	144			$\sim 4 \times 10^{-2}$	~ 0.3	$\sim 6 \times 10^{-2}$	$\sim 2.5 \times 10^{-1}$			

TABLE 14

Selectivity Constants K_{ij}^{pot} for Various Pressed Ag_2S Membranes[a]

K_{ij}^{pot}	Theoretical	Crytur	A	B	C	D
$K_{Ag^+-Cu^{2+}}$	9.0×10^{-8}	10^{-5}	10^{-6}	2×10^{-6}	10^{-5}	5×10^{-7}
$K_{Ag^+-Pb^{2+}}$	9.0×10^{-12}	10^{-6}	8×10^{-9}	2×10^{-9}	10^{-8}	5×10^{-11}
$K_{Ag^+-H^+}$	4.0×10^{-12}	9×10^{-6}	4×10^{-7}	3×10^{-6}	5×10^{-6}	6×10^{-12}

[a]Theoretical: ratio of solubility products according to Eq. (178) of Chapter 3. Crytur: commercial electrode (Czechoslovakia). A: Ag_2S precipitated from a solution containing 3% excess sulfur and purified by using carbon disulfide. B: Ag_2S precipitated from a solution containing 3% excess sulfur. C: Ag_2S precipitated from a solution containing 3% excess silver. D: Ag_2S precipitated from a solution containing 0–3% excess of either Ag or sulfur and purified by using HNO_3 and/or carbon disulfide.

found to be selective to Ag^+, Tl^+, and Hg^{2+} ions and has been used in titrations involving some halide ions.[155] The silver ion-selective Ag_2S membrane electrode has been used in the potentiometric determination of Ag in ZnS, CdS, ZnSe, and CdSe. A cell of the type

$$Ag|0.1\ N\ AgNO_3|Ag_2S\ \text{membrane}|Ag(x\ N)|\text{Sat. } KNO_3|\text{Sat. } KCl, AgCl, Ag$$

was used.[156] The electrode showed a Nernstian response for 10^{-3}–10^{-6} N Ag^+ at pH 5.0. Zinc nitrate and cadmium nitrate in excess (10^5-fold) did not interfere with the estimation of Ag^+ ions. In the pH range 2–3, Ag^+ concentration (0.003%) was determined with $\pm 7.5\%$ error.

K. SODIUM-SELECTIVE MEMBRANE ELECTRODES

The glass membrane electrode is the most popular Na ion-selective electrode. Recently Ammann *et al.*[157] prepared the ligand (XXVII) which when used as a membrane (solid or liquid) was selective to Na ions.[158] Two

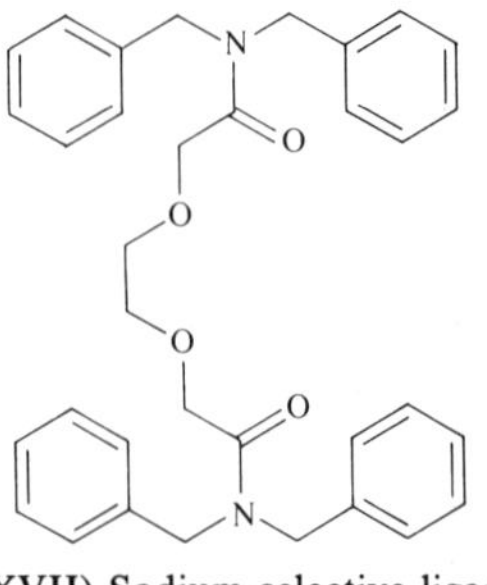

(XXVII) Sodium-selective ligand

types of electrodes (A and B) were prepared as follows:

Electrode A: Na ligand 2.4% (w/w)
Dibenzyl ether 67% (dielectric constant 4.0)
PVC 30.6%

Electrode B: 2.6% Na ligand
o-Nitrophenyl *n*-octyl ether 64.4% (dielectric constant 24)
PVC 33%

Membranes of 0.2 mm thickness were cast from these solutions and used as electodes. Electrode A gave a Nernstian response to Na^+ ions in the concentration range 10^{-1}–5×10^{-5} M with a slope of 57.1 mV/pNa. The response of electrode B was slightly less than Nernstian. Their selectivities to various ions are shown in Table 15.

The membrane formed of the solvent of low dielectric constant (electrode A) seems to have better selectivity than the membrane electrode made from using the solvent of higher dielectric constant.

TABLE 15

Selectivity Constants K_{ij}^{pot} of Na-Selective PVC Membrane Electrodes Containing the Ligand XXVII[a]

	K_{Na-M}^{pot}	
Cation	Electrode A	Electrode B
H^+	1	7.2×10^{-1}
NH_4^+	3.9×10^{-3}	6×10^{-3}
Li^+	3.6×10^{-2}	4.8×10^{-2}
Na^+	1	1
K^+	1.9×10^{-1}	2.1×10^{-1}
Rb^+	6.7×10^{-2}	9.5×10^{-2}
Cs^+	3.6×10^{-2}	7.3×10^{-2}
Mg^{2+}	3.5×10^{-4}	8.2×10^{-4}
Ca^{2+}	5.6×10^{-3}	3.4×10^{-1}
Sr^{2+}	1.2×10^{-2}	2×10^{-1}
Ba^{2+}	4.4×10^{-2}	8.1×10^{-1}
Zn^{2+}	1.1×10^{-2}	1.3×10^{-1}

[a]See text for a description of electrodes A and B.

L. THALLIUM-SELECTIVE MEMBRANE ELECTRODES

Thallium salts of molybdophosphoric (Tl–Mo–P) and tungstophosphoric (Tl–W–P) acids in epoxy resin supports have been used to construct electrodes selective to thallium(I) ions. An evaluation of these electrodes[159] yielded a calibration curve with a slope of 41 mV/decade concentration of Tl. In the pH range 1–4, the electrode potential increased and then remained constant from pH 4 to 6. The electrodes gave a temperature coefficient of 0.27 mV/deg, and the response time (to reach constant potential) was about 2 min. The selectivity constants of these electrodes are given in Table 16. The slopes of the calibration curves obtained in 25% methanol, ethanol, *n*-propanol, and acetone solutions were 39, 39, 40, and 38 mV/decade, respectively, for the Tl–Mo–P electrode and 40, 42, 44.7, and 46.3 mV/pTl respectively, for the Tl–W–P electrode. These electrodes can be used as indicator electrodes in precipitation titrations involving thallium with potassium bromide, potassium chromate, and sodium tetraphenyl borate.

TABLE 16

K_{ij}^{pot} of Tl–Mo–P and Tl–W–P Membrane Electrodes Selective to Tl^{+} Ions[a]

Interfering ion	Tl–Mo–P	Tl–W–P	Interfering ion	Tl–Mo–P	Tl–W–P
Li^{+}	10^{-2}	0.28	Ag^{+}	—	0.65
Na^{+}	10^{-2}	0.53	Mg^{2+}	10^{-3}	2×10^{-2}
K^{+}	10^{-2}	0.66	Ca^{2+}	10^{-3}	2×10^{-2}
Rb^{+}	6.5×10^{-2}	0.8	Sr^{2+}	10^{-3}	1.3×10^{-2}
Cs^{+}	10^{-2}	0.29	Ba^{2+}	10^{-3}	1.2×10^{-2}
NH_4^{+}	10^{-2}	0.57			

[a]Values from Coetzee and Basson.[159]

M. URANYL ION-SELECTIVE MEMBRANE ELECTRODES

Several uranyl organic phosphorus complexes incorporated into a polyvinyl matrix have been used as sensors for the uranyl ion. The membrane was formed from a solution of the uranyl complex of di(2-ethylhexyl) phosphoric acid or other organophosphorus compound in a suitable diluent, polyvinyl chloride, and tetrahydrofuran.[160] The best electrodes were considered to be formed from di(2-ethylhexyl) phosphoric acid in diamyl phosphonate (DAP), di(2-ethylhexyl)ethyl phosphonate (DEHEP), or tri(2-ethylbutyl) phosphate (TEBP) as the diluent. The electrode formed by using each of these diluents gave near-Nernstian slopes. Ionic interference to the response of these electrodes to uranyl ions is indicated in Table 17 by the selectivity constants K_{ij}^{pot}.

TABLE 17

Selectivity Constant K_{ij}^{pot} of Electrodes Formed from Uranyl Complex of Di-2-ethylhexyl Phosphoric Acid in PVC

		Diluent solvent		
Interfering ion (j)	Concentration (M)	Diamyl phosphonate (DAP)	Di(2 - ethythexyl)-ethyl phosphonate (DEHEP)	Tri(2 - ethylbutyl) phosphate (TEBP)
Ca^{2+}	10^{-2}	2.7×10^{-3}	9×10^{-4}	1.4×10^{-3}
Cd^{2+}	10^{-2}	2.7×10^{-3}	3.4×10^{-3}	1.3×10^{-3}
Mg^{2+}	10^{-2}	2.5×10^{-3}	1×10^{-3}	4×10^{-4}
Zn^{2+}	10^{-2}	6.4×10^{-3}	1.7×10^{-3}	7×10^{-4}
Cu^{2+}	10^{-2}	9.3×10^{-3}	1.9×10^{-3}	5×10^{-4}
Al^{3+}	10^{-2}	2.9×10^{-3}	5.5×10^{-3}	3.8×10^{-3}
Fe^{3+}	10^{-4}	0.42	0.15	0.48
Ni^{2+}	10^{-2}	6×10^{-3}	1.7×10^{-3}	3×10^{-3}

REFERENCES

1. G. J. Hills, *in* "Reference Electrodes" (D. J. G. Ives and G. J. Janz, eds.), p. 411. Academic Press, New York, 1961.
2. M. Cremer, *Z. Biol.* **47**, 562 (1906).
3. F. Haber and Z. Klemensiewicz, *Z. Phys. Chem.* **67**, 385 (1909).
4. J. W. Ross, Jr., *in* "Ion Selective Electrodes" (R. A. Durst, ed.), Chapter 2. Nat. Bur. Std. Spec. Publ. 314, Washington, D.C., 1969.
5. J. D. Czaban and G. A. Rechnitz, *Anal. Chem.* **45**, 471 (1973).
6. T. Anfalt and D. Jagner, *Anal. Chim. Acta* **56**, 477 (1971).
7. M. Mascini and A. Liberti, *Anal. Chim. Acta* **47**, 339 (1969); **51**, 231 (1970); **53**, 202 (1971).
8. J. Ruzicka and C. G. Lamm, *Anal. Chim. Acta* **53**, 206; **54**, 1 (1971).
9. J. Ruzicka, C. J. Lamm, and J. C. Tjell, *Anal. Chim. Acta* **62**, 15 (1972).
10. C. G. Lamm, E. H. Hansen, and J. Ruzicka, *Anal. Lett.* **5**, 451 (1972).
11. J. H. Wang and E. Copeland, *Proc. Nat. Acad. Sci.* **70**, 1909 (1973).
12. M. Sharp and G. Johannson, *Anal. Chim. Acta* **54**, 13 (1971).
13. P. Weidenthaler and E. Pelinka, *Collect. Czech. Chem. Commun.* **34**, 1482 (1969).
14. M. Sharp, *Anal. Chim. Acta* **59**, 137 (1972).
15. C. Liteanu and C. Hopirtean, *Rev. Roum. Chim.* **16**, 559 (1971).
16. H. Hirata and K. Higashiyama, *Talanta* **19**, 391 (1972).
17. A. G. Fogg, A. S. Pathan, and B. T. Burns, *Anal. Chim. Acta* **69**, 238 (1974).
18. G. I. Goodfellow and H. M. Webber, *Analyst* **97**, 95 (1972).
19. W. M. Baumann and W. Simon, *Helv. Chim. Acta* **52**, 2054 (1969).
20. R. E. Cosgrove, C. A. Mask, and I. H. Krull, *Anal. Lett.* **3**, 457 (1970).
21. D. Midgley and K. Torrance, *Analyst* **97**, 626 (1972).
22. J. G. Motalvo, Jr., *Anal. Chim. Acta* **65**, 189 (1973).
23. J. Ruzicka and E. H. Hansen, *Anal. Chim. Acta* **69**, 129 (1974).
24. J. Ruzicka, E. H. Hansen, P. Bisgaard, and E. Reymann, *Anal. Chim. Acta* **72**, 215 (1974).

25. E. H. Hansen, H. B. Filho, and J. Ruzicka, *Anal. Chim. Acta* **71**, 225 (1974).
26. U. Fiedler, E. H. Hansen, and J. Ruzicka, *Anal. Chim. Acta* **74**, 423 (1975).
27. M. L. Eagan and L. Dubois, *Anal. Chim. Acta* **70**, 157 (1974).
28. R. A. Llenado and G. A. Rechnitz, *Anal. Chem.* **46**, 1109 (1974).
29. T. R. Gilbert and A. M. Clay, *Anal. Chem.* **45**, 1757 (1973).
30. W. H. Evans and B. F. Partridge, *Analyst* **99**, 367 (1974).
31. I. Sekerka and J. F. Lechner, *Anal. Lett.* **7**, 463 (1974).
32. D. Midgley and K. Torrance, *Analyst* **98**, 217 (1973).
33. C. H. Sloan and G. P. Morie, *Anal. Chim. Acta* **69**, 243 (1974).
34. A. V. Gordievskii, V. S. Shterman, Ya. A. Syrchenkov, N. I. Savvin, A. F. Zhukov, and Yu. I. Urusov, *Zh. Anal. Khim.* **27**, 2170 (1972).
35. H. Hirata and K. Higashiyama, *Z. Anal. Chem.* **257**, 104 (1971).
36. H. Hirata and K. Date, *Bull. Chem. Soc. Japan* **46**, 1468 (1973).
37. M. Mascini and A. Liberti, *Anal. Chim. Acta* **64**, 63 (1973).
38. Analytical Methods Guide, 2nd ed., 1972, Orion Research, Massachusetts.
39. M. J. D. Brand, J. J. Militello, and G. A. Rechnitz, *Anal. Lett.* **2**, 523 (1969).
40. J. Ruzicka and E. H. Hansen, *Anal. Chim. Acta* **63**, 115 (1973).
41. H. J. C. Tendeloo, *J. Biol. Chem.* **113**, 333 (1936); *Rec. Trav. Chim.* **55**, 227 (1936).
42. H. J. C. Tendeloo and A. Krips, *Rec. Trav. Chim.* **76**, 703, 946 (1957).
43. H. J. C. Tendeloo and F. H. Van der Voort, *Rec. Trav. Chim.* **79**, 639 (1960).
44. P. Cloos and J. J. Fripaat, *Bull. Soc. Chim. Fr.* 423 (1960).
45. A. Shatkay, *Anal. Chem.* **39**, 1056 (1967).
46. H. P. Gregor and H. J. Schonhorn, *J. Amer. Chem. Soc.* **81**, 3911 (1959).
47. I. Langmuir and V. J. Schaefer, *J. Amer. Chem. Soc.* **59**, 2400 (1937).
48. R. Block, A. Shatkay, and H. A. Saroff, *Biophys. J.* **7**, 865 (1967).
49. E. A. Materova, A. L. Grekovich, and S. E. Didina, *Elektrokhimiya* **8**, 1829 (1972).
50. M. E. Thompson and J. W. Ross, Jr., *Science* **154**, 1643 (1966).
51. J. W. Ross, Jr., *Science* **156**, 1378 (1967).
52. F. A. Schultz, A. J. Petersen, C. A. Mask, and R. P. Buck, *Science* **162**, 267 (1968).
53. G. J. Moody, R. B. Oke, and J. D. R. Thomas, *Analyst* **95**, 910 (1970).
54. G. H. Griffiths, G. J. Moody, and J. D. R. Thomas, *Analyst* **97**, 420 (1972).
55. R. W. Cattrall and H. Freiser, *Anal. Chem.* **43**, 1905 (1971).
56. A. Ansaldi and S. I. Epstein, *Anal. Chem.* **45**, 595 (1973).
57. J. Ruzicka, E. H. Hansen, and J. C. Tjell, *Anal. Chim. Acta* **67**, 155 (1973).
58. F. W. Orme, *in* "Glass Micro Electrodes" (M.Lavallace, O. F. Schanne, and N. C. Hebert, eds.), p. 376, Wiley, New York, 1967.
59. G. A. Rechnitz and Z. F. Lin, *Anal. Chem.* **40**, 696 (1968).
60. G. A. Rechnitz and M. T. Hseu, *Anal. Chem.* **41**, 111 (1969).
61. A. Shatkay, *J. Phys. Chem.* **71**, 3858 (1967); *Biophys. J.* **8**, 912 (1968).
62. A. Shatkay, *Electrochim. Acta* **15**, 1759 (1970).
63. A. Shatkay, *Anal. Chem.* **40**, 457 (1968).
64. G. A. Rechnitz, *Anal. Chem.* **40**, 456 (1968).
65. M. S. Frant, *Anal. Chem.* **40**, 457 (1968).
66. C. J. Coetzee and A. J. Basson, *Anal. Chim. Acta* **57**, 478 (1971).
67. C. J. Coetzee and A. J. Basson, *Anal. Chim. Acta* **56**, 321 (1971).
68. T. Nomura and G. Nakagawa, *Bunseki Kagaku* **20**, 1570 (1971).
69. N. T. Savvin, V. S. Shterman, A. V. Gordievskii, and Ya,. A. Syrchenkov, *Zavod. Lab.* **37**, 1025 (1971).
70. N. I. Savvin, V. S. Shterman, O. A. Zemskaya, Ya. A. Syrchenkov, and A. V. Gordievskii, *Zh. Anal. Khim.* **23**, 1281 (1971).

71. L. F. Heerman and G. A. Rechnitz, *Anal. Chem.* **44**, 1655 (1972).
72. H. Hirata and K. Date, *Talanta* **17**, 883 (1970).
73. H. Hirata, K. Higashiyama, and K. Date, *Anal. Chim. Acta* **51**, 209 (1970).
74. J. Vesely, *Collect. Czech. Chem. Commun.* **36**, 3365 (1971).
75. L. Sucha and M. Suchanek, *Anal. Lett.* **3**, 613 (1970).
76. G. Johansson and K. Edstrom, *Talanta* **19**, 1623 (1972).
77. D. J. Crombie, G. J. Moody, and J. D. R. Thomas, *Talanta* **21**, 1094 (1974).
78. W. J. Blaedel and D. E. Dinwiddie, *Anal. Chem.* **46**, 873 (1974).
79. H. Wada and Q. Fernando, *Anal. Chem.* **43**, 751 (1971).
80. R. Jasinski, I. Trachtenberg, and D. Andrychuck, *Anal. Chem.* **46**, 364 (1974).
81. M. Mascini, *Anal. Chim. Acta* **56**, 316 (1971).
82. J. W. Ross, Jr. and M. S. Frant, *Anal. Chem.* **41**, 1900 (1969).
83. E. W. Baumann and R. M. Wallace, *Anal. Chem.* **41**, 2072 (1969).
84. G. A. Rechnitz and N. C. Kenny, *Anal. Lett.* **2**, 395 (1969).
85. G. A. Rechnitz and N. C. Kenny, *Anal. Lett.* **3**, 509 (1970).
86. E. Hakoila, *Anal. Lett.* **3**, 273 (1970).
87. J. Pick, *Kem. Kozlem.* **39**, 19 (1973).
88. J. Pick, K. Toth, and E. Pungor, *Anal. Chim. Acta* **61**, 169 (1972).
89. J. Pick, K. Toth, and E. Pungor, *Anal. Chim. Acta* **65**, 240 (1973).
90. M. Mascini and A. Liberti, *Anal. Chim. Acta* **53**, 202 (1971).
91. E. H. Hansen, C. G. Lamm, and J. Ruzicka, *Anal. Chim. Acta* **59**, 403 (1972).
92. E. H. Hansen and J. Ruzicka, *Talanta* **20**, 1105 (1973).
93. C. T. Baker and I. Trachtenberg, *J. Electrochem. Soc.* **118**, 571 (1971).
94. R. Jasinski and I. Trachtenberg, *J. Electrochem. Soc.* **120**, 1169 (1973).
95. R. Jasinski and I. Trachtenberg, *Anal. Chem.* **44**, 2373 (1972).
96. R. Jasinski and I. Trachtenberg, *Anal. Chem.* **45**, 1277 (1973).
97. R. W. Cattrall and P. Chin-Poh., *Anal. Chem.* **47**, 93 (1975).
98. H. Thompson and G. A. Rechnitz, *Chem. Instrum.* **4**, 239 (1973).
99. J. W. Ross, Jr. and M. S. Frant, *Anal. Chem.* **41**, 967 (1969).
100. W. Selig, *Mikrochim. Acta* 168 (1970).
101. W. Selig, *Microchem. J.* **15**, 452 (1970).
102. W. Selig, *Mikrochim. Acta* 564 (1970).
103. G. A. Rechnitz, *Science* **166**, 532 (1969).
104. M. Young, J. N. Driscoll, and K. Mahony, *Anal. Chem.* **45**, 2283 (1973).
105. G. Gardner and G. H. Nancollas, *Anal. Chem.* **42**, 794 (1970).
106. M. Mascini, *Analyst* **98**, 325 (1973).
107. G. A. Rechnitz and N. C. Kenny, *Anal. Lett.* **3**, 259 (1970).
108. H. Hirata and K. Date, *Anal. Chem.* **43**, 279 (1971).
109. M. Mascini and A. Liberti, *Anal. Chim. Acta* **60**, 405 (1972).
110. H. Hirata and K. Higashiyama, *Anal. Chim. Acta* **54**, 415 (1971).
111. H. Hirata and K. Higashiyama, *Bull. Chem. Soc. Japan* **44**, 2420 (1971).
112. H. Hirata and K. Higashiyama, *Anal. Chim. Acta* **57**, 476 (1971).
113. V. Majer, J. Vesely, and K. Stulik, *Anal. Lett.* **6**, 577 (1973).
114. E. H. Hansen and J. Ruzicka, *Anal. Chim. Acta* **72**, 365 (1974).
115. I. H. Krull, C. A. Mask, and R. E. Cosgrove, *Anal. Lett.* **3**, 43 (1970).
116. O. Kedem, M. Furmansky, E. Loebel, S. Gordon, and R. Bloch, *Israel J. Chem.* **7**, 87 (1969).
117. J. Pick, K. Toth, M. Vasak, E. Pungor, and W. Simon, *in* "Ion Selective Electrodes" (E. Pungor, ed.), p. 245. Akademai Kaido, Budapest, 1973.
118. J. Pick, K. Toth, E. Pungor, M. Vasak, and W. Simon, *Anal. Chim. Acta* **64**, 477 (1973).

119. L. A. R. Pioda, V. Stankova, and W. Simon, *Anal. Lett.* **2**, 665 (1969).
120. A. G. Fogg, A. S. Pathan, and D. T. Burns, *Anal. Lett.* **7**, 539 (1974).
121. U. Fiedler and J. Ruzicka, *Anal. Chim. Acta* **67**, 179 (1973).
122. M. S. Frant and J. W. Ross, Jr., *Science* **167**, 987 (1970).
123. G. Baum and M. Lynn, *Anal. Chim. Acta* **65**, 393 (1973).
124. R. W. Cattrall, S. Tribuzio, and H. Freiser, *Anal. Chem.* **46**, 2223 (1974).
125. M. D. Smith, M. A. Genshaw, and J. Greyson, *Anal. Chem.* **45**, 1782 (1973).
126. C. J. Pedersen, *J. Amer. Chem. Soc.* **89**, 7017 (1967).
127. *Chem. Eng. News* **48** (9), 26 (1970).
128. J. J. Christensen, J. O. Hill, and R. M. Izalt, *Science* **174**, 459 (1971).
129. C. J. Pedersen, *J. Amer. Chem. Soc.* **92**, 386 (1970).
130. C. J. Pedersen, *J. Amer. Chem. Soc.* **92**, 391 (1970).
131. H. K. Frensdorff, *J. Amer. Chem. Soc.* **93**, 600 (1971).
132. C. J. Pedersen and H. K. Frensdorff, *Angew. Chem.* **84**, 16 (1972).
133. M. R. Truter and C. J. Pedersen, *Endeavour* **30**, 142 (1971).
134. L. M. Thomassen, T. Ellingsen, and J. Ugelstad, *Acta Chem. Scand.* **25**, 3024 (1971).
135. J. L. Dye, M. G. DeBaker, and V. A. Nicely, *J. Amer. Chem. Soc.* **92**, 5226 (1970).
136. M. A. Bush and M. R. Truter, *Chem. Commun.* 1439 (1970).
137. M. R. Truter and D. Bright, *Nature (London)* **225**, 176 (1970).
138. J. Petranek and O. Ryba, *Anal. Chim. Acta* **72**, 375 (1974).
139. O. Ryba, E. Knizakova and J. Petranek, *Collect. Czech. Chem. Commun.* **38**, 497 (1973).
140. O. Ryba and J. Petranek, *J. Electroanal. Chem. Interfacial Electrochem.* **44**, 425 (1973).
141. M. Semler and H. Adametzova, *J. Electroanal. Chem. Interfacial Electrochem.* **56**, 155 (1974).
142. M. Mascini and F. Pallozzi, *Anal. Chim. Acta* **73**, 375 (1974).
143. E. Eyal and G. A. Rechnitz, *Anal. Chem.* **43**, 1090 (1971).
144. G. A. Rechnitz and E. Eyal, *Anal. Chem.* **44**, 370 (1972).
145. C. Liteanu, I. C. Popescu, and H. Nascu, *Rev. Roum. Chim.* **17**, 1651 (1972).
146. J. Vesely, O. J. Jensen, and B. Nicolaisen, *Anal. Chim. Acta* **62**, 1 (1972).
147. D. C. Muller, P. W. West, and R. H. Muller, *Anal. Chem.* **41**, 2038 (1969).
148. J. G. Frost, *Anal. Chim. Acta* **48**, 321 (1969).
149. F. J. Conrad, *Talanta* **18**, 952 (1971).
150. W. E. Bazzelle, *Anal. Chim. Acta* **54**, 29 (1971).
151. A. V. Gordievskii, V. S. Shterman, Ya. A. Syrchenkov, N. I. Savvin, A. F. Zhukov, and Yu. I. Urusov, *Zh. Ana. Khim.* **27**, 2170 (1972).
152. I. C. Popescu and C. Liteanu, *Rev. Roum. Chim.* **17**, 1629 (1972).
153. C. Liteanu, I. C. Popescu, and V. Ciovirnache, *Talanta* **19**, 985 (1972).
154. E. Schmidt and E. Pungor, *Magy. Kem. Foly.* **77**, 397 (1971).
155. C. Liteanu and H. Hopirtean, *Talanta* **19**, 971 (1972).
156. B. G. Lofis, N. I. Savvin, A. V. Vishnyakov, and A. V. Gordievskii, *Zavod. Lab.* **39**, 267 (1973).
157. D. Ammann, E. Pretsch, and W. Simon, *Helv. Chim. Acta* **56**, 1780 (1973).
158. D. Ammann, E. Pretsch, and W. Simon, *Anal. Lett.* **7**, 23 (1974).
159. C. J. Coetzee and A. J. Basson, *Anal. Chim. Acta* **64**, 300 (1973).
160. D. L. Manning, J. R. Stokely, and D. W. Magouyrk, *Anal. Chem.* **46**, 1116 (1974).

Chapter 8

LIQUID MEMBRANE ELECTRODES

The theoretical aspects of liquid membrane electrodes were discussed briefly in Chapter 3. In this chapter the construction and uses of some liquid membrane electrodes selective to anions and cations are described.

The membrane is formed by a layer of a suitable solvent which does not dissolve in the test solution. Some arrangements of the membrane between the test solution and the reference electrode solution are shown in Fig. 1.

Orion Research uses a porous flexible plastic membrane that is saturated with the liquid ion exchanger, and Corning Glass Laboratories uses a sintered glass filter. A number of other materials for holding the liquid have been tried.[1, 2] The solvent and/or the liquid saturating the membrane must be quite insoluble in water and must have a low vapor pressure. In addition if the liquid has sufficiently high viscosity, the membrane remains stable for long periods. Furthermore, if the dielectric constant of the liquid (whose molecular weight is generally high) is low, considerable ion association in the liquid membrane phase will take place. Some of these properties are possessed by substances with long hydrocarbon chains in their molecules. A high selectivity toward the ion whose activity is to be monitored by the liquid membrane electrode requires high stability of the ion complex which is influenced among other things, by the solvent.[3]

The history of liquid membranes goes back to Nernst and Haber who, with their colleagues, made a systematic study of the electrochemical behavior of liquid membranes. An account of the early work has been summarized by Sollner.[4] Sandblom and Orme[5] have discussed the different liquid membrane systems and transport phenomena arising across them.

Many substances have been used in a pure liquid state and/or in a suitable solvent to form liquid membrane electrodes. These substances are

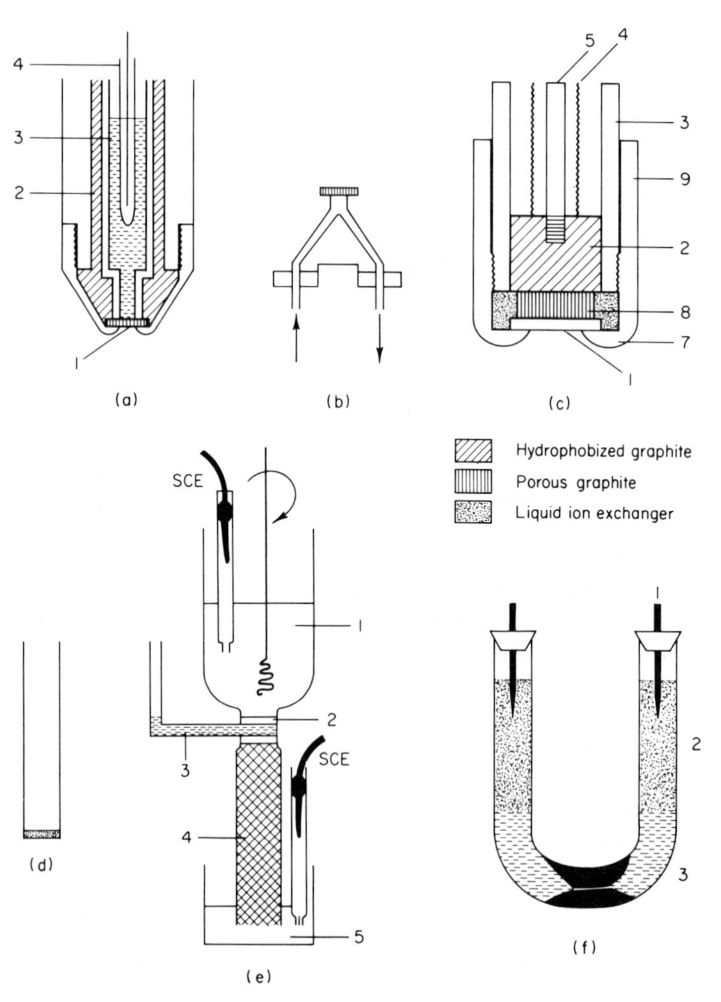

Fig. 1. Some arrangements of liquid membranes between the test solution and the reference electrode solution. (a) A liquid membrane ion-selective electrode: 1 is the membrane saturated with the liquid ion exchanger 2; 3 is the inner reference electrolyte containing the inner reference electrode 4. (b) A flow-through device that can be attached to the membrane electrode so that the test solution can be flowed over the surface of the membrane of the sensing device. (c) Construction of a Selectrode. Activated liquid state Selectrode: 1, sensitive surface; 2, cylinder pressed from graphite hydrophobized by Teflon; 3, Teflon tubing; 4, screening; 5, stainless steel contact; 7, cup; 8, a porous pellet; and 9, Teflon tube with threads. (d) An electrode with sintered glass filter that can hold the liquid ion exchanger to form the liquid membrane. (e) A titration cell arrangement after Covington and Thain[161]: 1 is the solution to be titrated; 2 is the glass frit which is saturated with the ion exchange liquid 3; 4 is the inner filling solution in agar; and 5 is the saturated KCl solution. SCE is standard calomel electrode. (f) Thick liquid membrane of an ion exchanger: 1, reference Ag–AgCl electrode; 2, aqueous electrolyte solution; and 3, liquid ion exchange membrane.

referred to in sections dealing with the particular cation- and/or anion-selective electrodes in whose construction they are used. The common feature of these compounds is their ability to bind certain small ions selectively either at a charged site of opposite sign (liquid ion exchanger) or at neutral sites of organic nature. These substances in the form of a liquid membrane generally separate two aqueous phases. At the membrane–solution interface, rapid exchange takes place between the free ions in the aqueous phases and the same ions bound to the organic groups in the membrane phase. The selectivity of the electrode depends primarlily on the selectivity of this ion exchange process. A number of liquid ion exchangers have been proposed for the formation of ion-selective electrodes.[6–8] Although these ion exchangers generally display a preference for cations (cation exchangers) or anions (anion exchangers) because they carry opposite electric charges, they exhibit poor selectivity to any *particular* cation or anion, respectively. In their ion preference, they follow the Hofmeister lyophilic series.[4] On the other hand, some macrocyclic compounds which are electoneutral show good selectivity to certain cations. Some features of these compounds are summarized below.

A. MACROCYCLIC COMPOUNDS

These compounds belong to cyclic depsipeptides (α-amino acids and α-hydroxyaliphatic acids alternately bound in a ring), macrotetralides (tetralactone of nonactinic acid and its derivatives), polyethers, and other substances. Most of these are electrically neutral and form 1 : 1 complexes with the alkali metal ions. There are also some compounds that carry a negative group (e.g., nigericin and monensin) and form complexes with alkali metal ions.

1. Depsipeptides

The most important member of this group is valinomycin (molecular weight 1111); it contains, per molecule,[9, 10] three molecules of L-valine, three molecules of D-valine, three molecules of L-lactic acid, and three molecules of D-α-hydroxyisovaleric acid as shown in Fig. 2. The valinomycin molecule thus has a 36-membered ring with numerous possibilities for undergoing conformational changes.[11] In solution it exists in three forms (A, B, and C) in equilibrium with one another (see Fig. 3). The state of equilibrium is highly dependent on the nature of the solvent. In nonpolar solvents, it exists predominantly in form C in which all the amide groups take part to form a rigid system of six intramolecular hydrogen bonds. Form B has only three hydrogen bonds and is flexible. Form A,

```
      D-Val—L-Lac—L-Val
     /                 \
D-Hov                   D-Hov
  |                       |
L-Val                   D-Val
  |                       |
L-Lac                   L-Lac
     \                 /
      D-Val—D-Hov—L-Val
```

Fig. 2. Structure of valinomycin: Val = valine, Lac = lactic acid, and Hov = α-hydroxyisovaleric acid.

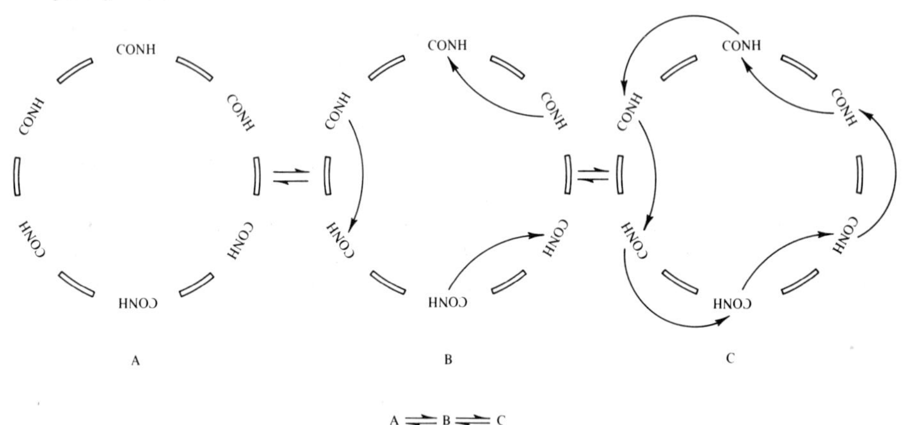

Fig. 3. Conformational equilibria of valinomycin.

with no hydrogen bonds, exists in a highly dynamic state containing an equilibrium mixture of energetically equivalent conformers. It is this state that undergoes interesting changes when the environment is altered. Complex formation with the alkali metal ion brings about a profound change in the ligand structure in which all the polar groups (carbonyl oxygen atoms) are directed toward the center of the molecule where in the cavity of the ligand the alkali metal ion is held by ion–dipole interactions while the nonpolar lipophilic groups encapsulate the whole structure (see Fig. 4). The cation is thus effectively shielded from the solvent and the anion by the ester groups, the hydrogen bonds, and the valyl isopropyl groups.[12, 13]

The lipophilic character of the complex cation surface is responsible for its stability and high solubility in neutral organic solvents. The large ligand volume and the polar character of the inside of the complex probably prevent ion pair formation between the complex and anions in a solvent of low dielectric constant. The stability of the potassium–valinomycin complex decreases with increasing polarity of the medium and no complex is detectable in water.

The size of the internal cavity in the valinomycin molecule is sufficient to accommodate Na^+ or K^+ ion without steric strain. Because of the small size (crystal radius) of the Na^+ ion, it is likely to be shifted toward the rim

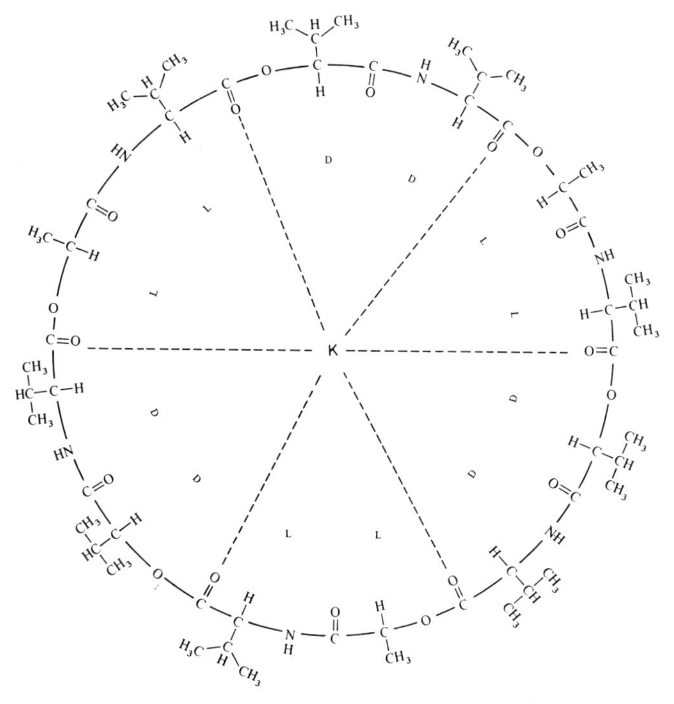

Fig. 4. Structure of potassium–valinomycin complex.

of the cavity, whereas the larger Rb and Cs ions are accommodated by an increase in the cavity size brought about by an elongation of the hydrogen bonds. The K^+ ion (radius 1.33 Å) just fits the cavity which is about 2.7–3.3 Å.[11]

Compounds in this category include the antibiotics, enniatins, which contain 18-membered rings with three molecules of D-α-hydroxyisovaleric acid and three molecules of either *N*-methylvaline (enniatin B) or *N*-methylisoleucine (enniatin A) as cyclic hexadepsipeptides[14, 15] (I, II), beauvericin[16] (III), and antamanide (IV). The latter compound, a decapeptide, shows significant selectivity to Na compared to K.[11, 17, 18]

D-Hov L-Met-Ile

$$\left[-O-\underset{\substack{| \\ CH(CH_3)_2}}{CH}-\underset{\substack{\| \\ O}}{C}-\underset{\substack{| \\ CH_3}}{N}-\underset{\substack{| \\ CH(CH_3)CH_2-CH_3}}{CH}-\underset{\substack{\| \\ O}}{C}- \right]_3$$

(I) Ennatin A

D-Hov L-Met-Val

$$\left[-O-\underset{\substack{| \\ CH(CH_3)_2}}{CH}-\underset{\substack{\| \\ O}}{C}-\underset{\substack{| \\ CH_3}}{N}-\underset{\substack{| \\ CH(CH_3)_2}}{CH}-\underset{\substack{\| \\ O}}{C}- \right]_3$$

(II) Ennatin (B)

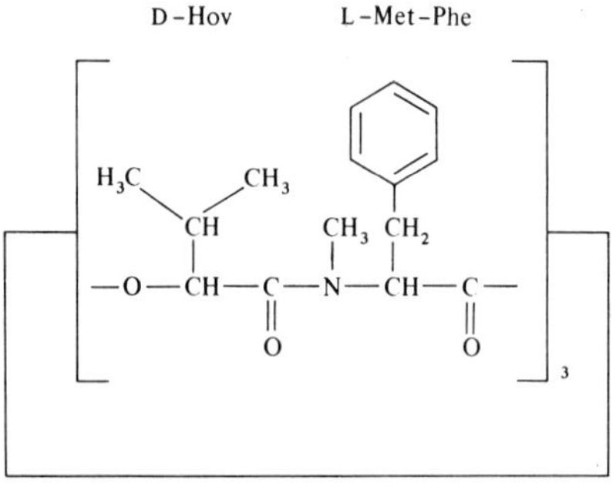

D-Hov L-Met-Phe

(III) Beauvericin

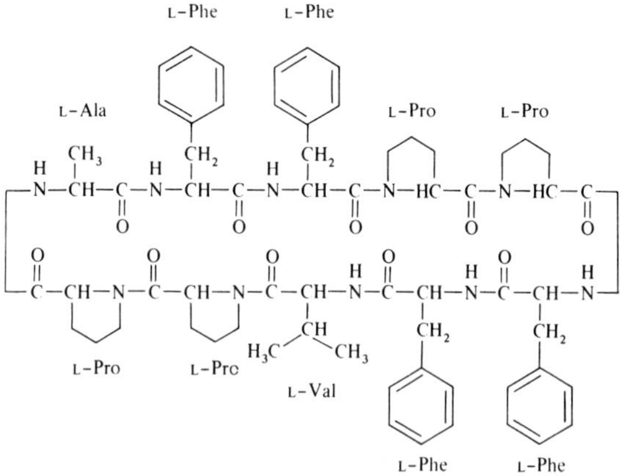

(IV) Antamanide

2. Macrotetralides

This category of antibiotics includes tetralactones derived from nonactinic acid (V). Substitutions are made in various ways to give the different nonactins (VI): for nonactin: $R_1 = R_2 = R_3 = R_4 = CH_3$; monactin: $R_1 = R_2 = R_3 = CH_3$, $R_4 = C_2H_5$; dinactin: $R_1 = R_2 = CH_3$, $R_3 = R_4 = C_2H_5$; trinactin: $R_1 = CH_3$, $R_2 = R_3 = R_4 = C_2H_5$; tetranactin: $R_1 = R_2 = R_3 = R_4 = C_2H_5$.

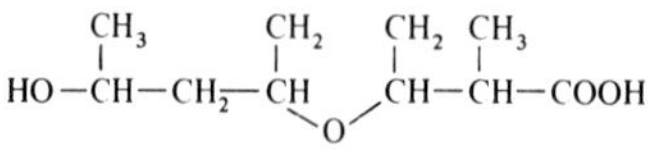

(V) Nonactinic acid

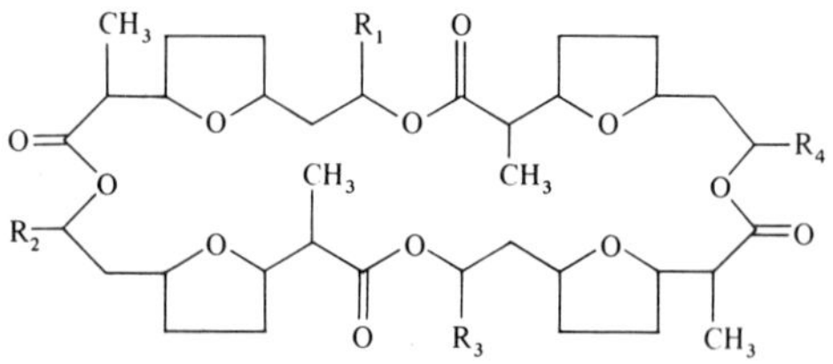

(VI) Different nonactins

These compounds bind cations with different ionic radii.

3. Miscellaneous Compounds

Polyethers have already been referred to in Chapter 7. The other compounds that have attracted attention are gramicidins, which are antibiotics possessing the peptide backbone. These (gramicidin A, B, C) are linear peptides (acyclic pentadecapeptide ethanolamines) of molecular weight 1000 with the N-terminal residue formylated and the C-terminal group masked by ethanolamine[19] (VII).

1 2 3 4 5 6 7 8 9
HCO—L-(X)-Gly-L-Ala-D-Leu-L-Ala-D-Val-L-Val-D-Val-L-Trp-

10 11 12 13 14 15
D-Leu-L-(Y)-D-Leu-L-Trp-D-Leu-L-Trp —$NHCH_2CH_2OH$

X = Val	Y = Trp (tryptophan)	Valine gramicidin A
X = Ile	Y = Trp	Isoleucine gramicidin A
X = Val	Y = Phe (phenylalanine)	Valine gramicin B
X = Ile	Y = Phe	Isoleucine gramicidin B
X = Val	Y = Tyr (tyrosine)	Valine gramicidin C
X = Ile	Y = Tyr	Isoleucine gramicidin C

(VII) Gramicidins

There is a cyclic decapeptide called gramicidin S (VIII) which is a dimer of the left-hand portion of tyrocidin A (IX).[20]

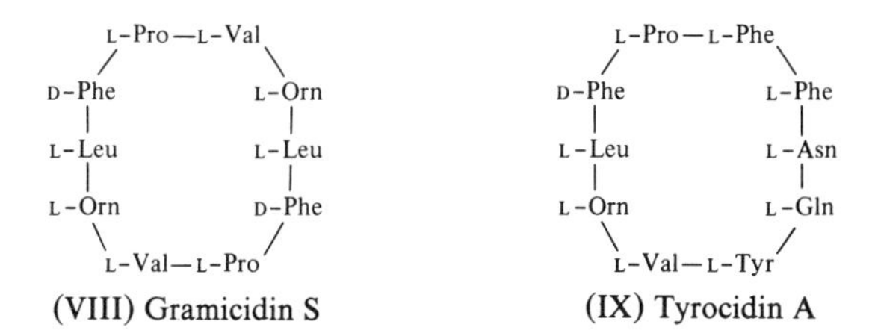

(VIII) Gramicidin S (IX) Tyrocidin A

Gramicidins A, B, and C, unlike valinomycin, do not form stable complexes with alkali metal ions, but are able to increase effectively the ionic permeability of artificial and biological membranes without manifesting any selectivity to any of the ions.[11, 21] In view of this, they are of little use in the fabrication of ion-selective sensors.

Unlike the compounds just discussed which are electroneutral, the members of the nigericin group (nigericin, monensin, grisorixin, X-206, X-537A) have a carboxyl group and thus at physiological pH are present as anions.[22] The structures of compounds of this group are shown in Figs. 5–8.

The X-ray analysis of the crystalline complexes of the antibiotics has been carried out and the following summary has been given by Simon and Morf.[22]

(a) The metal ion is coordinated by five to eight oxygen atoms.

(b) The external surface of the complexes is lipophilic, enhancing the lipid solubility of the carrier complexes.

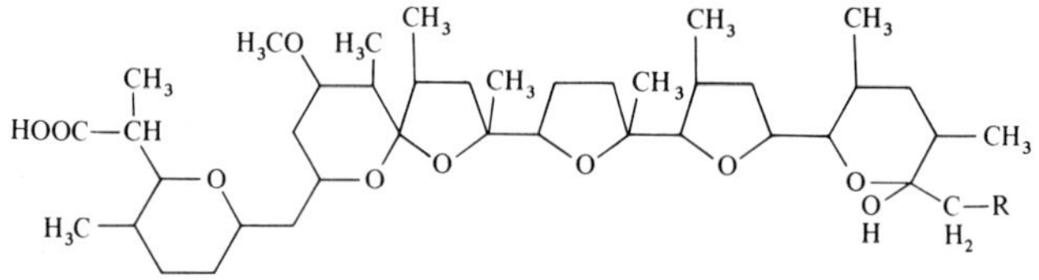

Fig. 5. Structures of nigericin (R = OH) and grisorixin (R = H).

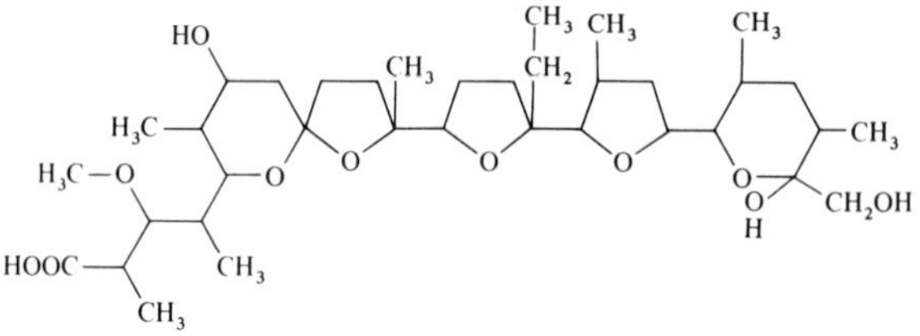

Fig. 6. Structure of monensin.

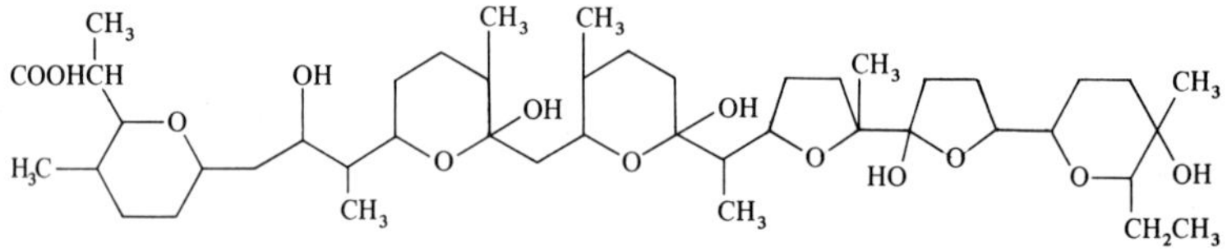

Fig. 7. Structure of the antibiotic X-206, belonging to the nigericin group.

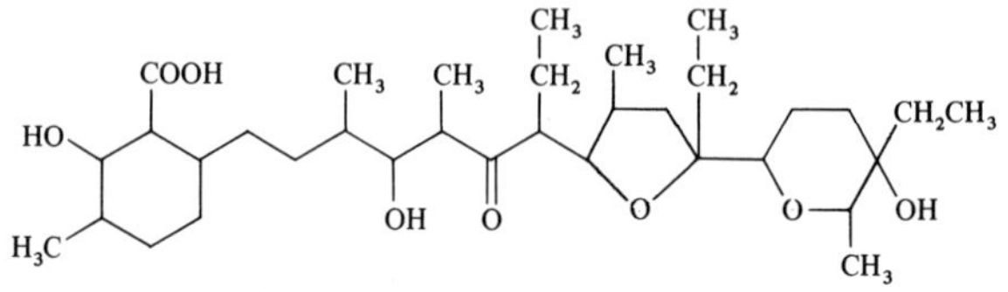

Fig. 8. Structure of the antibiotic X-537A, belonging to the nigericin group.

(c) The alkali metal cation is not hydrated.
(d) The alkali metal cation complexes of compounds of the nigericin group are electrically neutral.
(e) The alkali metal cation complexes of compounds of the valinomycin group are positively charged.
(f) The carrier molecules have sufficient flexibility to allow for a step-by-step substitution of the solvent molecules. This leads to a low activation energy barrier and fast complexation reaction rates.

The complex formation constants of macrocyclic compounds and alkali metal ions have been determined by applying a variety of physical techniques such as conductometry, nuclear magnetic resonance, potentiometry, optical rotatory dispersion, and relaxation spectrometry. Simon and Morf[22] have given a survey of stability constants for antibiotic molecules some of which are given in Table 1 along with values determined for other compounds. The stability of the complexes in aqueous solution is very low and so the values are obtained mainly for various nonaqueous solvents.

TABLE 1

Selected Stability Constants of Macrocyclic Compound Complexes with Alkali Metal Ions

Macrocyclic compound	log (stability constant) for the ion (liter/mole)[a]					Solvent	Temperature (°C)	Ref.
	Li^+	Na^+	K^+	Rb^+	Cs^+			
Valinomycin	—	—	6.08	—	—	Ethanol	25	23
	—	0	6.30	6.42	5.81	Ethanol	25	13
	< 0.70	0.67	4.90	5.26	4.42	Methanol	25	22
		1.08	> 3.90	—	—	Methanol	25	24
Enniatin A	—	—	3.08	—	—	Methanol	25	24
Enniatin B	—	3.11	3.57	3.60	3.34	Ethanol	25	13, 25
	—	2.38	2.92	—	—	Methanol	25	24
	1.28	2.42	2.92	2.74	2.34	Methanol	25	22
Beauvericin	2.0	2.48	3.49	3.54	3.54	Ethanol	—	26
Antamanide	—	3.40	2.40	—	—	Ethanol	—	27
Nigericin	—	4.38	5.18	—	—	Methanol	25	28
Monensin	—	5.85	4.98	—	—	Methanol	25	28
	—	—	4.40	—	—	Methanol	25	29
			4.58	—	—	Methanol	25	29
Nonactin	—	2.38	3.58	3.53	2.95	Methanol	30	22
	—	3.46	4.74	—	—	Ethanol	30	22
	—	3.71	3.71	—	3.04	Acetone	17	22, 30
	Complex with Ba^{2+} : $\log K_{Ba\text{-}X} = 1.72$					Methanol	30	31
	$= 2.43$					Ethanol		
Monactin	—	2.58	3.83	3.57	3.15	Methanol	30	22
	—	3.59	4.42	—	—	Ethanol	30	22
	—	2.70	—	—	—	Methanol	30	32
	Complex with Ba^{2+} : $\log K_{Ba\text{-}X} = 1.28$					Methanol	30	31
	$= 2.43$					Ethanol	30	31
Dinactin	—	2.94	3.70	3.70	3.28	Methanol	30	22
	—	3.63	—	—	—	Ethanol	30	22
Trinactin	—	3.64	—	3.77	3.40	Methanol	30	22
Polyethers	—	3.61	—	—	—	Ethanol	30	22
Dicyclohexyl-18-crown-6	0.6	1.5–1.85	2.18	—	1.25	Water	25	33
(isomer A)	—	—	2.02	1.52	0.96	—	—	34
	$\log K_{NH_4\text{-}X} = 1.33$; $\log K_{Ag\text{-}X} = 2.3$					Water	25	33
$\log K_{Tl\text{-}X} = 2.4$; $\log K_{Sr\text{-}X} = 3.24$; $\log K_{Ba\text{-}X} = 3.57$						—	—	34, 35
Dicyclohexyl-	—	4.08	6.01	—	4.61	Methanol	25	33
18-crown-6	—	1.2–1.6	1.63	—	0.9	Water	25	33

TABLE 1 (continued)

Macrocyclic compound	log (stability constant) for the ion (liter/mole)[a]					Solvent	Temperature (°C)	Ref.
	Li^+	Na^+	K^+	Rb^+	Cs^+			
(isomer B)	—	—	1.78	0.87	0.9	—	—	34
Log $K_{NH_4\text{-}X} = 0.8$; log $K_{Ag\text{-}X} = 1.59$						—	—	33
Log $K_{Tl\text{-}X} = 1.8$; log $K_{Sr\text{-}X} = 2.64$; log $K_{Ba\text{-}X} = 3.27$						—	—	34, 35
	—	3.68	5.38	—	3.49	Methanol	25	33
Dicyclohexyl-14-crown-4	—	2.18	1.38	—	—	Methanol	25	33
Cyclohexyl-18-crown-5	—	3.71	3.58	—	2.78	Methanol	25	33
Dibenzo-18-crown-6	—	4.36	5.00	—	3.55	Methanol	25	33
Dibenzo-21-crown-7	—	2.40	4.30	—	4.20	Methanol	25	33
Dibenzo-24-crown-8	—	—	3.49	—	3.78	Methanol	25	33
Dibenzo-30-crown-10		2.00	4.60	—	—	Methanol	25	33
Dibenzo-60-crown-20		—	3.90	—	—	Methanol	25	33

[a] K is the stability constant for the ion.

B. CATION-SELECTIVE LIQUID MEMBRANE ELECTRODES

Typical ion exchange compounds that can serve as liquid membranes are the quaternary ammonium base trioctylpropyl ammonium hydroxide or its salts (anion exchangers), and analogous secondary and tertiary amines and acidic compounds such as dinonylnaphthalenesulfonic acid or monodioctylphenylphosphoric acid (cation exchangers).

Some solid membrane electrodes selective to Cu^{2+}, Pb^{2+}, F^-, and Cl^- have been found to have greater selectivity than the corresponding liquid membrane electrodes. The Orion 92-32 electrode (liquid membrane) responded alike to Ca^{2+}, Mn^{2+}, Co^{2+}, Ni^{2+}, Cu^{2+}, and Zn^{2+} ions[36] and could be used to determine the solubility products of hydroxides, sulfides, and halides of these metal ions.

The dinonylnaphthalenesulfonic acid (DNNS) membrane system has been evaluated for its response to various cations.[37] The liquid membrane in suitable ionic form was used as an electrode to sense the various cations. The results are given in Table 2. The theoretical slopes (according to the Nernst equation) for di-, tri-, and tetravalent ions at 23°C are 29.38, 19.58,

TABLE 2

Responses of DNSS Liquid Membrane to Various Ions[a]

M^{z+}/DNNS[b]	pH	Slope (mV/log $a_{M^{z+}}$)	Concentration range for linear response(M)	Useful molar concentration range(M)
$CaCl_2$/DNNS	6.0	17.8	$1.0–4 \times 10^{-2}$	1.0×10^{-3}
$CaCl_2$/DNNS	4.0	16.7	$1.0–2 \times 10^{-2}$	$1.0–10^{-3}$
$NiCl_2$/DNNS	4.0	16.7	$1.0–10^{-3}$	$1.0–10^{-4}$
$CrCl_3$/DNNS	4.0	20.0	$0.3–10^{-3}$	$0.5–10^{-4}$
$LaCl_3$/DNNS	4.0	20.0	$0.2–4 \times 10^{-4}$	$0.5–5 \times 10^{-4}$
$La(ClO_4)_3$/DNNS	4.0	21.4	$0.2–10^{-4}$	$0.5–10^{-4}$
$ThCl_4$/DNNS	3.0	15.0	$0.1–10^{-3}$	$0.1–10^{-4}$

[a]Measured at 23°C.
[b]DNNS = dinonylnaphthalenesulfonic acid.

and 14.69, respectively. Thus the tetra- and trivalent ions gave a Nernstian response, whereas the divalent ions gave a sub-Nernstian response, although it was linear. The interferences of other anions, ClO_4^-, NO_3^-, Cl^-, and cations with the response of the electrode to La^{3+} were also evaluated. Even though there was little interference from anions, the cations tested affected the response of the membrane electrode. Thus the DNNS cation exchanger, like any other solid or liquid cation exchanger, showed no selectivity to any specific cation. However, liquid membranes formed from benzene solutions of trilauryl ammonium and tetraheptyl ammonium salts of zinc tetrachloride and tetrachloropalladium have been found to be sensitive and selective to Zn^{2+} and Pd^{2+} ions[38] and so these electrodes have been used in the estimation of those ions. Similarly, metal chelates used as liquid membrane electrodes have been tested for their responses to metal ions.[39] Some of these responses to ions in equilibrium with the organic reagent are shown in Table 3.

Solutions of metal dithizonates in solvents such as carbon tetrachloride, chloroform, benzene, and xylene, coated on a carbon rod, and previously rendered hydrophobic by treatment with the solvent, have been used to prepare electrodes selective to H^+, Hg^{2+}, Ag^+, Cu^{2+}, Pb^{2+}, and Zn^{2+} ions. The theory and the characteristics of these electrodes have been discussed.[40] Solutions of bis(*O*,*O*′-diisobutyldithiophosphato)nickel(II), bis-(*O*,*O*′-diisobutyldithiophosphato) cadmium(II), and bis(*O*,*O*′-diisobutyldithiophosphato)lead(II) in chlorobenzene have been used as liquid membranes.[41] These respond selectively to Ni^{2+}, Cd^{2+}, and Pb^{2+} ions, respectively, over the concentration range $10^{-1} - 10^{-4}$ M in the presence of alkali and alkaline earth ions. The lead electrode has been

TABLE 3

Responses of Metal Chelates Used as Liquid Membrane Electrodes[a]

Ion (M^{z+})	Organic reagent	pH range	Slope (mV/pM unit)
K^+	Potassium picrylaminate	5–12	50
Cs^+	Cesium tetraphenyl borate	9.5–12	59
H^+	Quinhydrone	1.0–5.5	59
		5.5–8.5	29
Ag^+	Silver dithizonate	2.0–12	~59
Cu^{2+}	Copper dithizonate (secondary)	4.5–12	~59
	Copper dithizonate (primary)	3.0–4.5	29
	Copper diethyldithiocarbamate	3.0–12	29
Pb^{2+}	Lead dithizonate	7.0–12	~29
Cd^{2+}	Cadmium dithizonate	6.5–12	~29
Zn^{2+}	Zinc dithizonate	7.5–12	~23

[a]Data from Ruzicka and Tjell.[39]

found to be highly selective in solutions containing H^+ and a number of heavy metal ions.

In very early studies, Ilani[42] found that just the solvents alone (e.g., butanol, *n*-octanol, toluene, and chloroform) saturating a Millipore filter acted as liquid membranes selective to K^+ ions over Na^+ ions, toluene being the one exhibiting the highest selectivity to K^+ over Na^+. Other solvents, such as bromobenzene, *n*-butyl bromide, and ethylene chloride, have been used.[43] Bromobenzene solvent membrane selectivity to various cations decreased in the order: quinine > tetraethyl ammonium > Cs = K > acetylcholine > NH_4 > Na = choline > Li.[44] Glass frit, ultrafilter, parchment paper, and PVC in the place of Millipore filter and water-immiscible organic solvents have been evaluated as membrane electrodes, particularly in acid–base titrations.[2, 45–47] A similar approach has been taken by Higuchi *et al.*[48] to develop electrodes sensitive to organic ions. A typical cell used for measuring the potentials is

$$\begin{array}{l|l||l||l|l} \text{Saturated calomel electrode} & 0.1\,M\ \text{KBr},\ 0.1\,M\ \text{KH}_2\text{PO}_4 & \text{Plastic membrane} & \text{R}_4\text{N}^+ \text{test solution} & \text{Saturated calomel electrode} \end{array} \quad (1)$$

Any plastic matrix of limited hydrophilicity is considered suitable. A PVC membrane plasticized with *N*, *N*-dimethyl oleamide acted as an electrode and gave a Nernstian slope for tetrabutyl ammonium ion, whereas the same electrode in KBr solutions showed a completely nonlinear response. The electrode is sensitive to H^+ ions with a slope of 56 mV/pH. However, PVC plasticized with dioctylphthalate showed excellent response to organic cations with slopes (in millivolts per decade change in concentration) of 57.5 (tetrapropyl ammonium), 58.5 (tetrabutyl ammonium), 59.5 (tetrapentyl ammonium), and 58.5 (tetrahexyl ammonium). In the case of tetrabutyl ammonium (TBA^+) ion, the selectivities for the two membrane electrodes mentioned above were $K_{TBA^+-K^+} = 5.3 \times 10^{-5}$, $K_{TBA^+-H^+} = 7.8 \times 10^{-2}$ for the PVC–amide membrane and $K_{TBA^+-K^+} = 1.4 \times 10^{-6}$, $K_{TBA^+-H^+} = 4.3 \times 10^{-7}$ for the PVC–DOP membrane. These electrodes are considered potentially useful in titrimetric analyses. Another electrode showing high selectivity for acetylcholine ion over Na^+, K^+, and NH_4^+ has been constructed and is commerically available.[49] The liquid membrane consists of a 5% solution of acetylcholine tetra(*p*-chlorophenyl) borate in either 3-*o*-nitroxylene, dibutylphthalate, or tri(2-ethylhexyl)phosphate.[50] This liquid membrane showed rapid Nernstian response to acetylcholine (Ach) ion activity from 10^{-1} to 10^{-5} *M*.[51] It also showed slightly less than ideal Nernstian response to choline ion (Ch^+). The selectivity constants (K_{Ach-M}) were 1×10^{-4} (M = Na), 1×10^{-3} (NH_4^+, K^+), and 6.6×10^{-2} (Ch^+). The response of the electrode to a series of alkyl esters of choline ranging from acetyl- to benzoylcholine has been examined.[52] The selectivity ratios were calculated by the relationship

$$\log K = \frac{E_2 - E_1}{2.303RT/F} \tag{2}$$

where E_2 is the observed potential of a 10^{-1} *M* solution of choline derivative, and E_1 is that of a 10^{-1} *M* solution of choline. In general the logarithm of the selectivity ratio *K* for choline esters increased linearly as the total number *n* of carbon atoms in the choline derivative increased. The data have been described by the empirical equation

$$\log K = 0.537 + 0.159n \tag{3}$$

The values of *K* were 1 (choline), 37.6 (acetylcholine), 121 (propionylcholine, acetyl β-methylcholine), 407 (butyrylcholine), 1550 (valerylcholine), and 10,000 (benzoylcholine). In addition, the selectivity ratios for some of these choline esters have been determined using two other solvents, dibutylphthalate and tri(2-ethylhexyl)phosphate, with acetylcholine tetra(*p*-chlorophenyl) borate forming the liquid membrane.[50] The results given above for the selectivity ratio were obtained using 3-*o*-nitroxylene as the solvent. The selectivity ratios for various esters determined by the electrode using different solvents are shown in the accompanying table.

Solvent	Choline	Acetylcholine	Propionylcholine	Butyrylcholine
Tri(2-ethylhexyl)phosphate	1	1.08	3.36	10.2
Dibutyl phosphate	1	6.85	18.2	57.6
3-*o*-Nitroxylene	1	37.6	121	1550

The data indicate that the solvent has a strong effect on the selectivity of the electrode. The selectivity has been shown to be related to the free energy of transfer of organic ion from an aqueous to a hydrophobic environment and to obey the relation

$$\log K_{ij} = (\Delta G_{t,ji}^{w\to e} - \Delta G_{t,i}^{e\to s})/23.05S \qquad (4)$$

where $\Delta G_{t,ji}^{w\to e}$ is an additive, constitutive free energy of transfer term available from the distribution coefficients of amino acids, $\Delta G_{t,i}^{e\to s}$ is a solvent-adjusted term, and S is the Nernst slope.

The free energies of transfer (ΔG_t) of amino acids going from water to ethanol have been computed.[53] That is,

$$\Delta G_t^{w\to s} = \Delta G_t^{w\to e} + \Delta G_t^{e\to s} \qquad (5)$$

where w stands for water, s for solvent, and e for ethanol.

Then the difference in ΔG_t between solutes j and i is

$$\Delta G_{t,j}^{w\to s} - \Delta G_{t,i}^{w\to s} = (\Delta G_{t,j}^{w\to e} - \Delta G_{t,i}^{w\to e}) - (\Delta G_{t,i}^{e\to s} - \Delta G_{t,j}^{e\to s}) \qquad (6)$$

Thus $\Delta G_{t,j}^{w\to e} - \Delta G_{t,i}^{w\to e} = \Delta_{t,ji}^{w\to e}$ evaluated for the choline esters are given in Table 4.

TABLE 4

Estimated Difference of Free Energy of Transfer between Choline and Choline Esters[a]

Choline ester	$\Delta G_{t,ji}^{W\to e}$ (cal/mole)
Acetylcholine	1460
Propionylcholine	2190
Butyrylcholine	2920
Valerylcholine	3550
Isobutyrylcholine	2420
Acetyl β-methylcholine	2190

[a] The side chain contributions were CH_3–, –CH_2–, –C(=O)– = + 730 cal/mole and isobutyl = +2420 cal/mole.

According to Eq. (4), a plot of $\log K_{ij}$ vs. $\Delta G_{t,ji}^{w\to e}$ should give a straight line from whose intercept $\Delta G_{t,i}^{e\to s}$ for each solvent may be evaluated. The slope should be equal to the constant $1/23.05S$, i.e., 0.75×10^{-3}. The experimentally determined values for $\Delta G_{t,i}^{e\to s}$ and the slope for the three solvents used to form the electrode were $\Delta G_{t,i}^{e\to s} = 328$ cal/mole (3-*o*-nitroxylene), -70 cal/mole (dibutylphthalate), and -1000 cal/mole [tri-(2-ethylhexyl)phosphate]; and slopes of 0.63×10^{-3}, 0.61×10^{-3}, and

0.65×10^{-3}, respectively. These results are considered consistent with the relationships given in Chapter 3 (see Sandblom *et al.*[54]).

The polyvinyl chloride membrane has been rendered electroactive by treating it with a solution of acetylcholine tetra-*p*-chlorophenyl borate in phthalate ester which acted as a plasticizer.[55] This membrane used as an electrode showed Nernstian response to choline and choline esters and so could be used like the other commercially available liquid membrane electrodes described above for assaying the activity of acetylcholinesterase.[56] However, the PVC membrane electrode responses were considered superior to those of the commercial membrane electrodes.[55]

Organophosphate pesticides exhibit acetylcholinesterase activity and so the acetylcholine electrode has been used in the analysis of organophosphate pesticides.[57] Furthermore, the electrode has been used to assay the cholinesterase activity of blood-derived fractions (whole blood, serum, and erythrocytes[58]; also see Chapter 11).

1. Potassium-Selective Membrane Electrode

It is seen from Table 1 that macrocyclic compounds can discriminate between similar ions, and so serve as good candidates for membrane components in electrode systems of high selectivity. Originally nonactin and its homologs were used in the preparation of membrane electrodes.[59] However valinomycin gave an electrode analytically useful for K^+ and discriminated Na^+ by a factor of about 5000.[60] This electrode contained a Millipore filter that was saturated with a solution of valinomycin[61–63] and is produced by Orion Research[64] and Philips.[65]

The electrode response has been found to be Nernstian in the potassium concentration range 1.0–10^{-6} M,[66] whereas others find it to be in the range 10^{-1}–10^{-4} [67] and 10^{-1}–10^{-5} M with a slope of 58.3 mV/decade.[62] The selectivity of the electrode to K^+ over other ions is shown in Table 5. Lal and Christian[67] found the selectivity ratio K/Na to depend on the relative concentrations of Na and K. Iodide, hydroxide, chromate, and oxalate interfered significantly with the response of the electrode. Tetraphenyl borate gave large potential shifts and consequently the electrode has limited usefulness in its presence. Similarly, Hammond and Lambert[69] investigated the responses of the electrode to various inorganic (see Table 5 for selectivity constants) and organic ions and found the membrane-soluble tetraphenyl borate to cause interference with the response of the electrode. Also cetyltrimethyl ammonium bromide, which is a positively charged surfactant, caused interference. But the negatively charged surfactant sodium dodecyl sulfate, the nonionic surfactant "Tween 80," and the positively charged nonsurfactant tetramethyl ammonium bromide, did not cause any interference. These facts indicate that the compound

TABLE 5

The Selectivity Constant K_{ij}^{pot} for Potassium-Selective Liquid Membrane Electrodes

Substance	Solvent forming the membrane	K_{ij}^{pot}, interfering ion j									Ref.
		K^+	Li^+	Na^+	Rb^+	Cs^+	NH_4^+	Tl^+	Ag^+	H^+	
Valinomycin (0.009 M)	Diphenyl ether	1	2.0×10^{-4}	2.5×10^{-4}	1.9	0.4	1.0×10^{-2}		2.0×10^{-9}	5.0×10^{-5}	62
	Orion	1	3.0×10^{-2}	9.0×10^{-2}	2.9	0.5	5.0×10^{-2}	9.0×10^{-2}	2.0×10^{-3}	4.0×10^{-2}	67
	Orion	1	0	1.0×10^{-4}	—	1.0	2.0×10^{-2}				66
	Orion	1	$\sim 6.6 \times 10^{-5}$	$\sim 1.0 \times 10^{-4}$	2.5	—	2.0×10^{-2}				68
Valinomycin	Diphenyl ether		2.5×10^{-4}	2.2×10^{-4}	1.8	0.4	1.4×10^{-2}			5.8×10^{-5}	69
Nonactin (0.3 M)	50% Nujol, 50% 2-octanol	1	5.6×10^{-4}	6.7×10^{-3}	0.42	3.1×10^{-2}	2.5			1.8×10^{-2}	61
Nonactin (0.04 M)	Diphenyl ether			1.0×10^{-2}							61
Monactin (0.04 M)	50% Nujol, 50% 2-octanol			2.0×10^{-2}							61, 62
Monactin (0.03 M)	Diphenyl ether			8.3×10^{-3}							61, 62
72% Nonactin + 28% monactin (0.3 M)	50% Nujol, 50% 2-octanol			3.1×10^{-2}							61, 62
Saturated solution of nonactin (72%) + monactin (28%)	Tris(2-ethylhexyl) phosphate	1.2×10^{-1}	4.2×10^{-3}	2.0×10^{-3}	4.3×10^{-2}	4.8×10^{-3}	1			1.6×10^{-2}	70

Substance	Solvent forming the membrane	K_{ij}^{pot}, interfering ion j									Ref.
		Mg^{2+}	Ca^{2+}	Sr^{2+}	Ba^{2+}	Fe^{2+}	Mn^{2+}	Cu^{2+}	La^{3+}	Al^{3+}	
Valinomycin	Diphenyl ether	2.0×10^{-4}	2.5×10^{-4}		6.0×10^{-5}	4.0×10^{-2}					62
(0.009 *M*)	Orion	1.0×10^{-3}	2.6×10^{-2}	2.6×10^{-4}			3.5×10^{-4}	1.3×10^{-3}	5.0×10^{-5}	5.0×10^{-4}	67
	Orion	$> 2.0 \times 10^{-4}$	$> 2.0 \times 10^{-4}$								66
	Orion										68
Valinomycin	Diphenyl ether		2.4×10^{-4}								69
Nonactin (0.3 *M*)	50% Nujol, 50% 2-octanol										61
Nonactin (0.04 *M*)	Diphenyl ether										61
Monactin (0.04 *M*)	50% Nujol, 50% 2-octanol										61, 62
Monactin (0.03 *M*)	Diphenyl ether										61, 62
72% Nonactin + 28% monactin (0.3 *M*)	50% Nujol, 50% 2-octanol										61, 62
Saturated solution of nonactin (72%) + monactin (28%)	Tris(2-ethylhexyl) phosphate		1.7×10^{-4}								70

causing interference with the response of the K-selective liquid membrane electrode must possess a net positive charge together with surface-active properties.

The valinomycin-based K-selective electrode has been used in the estimation of K^+ ion directly in solutions,[62] in seawater by using the standard addition technique,[71] in natural and waste waters on an automated continuous flow system with on-line minicomputer and printer,[72] and in blood serum.[73, 74] It has been used in the measurement of activity coefficients of KCl in mixed electrolyte solutions (KCl + NaCl)[75] and to measure the activity of K^+ ions in equilibrium with adenosine triphosphate. The latter measurement enabled Rechnitz and Mohan[76] to derive a value of 219 for the formation constant K_f of the potassium–adenosine triphosphate complex ($KATP^{3-}$). This value is nearly 20 times greater than the values (10–14) derived indirectly. These K activity measurements were carried out at pH 9.2 by adding increments of K_2H_2ATP ($\sim 10^{-2}$ *M*) and 0.05 *M* KOH into 50 ml of distilled water.[77] The free K^+ ion concentration was determined from the measured cell emf. Assuming that the ligand was present as ATP^{4-} or $KATP^{3-}$, the terms in the equation

$$K_f = \frac{(KATP^{3-})f_3}{(ATP^{4-})(K^+)f_1 f_2} \tag{7}$$

corresponding to the reaction

$$K^+ + ATP^{4-} \rightleftharpoons KATP^{3-} \tag{8}$$

were evaluated from

$$T_m = 2T_{ATP} + T_{base} \tag{9}$$

$$(KATP^{3-}) = T_m - (K^+) \tag{10}$$

$$(ATP^{4-}) = T_{ATP} - (KATP^{3-}) \tag{11}$$

where T_{ATP} and T_{base} are the total ligand and total base added and T_M is the total potassium. The activity coefficients f_1, f_2, and f_3 were calculated by using the Davies equation [Eq. (111) of Chapter 2].

Eyal and Rechnitz[68] extended this approach to determine the formation constants of valinomycin with K^+, NH_4^+, Rb^+, and Na^+ ions in order to correlate the ability of valinomycin to form complexes with its ability to sense K^+ ions selectively over NH_4^+, Rb^+, and Na^+ ions when it was in the form of a liquid membrane. They determined the formation constants as a function of water content of the medium. These results are given in Table 6 along with the selectivity ratios determined by the valinomycin-based electrode. According to expectations the results show that the formation constant ratios determine the selectivities of the membrane

TABLE 6

Comparison of Complex Formation and Electrode Selectivity in Water–Methanol Mixtures

Mole fraction water	Formation constants (K_f) of valinomycin				Ratio $K_f(K^+)/K_f(M^+)$	Selectivity ratio of electrode $(K^+)/(M^+)$
	K^+	NH_4^+	Rb^+	Na^+		
0.491	164		436		0.38	
0.360	674		1530		0.44	0.4
				< 15	> 45	
0.284	1290	19.9			~ 65	50
0.200	2830	40			71	
			6040		0.47	
				< 75	> 38	10^4
0.106	7260	77			94	
				< 9	> 807	

electrode. This means that the transport through the membrane must involve a carrier mechanism which is dependent mainly on the formation of the complex between valinomycin and the metal ion. Experiments carried out on "frozen" electrodes (see Table 7) showed that the mechanism of transport involved was of the mobile site type in which valinomycin acted as a carrier for the K^+ ions. A channel or pore-type mechanism of transport was excluded because the results of Table 7 show that at low temperature, when diphenyl ether was frozen, the selectivity of the electrode was destroyed.

The studies have been extended[78] to other liquid membranes formed from some polyethers (see Chapter 7). The membranes were formed by

TABLE 7

Results of "Freezing" Experiments[a]

Membrane system	Temperature (°C)	Calibration (mV/decade)	Selectivity ratio	
			K^+/Na^+	K^+/NH_4^+
Millipore saturated with valinomycin in diphenyl ether (liquid)	25	55	~ 10,000	~ 50
Millipore saturated with valinomycin in diphenyl ether (frozen)	5	9–13	~ 2	~ 1
Millipore with diphenyl ether (liquid)	25	55	~ 500	~ 50
Millipore with diphenyl ether (frozen)	5	9–13	~ 1	~ 1

[a] Data from Eyal and Rechnitz.[68]

dissolving the polyethers in nitrobenzene and were used in the construction of electrodes whose selectivity constants (K_{ij}^{pot}) are given in Table 8. The results given in Table 9 show that the complex formation quotients follow the selectivity ratios. Similar conclusions about the mechanism of ion transport by macrocyclic compounds were reached by Simon and co-workers who performed transport experiments by applying an electric field to a membrane system

Anode|Solution (0.05 M KCl + 0.05 M Na Cl or LiCl)|Membrane|Solution (0.1 M KCl)|Cathode

in which a PVC membrane impregnated with either macrotetralide antibiotic (nonactin + monactin)[79] or valinomycin[80] was used. In both cases selective transport of K^+ ion was observed.

Using ^{14}C-labeled macrocyclic compounds, it was shown that in the presence of the electric field the transport of K^+ was accompanied by an equivalent transport of the macrocyclic compound.[79-81] This is consistent with the formation of a 1 : 1 complex of K^+ with valinomycin or macrotetralide. That nigericin and monensin also facilitate transport of K^+ ion has been demonstrated.[82]

The effects of various oil-soluble anions on the response of the valinomycin-based K-selective electrode have been investigated both

TABLE 8

Selectivity Constants K_{ij}^{pot} of Liquid Membrane Electrodes Formed from Polyether Crown Compounds in Nitrobenzene[a]

Metal ion	Dicyclohexyl-18-crown-6	Dibenzo-18-crown-6	Benzo-15-crown-5	Dibenzo-30-crown-10
Orion electrode body used to form the electrode				
K^+	1	1	1	1
Rb^+	0.37–0.34[b]	0.37–0.32[b]	0.33–0.31[b]	1.2–1.1[b]
NH_4^+	0.19–0.15	0.07–0.06	0.055–0.049	0.22–0.20
Cs^+	0.13–0.12	0.29–0.25	0.144–0.137	0.48
Na^+	0.05–0.03	0.043–0.040	0.031–0.027	0.021–0.015
Corning electrode body used to form the electrode				
Rb^+	0.32–0.29			
NH_4^+	0.16–0.14			
Cs^+	0.11–0.07			
Na^+	0.0145–0.0139			
Sr^{2+} (K_{Ba-Sr})	0.454–0.417			

[a] Data from Rechnitz and Eyal.[78]
[b] Reciprocal of the values given by Rechnitz and Eyal.[78]

TABLE 9

Metal Complex Formation Constants K_f of Macrocyclic Compounds (Polyethers) and the Selectivity Ratios K_{ji} [a]

Polyether	Metal ion	K_f (liter/mole)	$K_f(K^+)/K_f(M^{z+})$	Selectivity ratios K^+/M^{z+}: Orion body	Corning body
Dicyclohexyl-18-crown-6	K^+	104.7[b]	1.	1	1
	Rb^+	33.1[b]	3.2	2.7–2.9	3.1–3.4
	NH_4^+	21.4[b]	4.9	5.3–6.5	6.1–7.4
	Cs^+	9.1[b]	11.4	7.9–8.6	9.1–14
	Na^+	31.6–70.8[c]	3.1	21–29	69–72
	Ba^{2+}	3715.[b]	$K_f(Ba)/K_f(Sr)$		Ba/Sr
	Sr^{2+}	1738.	2.1		2.2–2.4
Dibenzo-18-crown-6	K^+	74.1[d]			
	Rb^+	22.4[d]	3.3	2.7–3.1	
Benzo-15-crown-5	K^+	9.3[d]			
	Rb^+	2.9[d]	3.2	3.0–3.2	
Dibenzo-30-crown-10	K^+	22.4[d]			
	Rb^+	36.3	0.62	0.85–0.89	

[a] The values were determined by using the liquid membrane electrodes formed from nitrobenzene solutions of polyethers.
[b] From Izatt *et al.*[34]; the solvent used was water.
[c] From Frensdorff[33]; the solvent used was water.
[d] From Rechnitz and Eyal[78]; the solvent used was 50% water and 50% tetrahydrofuran.

theoretically and experimentally by Boles and Buck,[83] and Simon and co-workers.[84, 85] The responses of the electrode to potassium salts of picrate, thiocyanate, propionate, cyclohexane butyrate, benzene sulfonate, and benzoate at room and elevated temperatures have been discussed in terms of existing theories based on anion solubility, mobility, and ion pairing.[83] Similarly, equations have been given by Morf *et al.*[84] to reduce the interference of cation response by lipid-soluble anions in the sample solution. It has already been mentioned that the tetraphenyl borate anion interfered seriously with the response of the K-selective liquid membrane electrode to K^+ ions. However, this interference was reduced or eliminated by incorporation of the anion into a less polar membrane phase so as to influence or shift the extraction equilibrium of the system in accordance with the theoretical principles enunciated[84] to a state in which the electrode response to cations improved. This was accomplished in the case of K^+- and Ca^{2+}-selective electrodes in which the ligands [valinomycin and Ca ligand (see later)], and tetraphenyl borate were incorporated into a membrane formed from polyvinyl chloride dissolved in 2-nitro-*p*-cymene (to form a K-selective membrane) and *o*-nitrophenyloctyl ether (to form a Ca-selective membrane).[84, 85]

A very interesting type of microelectrode (of tip diameter 0.5–1.0 μm) containing a liquid ion exchanger (Corning Code 477317) selective to K^+ ions has been constructed.[86, 87] Other liquid ion exchangers (e.g., Corning 477315 selective to Cl^-) also may be used to form microelectrodes. These electrodes are being developed to monitor the activity of K^+ and Cl^- ions in cellular preparations.

2. Ammonium-Selective Membrane Electrode

In Chapter 7 the solid membrane electrode selective to NH_3 and/or NH_4^+ ions was described. The results given in Table 5 show that the liquid membrane [saturated solution of nonactin (72%) and monactin (28%) in tris(2-ethylhexyl)phosphate] acts as an electrode selective to ammonium ions exhibiting the selectivity sequence $NH_4^+ > K^+ > Rb^+ > H^+ > Cs^+ > Li^+ > Na^+ > Ca^{2+}$. Furthermore, the selectivity of NH_4^+ over K is 10 while that[70] over Na is 500. In the concentration range 10^{-1}–10^{-5} *M*, a Nernstian response was observed. The electrode is manufactured by Philips.[88] It has been used in automatic potentiometric determinations of ammonia in boiler feed waters.[89]

3. Barium-Selective Membrane Electrode

The neutral carrier complex of a polyethylene glycol derivative which contains 12 ethylene oxide units per Ba^{2+} and 2 moles of tetraphenyl borate ion [i.e., Igepal CO 880 (nonylphenoxypolyethylene ethanol) + $BaCl_2$ + excess sodium tetraphenyl borate, precipitate dissolved in *p*-nitroethylbenzene] has been used as a liquid membrane electrode[90] which showed a Nernstian response in the concentration range 10^{-1}–10^{-5} *M* Ba^{2+}. Most common anions (Cl^-, Br^-, I^-, NO_3^-, and SO_4^{2-}) had no effect on the response of the electrode, whereas copper ions poisoned the electrode.[91] The electrode had a selectivity constant K^{pot}_{Ba-M} of $< 10^{-4}$ for Ca^{2+}, Mg^{2+}, Ni^{2+}, Co^{2+}, Zn^{2+}, and Fe^{2+}; 2×10^{-2} for Sr^{2+}; 8×10^{-3} for K^+; 2×10^{-4} for H^+, Li^+, and Na^+; and 6×10^{-4} for NH_4^+.[91] The electrode could be used in the potentiometric estimation of the SO_4^{2-} ion and of other anions that formed precipitates with Ba^{2+} ions.

4. Calcium-Selective Membrane Electrode

In the early stages of development of ion-selective sensors, ion-site salts derived from diesters of phosphoric acid were used.[92] Electrodes made from diesters with hydrocarbon chains in the C_8–C_{16} range showed good selectivity for calcium ions in the presence of Na ions.[3] The Orion calcium electrode has a cellulose membrane that is saturated with the calcium salt of the ion exchange liquid formed from a solution of didecylphosphoric acid in di-*n*-octylphenyl phosphonate. The membrane is held in a holder (see Fig. 1a). The inner reference electrolyte and electrode are a solution of 0.1 *M* $CaCl_2$ and a Ag–AgCl wire, respectively. A linear response to Ca^{2+} in the presence of a constant amount of other ions (Na, K, and Mg) has been noted[93] over the concentration range 10^{-1}–10^{-5} *M*, the electrode behaving completely in a Nernstian manner. Below 10^{-6} *M*, the calibration curve formed a plateau and became independent of Ca^{2+} ion activity. The lower limit of detection is determined by the small solubility of the calcium phosphate ester salt in the aqueous phase.

The selectivity constants K_{ij}^{pot} determined by a number of investigators are given in Table 10. Also included in the table is the response of the beryllium electrode which was prepared by converting the ion exchanger into the beryllium form by treating the ion exchanger with a beryllium ion containing solution. This electrode has been used to measure the rates of reaction of beryllium ions with EDTA and nitrilotriacetic acid.[97]

The dynamic response of the Ca-selective liquid membrane electrode has been evaluated by measuring the time (t_{95}) required for the electrode to reach 95% of its equilibrium (steady state) potential after a rapid change in ion activity.[94] In pure Ca^{2+} ion solutions, the t_{95} was 2.25 sec, whereas in the presence of interfering ions, t_{95} was 4 (Mg^{2+}), 8.5 (Sr^{2+}), 22.4 (Ba^{2+}), 2.3 (H^+), and 2.2 [$(CH_3)_4N^+$] sec.

Selectivity isotherms for the Ca^{2+} + Na^+ system have been established over the ionic strength range 0.03–6 *M*.[98] A relatively strong base, cyclohexyl ammonium ion, has been found to interfere with the selectivity of the electrode,[99] although choline, triethanolamine, and tri(hydroxymethyl)aminomethane have no effect.

The electrode has been used to monitor the complexometric titrations in which EGTA [ethyleneglycol-bis(2-aminoethyl ether)tetraacetic acid] and DCTA (1, 2 -diaminocyclohexanetetraacetic acid) are used as the titrants.[100, 101] Similar complexometric titrations in which other derivatives of acetic acid are used have also been followed by using the Ca-selective electrode.[102–104] The electrode has been used in a number of studies involving the potentiometric estimation of Ca^{2+} ions[3, 102, 105, 106] and the determination of Ca in bentonites,[107, 108] soil,[109] seawater,[93, 110, 111] detergents,[112] and in the study of $MgSO_4$ and $CaSO_4$ association in seawater.[113]

In order to obtain accurate results for the amount of Ca^{2+} ions in water samples in which the Ca electrode is used, use of a constant complexation buffer (CCB) has been recommended.[114] CCB is made up as follows: 40.4 g of KNO_3, 3.6 g of disodium iminodiacetate, 160 ml of aqueous 0.5 *M* acetylacetone solution, 2 ml of 10 *M* ammonia, and 1.07 g of NH_4Cl dissolved and made up to 1 liter.

TABLE 10 Selectivity Constants K_{ij}^{pot} of the Ca^{2+}- and M^{2+}-Selective Liquid Membrane Electrodes

Interfering ion	Ca^{2+}-selective electrode$(RO)_2PO_2^-$ in dioctylphenyl phosphonate				M^{2+}-selective electrode$(RO)_2PO_2^-$ in mixed alcohol–phosphonate		$(RO)_2PO_2^-$ in Be form in mixed solvent	
Be^{2+}								1
Ca^{2+}	1	1	1	1	1		1	6.3×10^{-2}
Zn^{2+}	3.20	3.0×10^{-3}					3.50	
Fe^{2+}	0.80						3.5	—
Pb^{2+}	0.63	1.8						
Cu^{2+}	0.27	0.13					3.1	
Ni^{2+}	8.0×10^{-2}						1.35	
Sr^{2+}	2.0×10^{-2}	2.6×10^{-2}					0.54	
Mg^{2+}	1.0×10^{-2}	4.0×10^{-2}			1.8×10^{-2}	1	1.0	2.7×10^{-2}
Ba^{2+}	1.0×10^{-2}	9.0×10^{-3}					0.94	
H^+	10^7	2.0×10^4	18[a]					
Na^+	1.6×10^{-3}	1.0×10^{-3}		$\leqslant 6 \times 10^{-3}$[a]	1.0×10^{-2}	2.5×10^{-2}	1.0×10^{-2}	1.3×10^{-3}
Li^+				0.33[a]	0.156	0.124		2.9×10^{-2}
K^+				$\leqslant 6 \times 10^{-3}$[a]	6.0×10^{-3}	1.8×10^{-2}		5.0×10^{-4}
Rb^+				$\leqslant 6 \times 10^{-3}$[a]				
Cs^+				$\leqslant 6 \times 10^{-3}$[a]				
NH_4^+				$\leqslant 6 \times 10^{-3}$[a]				
$(CH_3)_4N^+$		6.0×10^{-4}			6.7×10^{-3}	1.6×10^{-2}		1.8×10^{-3}
$(C_2H_5)_4N^+$					5.1×10^{-3}	0.156		1.0×10^{-3}
$(C_3H_7)_4N^+$					2.72	7.4		0.432
$C_6H_5(CH_3)_3N^+$					$\sim 5.7 \times 10^3$	$\sim 1.6 \times 10^4$		1.4×10^2
Ref.	3	94[b]	95	96	97[c]	97	3	97

[a] The values of K_{ij}^{pot} were derived from the equation $E = \frac{RT}{F} \ln \left[1 + K_{ij}^{pot} \frac{a_{i+}}{a_{Ca^{2+}}^{1/2}} \right]$

[b] Continuous analysis. [c] Rapid mixing continuous flow system.

The use of the electrode in clinical practice has been reviewed by Moore[115] and others.[116–118] As an indicator electrode, it has been used in the titration of calcium with EDTA[119] and nitrilotriacetic acid.[120] It has been used to estimate the solubility and dissociation of calcium sulfate dihydrate[121, 122] and $CaCO_3$.[123] Also calcium molybdate solubility in water has been determined.[124] This is based on the assumption that the activity of Ca^{2+} ($a_{Ca^{2+}}$) measured with the Ca-selective electrode is equal to the activity of the molybdate ion. Thus using the relation $S_p = a_{Ca^{2+}}\, a_{MoO_4^{2-}}$, the value for the solubility product S_p is obtained. From the relation $\log S_p = -(2066/T) - 1.704$, where T is the absolute temperature, values for S_p at different temperatures have been calculated. These values are ($S_p \times 10^9$) 5 at 40°C, 3 at 30°C, and 1.8 at 20°C.

The electrode has been used in the determination of the stability constants of calcium complexes with several polyphosphates[125] and in the estimation of Ca^{2+} ion activities in aqueous electrolyte solutions[126–130] and in serum,[116, 131–138] blood, brain, and spinal fluid,[139, 140] gastric juices,[141] milk,[142, 143] and animal fodder.[144] It has been used to estimate the free calcium present in equilibrium with calcium chelates such as nitrilotriacetic acid, tripolyphosphoric acid, etc.,[145] in calcium–albumin aggregates,[146] venous human blood,[117, 147, 148] and other systems.[149]

The electrode has been used to follow the currents passing through the electrode on polarization with slow triangular pulses[150] and in the determination of activity coefficients of $CaCl_2$ in two aqueous systems, $CaCl_2$–$MgCl_2$ and $CaCl_2$–$SrCl_2$, over the range of ionic strength 0.1–6.0 mole/kg at 25°C[151] and in the system $CaCl_2$–NaCl over the ionic strength range 0.03–0.7 mole/kg at 25°C.[152]

Following extensive studies involving the complexation of alkali metal ions with electroneutral antibiotics (described earlier, see Table 1), Simon and co-workers[22, 153, 154] established the principles that regulate the characteristics of ligands which show selectivity to alkali and/or alkaline earth metal ions. Some of these characteristics of electrically neutral ligands which can act as carriers to metal ions are[153, 154] the following:

(1) A carrier must have both polar and nonpolar groups.

(2) Among polar groups, there must be preferably 5 to 8 but not more than 12 coordinating sites such as oxygen.

(3) All cations should accept the same given number of coordinating groups.

(4) The neutral ligand (i.e., carrier) should be capable of undergoing conformation with a cavity surrounded by polar groups suitable for taking up cations. The nonpolar groups of the ligand should form a lipophilic shell around the coordination sphere.

(5) High selectivity can be obtained provided the coordination sites of the ligand are able to form a rigid arrangement around the cavity. This rigidity can be increased by formation of hydrogen bonds or other bridged structures. The preference of the ligand is to that ion which fits the existing cavity.

(6) Despite the rigidity in arrangement of coordination sites dealt with under (5), the ligand should be flexible enough to allow for a fast exchange. This is achieved by a stepwise substitution of the solvent molecules by the ligand groups. Thus a compromise should be found between rigidity (item 5) and rate of exchange (item 6).

(7) The overall size of the ligand should be small and it should be lipid soluble.

(8) Electrically neutral Li- or Na-specific carriers should contain no more than six coordination sites.

(9) The ligands designed for large alkaline earth cations should be as small as possible in solvents of high dielectric constant.

Based on these considerations, it has been predicted that the chances of finding a suitable carrier for the Mg^{2+} ion are very small. Also, ligands have been synthesized which possess the above-mentioned characteristics and show a preference for divalent cations[154–156] [see (X), which is the structure of synthetic ligands specific for alkaline earth cations].

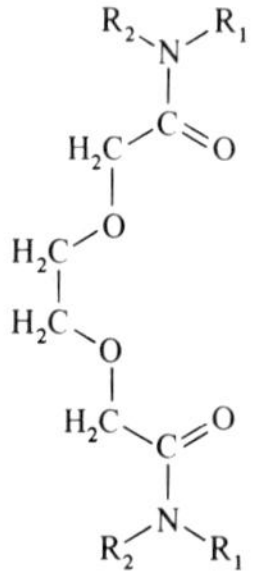

(i) $R_1 = -CH_3$, $R_2 = -(CH_2)_{10}-COO-CH_2-CH_3$
(ii) $R_1 = -CH_3$, $R_2 = -(CH_2)_6-CH_3$
(iii) $R_1 = R_2 = -CH_2-CH_2-CH_3$
(iv) $R_1 = -CH_2-CH_2-CH_3$, $R_2 = -CH_2-C-(CH_3)_3$
(v) $R_1 = -CH_3$, $R_2 = CH_2-COO-CH_2-CH_3$
(vi) $R_1 = R_2 = -C_6H_5$

(X)

Compound (i) has been used with solvents of different dielectric constant; the selectivity of the liquid membrane (saturating filter paper) electrodes so formed for Ca^{2+} over Na^+ has been determined.[157, 158] The values for K^{pot}_{Ca-Na} were ~ 14.1 (dibutylsebacate solvent, $\epsilon \approx 5$), ~ 2.5 [tris(2-ethylhexyl)phosphate, $\epsilon \approx 9$], ~ 2.0 (1-decanol, $\epsilon = 9$), 0.22 (acetophenone, $\epsilon \approx 17$), $\sim 5.6 \times 10^{-2}$ (2-nitro-*p*-cymene, $\epsilon \approx 20$), $\sim 2.0 \times 10^{-2}$ (nitrobenzene, $\epsilon \approx 35$), and $\sim 5.6 \times 10^{-3}$ (*p*-nitroethyl benzene, $\epsilon \approx 21$). These values were determined from emf measurements using 10^{-1} *M* $CaCl_2$ and 10^{-1} *M* NaCl sample solutions.[154] The highest selectivity is realized with the solvent of fairly high dielectric constant. The selectivity constants of the liquid membrane electrode prepared from ligand (i) in *p*-nitroethyl benzene solvent (10–20% by weight of ligand in solvent) are given in Table 11 along with other values derived for the same ligand but held in a polyvinyl chloride matrix. The PVC membrane was prepared from a solution of 3% ligand (i) and 30% PVC in 67% *o*-nitrophenyl octyl ether. The values for K^{pot}_{Ca-M} were calculated from emf measurements using pure solutions of $CaCl_2$ and other metal ion solutions according to the equation

$$\log K^{pot}_{Ca-M} = \frac{(\text{emf}_2 - \text{emf}_1)2F}{2.303RT} - \log a^{2/z}_{M^{z+}} + \log a_{Ca^{2+}}$$

where emf_2 is the potential measured with the interfering ion M^{z+} and emf_1 is that measured with pure $CaCl_2$ solution.

TABLE 11

Selectivity Constants K^{pot}_{Ca-M} of Ca-Selective Liquid and Solid Membrane Electrodes

Interfering ion	Liquid membrane electrode (10^{-1} *M* solution used)[a]	Solid PVC membrane electrode (10^{-2} *M* solution used)[b]
Ca^{2+}	1	1
Li^+	2.3×10^{-3}	6.0×10^{-2}
Na^+	5.7×10^{-3}	3.0×10^{-1}
		1.0×10^{-1} (in 10^{-1} *M*)
K^+	7.3×10^{-2}	1.0×10^{-1}
		4.0×10^{-2} (in 10^{-1} *M*)
Rb^+	1.6×10^{-1}	3.0×10^{-2}
Cs^+	5.2×10^{-2}	1.0×10^{-2}
H^+	4.1×10^{-2}	
NH_4^+	1.7×10^{-1}	1.0×10^{-1}
Mg^{2+}	3.0×10^{-5}	2.0×10^{-4}
		2.0×10^{-4} (in 10^{-1} *M*)
Sr^{2+}	1.0×10^{-2}	1.0×10^{-1}
Ba^{2+}	8.0×10^{-2}	9.0×10^{-1}
Cu^{2+}	4.0×10^{-3}	2.0×10^{-3}
Zn^{2+}	1.0×10^{-3}	6.0×10^{-4}
UO_2^{2+}	6.4×10^{-3}	
Al^{3+}	3.5×10^{-4}	
Ce^{3+}		2.0×10^{-2}

[a] Values from Ammann *et al.*[158]
[b] Values from Morf *et al.*[154]

Although the selectivities and the sensitivities (Nernstian response in the concentration range 10^{-1}–10^{-5} M Ca^{2+}) for both the liquid and solid membrane electrodes are comparable, the solid electrode is considered superior to the liquid electrode in that the former has a longer lifetime and greater stability.[154] The other ligands (ii)–(vi) of (X) have also been used with PVC to form the membrane electrodes whose responses are given in Table 12.

TABLE 12

Selectivity Behavior of Ligands (i)–(vi) of Basic Structure (X)[a]

Cation	($E_M - E_{Ca}$) mV for ligand					
M^{z+}	(i)	(ii)	(iii)	(iv)	(v)	(vi)
Ca^{2+}	0	0	0	0	0	0
Mg^{2+}	− 110	− 112	− 101	− 111	− 95	− 81
Ba^{2+}	− 1	− 3	− 40	− 50	− 65	+ 73
Na^+	− 73	− 77	− 91	− 103	− 92	+ 13
K^+	− 82	− 86	− 82	− 75	− 66	+ 3
Cs^+	− 103	− 109	− 103	− 107	− 89	− 34

[a]The emf responses of sensors (PVC membranes) to 10^{-12} M solutions of different cations referred to Ca^{2+} response.

The electrochemical activity indicated in Table 12 shows that the ester groups in ligand (i) are not involved in conferring Ca selectivity to the membrane electrode since the membrane electrode with ligand (ii) has properties similar to those of the membrane formed from ligand (i). Similarly, the *N*-alkyl groups [ligands (iii) and (iv)] also are of little importance in conferring selectivity to Ca^{2+} ions. The behavior of the ligands (i)–(v) shows that the same sites in all the ligands are involved in complexation with the Ca^{2+} ion. Substitution of *N*-alkyl groups by phenyl groups [ligand (vi)] leads to a preference for the larger Ba^{2+} ion. Probably a larger cavity between the coordination sites due to steric interactions among phenyl groups exists; this leads to a loss of discrimination of monovalent ions. All these ligands probably form 1 : 2 complexes with the Ca^{2+} ion and crystallize. The Ba^{2+} complex of ligand (vi) with 1 : 2 stoichiometry has been obtained.[154]

5. Copper-Selective Membrane Electrode

Orion Research puts out both the liquid and solid membrane electrodes selective to Cu^{2+} ions. The liquid membrane has an ion exchange resin of skeleton $R–S–CH_2COO^-$ saturating a cellulose membrane.[3] The selectivity constants K^{pot}_{Cu-M} of this electrode are as follows: Na^+ and K^+ : 5 ×

10^{-4}; Mg^{2+}, Ca^{2+}, Ba^{2+}, and Sr^{2+}: $\sim 10^{-3}$; Zn^{2+} and Ni^{2+}: $\sim 10^{-2}$; H^+: 10; and Fe^{2+}: 1.4×10^2. The anions Cl^-, Br^-, ClO_4^-, and $C_2O_4^{2-}$ interfered with the electrode (Orion 92-29) which has been used by Rechnitz and Lin[159] to study the formation complexes of Cu(II) with glycine (log $K_{f(1)} = 8.20$; log $K_{f(2)} = 14.96$), glutamic acid (log $K_{f(1)} = 8.48$; log $K_{f(2)} = 14.55$), tham (trishydroxymethylamino methane, log $K_f = 4.97$), and acetate (log $K_f = 2.04$).

6. Iron-Selective Membrane Electrode

Hemin incorporated into silicone rubber was tried as an electrode selective to Fe^{3+} ions. This gave results that were not as good as those obtained with a liquid membrane electrode that was formed by dissolving hemin in nitrobenzene.[160] The solution was used in the body of an Orion liquid electrode saturating a porous partition to form the electrode. This was used in a titration cell of the type suggested by Covington and Thain[161] (see Fig. 1e) to study its responses which were Nernstian in the concentration range 10^{-1}–10^{-4} M in the presence of NO_3^-, SO_4^{2-}, or Cl^- ions. The electrode could be used only in acid solutions since buffers could not be used because of the formation of complexes with Fe^{3+} ion. The electrode selectivity was $K_{Fe\text{–}Co} = 2 \times 10^{-3}$ and $K_{Fe\text{–}Cu} = 3.5 \times 10^{-3}$. The electrode could be used to detect end points in precipitation and complexometric titrations. Titration of Fe^{3+} with 8-hydroxyquinoline using the electrode gave three potential jumps corresponding to the successive formation of Fe(III)–oxine complexes.

7. Lead-Selective Membrane Electrode

Like the copper(II)-selective liquid membrane, a liquid membrane containing an ion exchange resin of matrix R–S–CH_2COO^-, has been marketed by Orion Research. This electrode has been found to have the following values for the selectivity constants[3] $K_{Pb\text{–}M}$: 2.6 for M = Cu^{2+}; 8×10^{-2} for Fe^{2+}; 3×10^{-3} for Zn^{2+}; 5×10^{-3} for Ca^{2+}; 7.0×10^{-3} for Ni^{2+}; and 8×10^{-3} for Mg^{2+}.

Lal and Christian[162] made a detailed study of this liquid membrane electrode and found it to respond well to monovalent ions. The selectivity ratios found for different ions as a function of concentration are given in Table 13. These results show that the relative selectivity to lead ions increases with an increase in the concentration of the interfering ions.

The electrode has been used in the titration of alkali metal ions, thallium, silver, and ammonium ions with sodium tetraphenyl borate as well as lead with various solutions of Na_2CrO_4, K_2SO_4, K_2CrO_4, $K_4Fe(CN)_6$, $K_4P_2O_7$, and Na_2WO_4.[162] The electrode has been used in the

TABLE 13

Selectivity Ratios for Different Ions at Different Concentrations Determined for a Pb-Selective Liquid Membrane Electrode

Interfering ion	K_{ij} at concentrations (M) of:				
	10^{-5}	10^{-4}	10^{-3}	10^{-2}	10^{-1}
H^+	1.9×10^7	4.2×10^5	2.9×10^5	1.0×10^3	1.1
Ag^+	2.2×10^6	1.7×10^5	8.6×10^4	5.6×10^2	9.0×10^{-1}
Na^+	8.9×10^6	5.6×10^4	2.6×10^2	2.0	3.0×10^{-2}
K^+	8.9×10^6	3.5×10^4	3.0×10^2	3.0	8.6×10^{-3}
Cu^{2+}	1.2×10^3	1.8×10^2	1.4×10^2	2.6	2.2×10^{-1}
Ca^{2+}	2.5×10^1	2.6	3.4×10^{-1}	3.5×10^{-2}	5.0×10^{-3}
Mg^{2+}	1.7×10^3	1.7×10^2	1.7×10^1	6.9×10^{-1}	9.2×10^{-2}
Ni^{2+}	8.9	1.3	2.8×10^{-1}	1.1×10^{-2}	2.9×10^{-3}
Zn^{2+}	1.0×10^1	3.0	2.4×10^{-1}	3.0×10^{-2}	2.3×10^{-3}

potentiometric determination of end points in the direct titration of sulfate with standard lead solutions.[163] From a practical standpoint, the liquid membrane electrodes of the heavy metals are now almost obsolete due to the recent advent of solid state membrane electrodes (see Chapter 7) which possess better selectivities than the liquid systems.

8. Molybdenum-Selective Membrane Electrode

Bistetraethylammonium pentathiocyanato-oxomolybdate(V) in a mixture of nitrobenzene and *o*-dichlorobenzene (2 : 3) has been used to form a liquid membrane electrode. The salt is insoluble in *o*-dichlorobenzene but is soluble in nitrobenzene which is not well absorbed by the lightly cross-linked natural rubber membrane. So this mixture of solvents is used to form a 2.5% (w/v) solution of the salt above.[164] Unlike the other ion-selective membrane electrodes, this electrode senses the amount of molybdenum present in the complex anion and not the molybdate ion. A Nernstian response to the molybdenum present as pentathiocyanatomolybdate(V) in the concentration range 10^{-2}–5×10^{-8} M has been noted. Metal ions such as iron, vanadium, tungsten, rhenium, and niobium which form thiocyanate complexes interfered with the response of the electrode.

9. Selenium-Selective Membrane Electrode

The active ingredient is a liquid membrane consisting of a saturated solution of 3,3′-diamino benzidine in hexane. The characteristics of this electrode have been determined by Malone and Christian.[165] At pH 2.5, the electrode gave a linear response to selenium(IV) up to 10^{-4} M with a slope of approximately 60–65 mV per tenfold change in selenium concentration. Between 10^{-4} and 10^{-3} *M*, a sudden change in potential which was analytically useful occurred. The interferences of other ions with the response of the electrode to selenium(IV) at a concentration of 10^{-4} *M* have been determined. In view of the multivalency of the mixed solution, precise values for K_{ij}^{pot} could not be calculated in the usual way. However, the selectivity ratios, as an approximate guide to relative selectivities, were obtained from the mixed potential readings in the presence of selenium(IV) assuming a slope of 60 mV for all ions using the equation

$$\text{mV change} = 240 + 60 \log\left[1 \times 10^{-4} + K_{ij}^{pot}\ (\text{interfering ion})\right]$$

The various ions that interfered with the response of the electrode followed the sequence V(V) > Te(IV) > Sb(III) > Mo(VI) > ClO_4^- > Ag^+ > As(III) > Na^+ > Br^- > Hg(II) > Cl^- > SO_4^{2-}.

10. Thallium- and Antimony-Selective Membrane Electrodes

Hexachloroantimonate(V) and tetrachlorothallium(III) salts of Sevron Red L, Sevron Red GL, Flavinduline O, and Phenazinduline O dissolved in *o*-dichlorobenzene and saturating a natural rubber membrane have been used as liquid membrane electrodes.[166] Calibration of the electrodes was done with saturated solutions of hexachloroantimonate and tetrachlorothallate in 2 *M* HCl. A full Nernstian response in the range 10^{-8}–10^{-2} *M* (Sb) and 10^{-6}–10^{-2} *M* (Tl) was observed. They were found to be suitable for determining Sb or Tl. However, the antimony electrode responded to both Sb(III) and Sb(V).

11. Zinc-Selective Membrane Electrode

The basic dye salt of Brilliant Green tetrathiocyanatozincate(II) was formed by adding stoichiometric amounts of zinc acetate and potassium thiocyanate to excess Brilliant Green. The precipitate was filtered, washed, and dried. The required quantity of the salt was dissolved in *o*-dichlorobenzene to form a 10^{-3} *M* solution. This solution, saturating a lightly cross-linked natural rubber membrane, served as the electrode[167] which responded to solutions of zinc (10^{-4}–10^{-1} *M*) containing a 20-fold excess of thiocyanate. A slope of 29.5 mV/decade concentration, which is expected of $Zn(SCN)_4^{2-}$, was observed. This electrode determined the amount of zinc present in the anion complex. Heterogeneous solid membrane electrodes have also been constructed using this salt. The basic dye salt, Silastomer 72, and a catalyst were mixed and the paste was compressed between glass plates and cured for 24 hr to form the silicone rubber membrane which was used as the electrode.

A 2% (w/v) solution of the dye salt in bromonaphthalene and special grade powdered graphite were used in the formation of the carbon paste electrode.[168] These electrodes gave steady responses after 10–15 min, whereas the liquid membrane electrode gave a response within 1 min.

C. ANION-SELECTIVE LIQUID MEMBRANE ELECTRODES

The possibility of employing a wide variety of organic solvent extraction (ion association and chelation) systems for use as ion-selective electrodes has been explored by Coetzee and Freiser.[169] Organic and inorganic salts of Aliquat 336S (tricaprylmethyl ammonium ion) dissolved in 1-decanol [10% (v/v)] functioned effectively as an organic phase component of the membrane electrode. The barrels and cellulose membrane of the Orion calcium electrode could be used to make a compact membrane electrode unit.

Some 16 electrodes were tested in pure solutions of the appropriate salt over a concentration range of 10^{-1}–10^{-5} M. Equilibrium potentials were observed quickly (20 sec to 1 min) and the values were highly reproducible. In every case, the potential response was linear with log(concentration) or activity from 10^{-1} to 10^{-3} M. The electrode response was independent of the decanol concentration used. The responses of the various electrodes and their interferences by different anions are shown in Table 14.

The values of selectivity constants given in Table 14 show that the interference of an ion j with the ion i to which the electrode is selective seems to be related to the relative extractability of the quaternary ammonium salts into the organic or membrane phase. The least interference arises when the ion i is highly extractable and the ion j is least extractable. The selectivities of the different electrodes in general do not seem to be very good.

Other quaternary ammonium salts could be used in place of Aliquat 336S. For example, tetraheptyl ammonium iodide in 1-decanol, on mixing with an aqueous solution of sodium salicylate, exchanges iodide for salicylate to form tetraheptyl ammonium salicylate in the organic phase. This acts as a liquid membrane electrode selective to salicylate ion. However, benzoate and p-hydroxybenzoate cause serious interferences. Consequently in the absence of these ions, the electrode has been used to follow the electroreduction of salicylic acid.[171] Similarly, Herman and Rechnitz,[171a] using a solution of Aliquat 336 in trifluoroacetyl-p-butylbenzene, constructed a liquid membrane electrode selective to carbonate ions in the presence of bicarbonate. The electrode showed a Nernstian response in the concentration range 10^{-2}–10^{-6} M. The selectivity constants K_{ij}^{pot} were

TABLE 14

Responses and the Selectivity Constants K_{ij}^{pot} of Anion-Responsive Liquid Membrane Electrodes[a]

Liquid electrode for ion i	Slope (mV/pX^-)	K_{ij}^{pot} for j ions				
		Cl^-	NO_3^-	SO_4^{2-}	Benzoate	Other anions
Perchlorate	59.2	7.9×10^{-3}	3.2	1.0×10^{-2}	—	ClO_3^- = 0.16; PO_4^{3-} = 5×10^{-3}
Thiocyanate	58.0	$< 3.2 \times 10^{-4}$	6.3×10^{-2}	7.9×10^{-3}		ClO_4^- = 0.54; I^- = 0.28
Nitrate	57.0	0.23		2.5×10^{-3}		ClO_3^- = 0.89; $NO_2^- \approx 0.5$
Iodide	59.0	$\sim 10^{-2}$	0.19	6.3×10^{-2}		$Br^- \approx 0.1$
Bromide	59.0	~ 0.16	0.85	3.2×10^{-2}		$I^- \approx 2 \times 10^{-2}$
Chloride	56.0		2.14	4.0×10^{-2}		$Br^- \approx 3.2$
Sulfate	36.0	39.8	10^3			$S_2O_3^{2-}$ = 0.79
Oxalate	40.0	~ 10		~ 0.32		Acetate = 1.0; benzoate = 31.6
Formate	53.0	1.32			5.0	Acetate = 0.46; propionate = 1.1
Acetate	53.0	1.95	5.0	0.16	5.0	Formate = 0.96; $C_2O_4^{2-} \approx 1.6$
Propionate	57.5	1.45			3.2	Formate = 0.63; acetate = 0.7
Benzoate	58.6	~ 0.5	0.1			
m-Toluate	58.0	~ 0.1		10^{-2}	0.48	
p-Toluate	57.0	4.0×10^{-2}		1.6×10^{-2}	0.74	Acetate $< 10^{-3}$; propionate = 1.05
p-Toluene sulfonate	57.0	3.2×10^{-2}	0.2		~ 0.1	Acetate $< 10^{-3}$; ClO_4^- = 2.5
Salicylate	56.0	$< 10^{-3}$	5.0×10^{-2}	6.3×10^{-3}		

[a] Values from Coetzee and Freiser.[169, 170]

determined to be 1.9×10^{-4} for CO_3^{2-}–Cl^-, 1.5×10^{-4} for CO_3^{2-}–SO_4^{2-}, and 2.6×10^{-4} for CO_3^{2-}–HPO_4^{2-}. The electrode has been used in the determination of CO_2 levels in human serum samples.[171b]

In early work, a chloride electrode using dimethyldistearyl ammonium cation was constructed and evaluated.[3] The selectivity constants K_{ij}^{pot} were as follows for the various interfering ions: ClO_4^-, 32; I^-, 17; NO_3^-, 4.2; Br^-, 1.6; OH^-, 1.0; acetate$^-$, 0.32; HCO_3^-, 0.19; SO_4^{2-}, 0.14; and F^-, 0.10. Srinivasan and Rechnitz[172] also evaluated the chloride electrode and found the selectivity constants to be 4.14 for NO_3^-, 2.86 for Br^-, and 23.1 for I^- in 0.1 *M* solutions.

An interesting type of liquid membrane electrode based on extraction of halides from a $LiNO_3$–KNO_3 eutectic at 160°C by tetraoctyl phosphonium nitrate has been described.[173] The electrode showed a selective linear response to iodide ions without interference from bromide and chloride ions.

Another interesting series of liquid membrane electrodes selective to amino acids has been constructed by Matsui and Freiser.[174] The organic phase consisted of either a 10% (v/v) or 1% (v/v) (for leucine or phenylalanine) decanol solution of Aliquat 336S which was converted to the amino acid anion salt by repeated equilibration of a 0.1 *M* aqueous solution of the corresponding K or Na salt. A Millipore Teflon membrane saturated with the organic phase was used to form the liquid membrane electrode. As indicated above, amino acid extractability into the organic phase determined the feasibility of preparation of that electrode. Anions such as polar glycine and alanine which are poorly extractable were not considered for electrode formation, whereas readily extractable ones such as tryptophan, phenylalanine, leucine, methionine, valine, and glutamic acid were used to form electrodes whose responses at pH-10.5 were followed over the concentration range 10^{-1}–10^{-5} *M*.

The tryptophan and phenylalanine electrodes responded in pure solutions in the concentration range 10^{-1}–$10^{-2.7}$ *M* linearly with a slope of 55.0 mV per log *C*, whereas the leucine electrode gave a slope of 57.0 in the same concentration range. Methionine, valine, and glutamic acid (dianion) electrodes responded linearly in the concentration ranges 10^{-1}–$10^{-2.5}$, 10^{-1}–$10^{-2.3}$, and 10^{-1}–$10^{-3.3}$ *M* with slopes of 52.6, 53.0, and 23.8 mV/decade (of concentration), respectively. The interferences caused by other amino acid anions and some inorganic ions have been determined for all six electrodes. Some of the values determined for the leucine and phenylalanine electrodes are listed in Table 15.

In order to study the quantitative relations governing extractability of ion pairs into the organic phase and selectivity exhibited by the electrodes constructed from such extraction systems, James *et al.*[175] carried out some

TABLE 15

Selectivity Constants K_{ij}^{pot} of Amino Acid-Selective Liquid Membrane Electrodes

Interfering anion (j)	Leucine electrode	Phenylalanine electrode
Leucine	—	0.4
Phenylalanine	1.59	
Isoleucine	0.63	0.40
Valine	0.25	0.16
Alanine	6.3×10^{-2}	5.0×10^{-2}
Methionine	3.2×10^{-2}	0.20
Glycine	6.3×10^{-2}	4.0×10^{-2}
Serine	3.2×10^{-2}	4.0×10^{-2}
Aspartic acid	1.0×10^{-2}	2.0×10^{-2}
Glutamic acid	1.3×10^{-2}	1.59×10^{-2}
HPO_4^{2-}	4.0×10^{-4}	1.0×10^{-2}
CO_3^{2-}	5.0×10^{-3}	1.26×10^{-2}
SO_4^{2-}	6.3×10^{-3}	2.5×10^{-2}
Cl^-	0.5	0.63
NO_3^-	1.6	1.26

model experiments using some of the organic ions and amino acids mentioned above. The distribution of tetra-*n*-hexyl ammonium iodide (THAI) between an organic solvent and aqueous phases was studied using carrier free ^{131}I. THAI (3.3×10^{-3} M) in organic solvent and ^{131}I in the aqueous phase were shaken together and the distribution of ^{131}I between the organic phase and the aqueous phase was measured. A competitive extraction using an organic anion (R^-) in the system was also studied.

THAI distribution between the organic and aqueous phases involves two equilibria:

$$Q^+ + I^- \overset{K_{IP}}{\rightleftharpoons} Q^+, I^- \tag{12}$$

Then there is the distribution of the ion pair between the organic (O) and the aqueous phases:

$$Q^+, I^- \overset{K_D}{\rightleftharpoons} Q^+I^-_{(O)} \tag{13}$$

Thus,

$$K_{IP}K_D = \frac{(Q^+\ I^-)_O}{(Q^+)(I^-)} \tag{14}$$

But due to the electroneutrality condition, $(Q^+) = (I^-)$. Thus,

$$K_{IP}K_D = \frac{(Q^+I^-)_O}{(I^-)^2} \tag{15}$$

The solvent extraction exchange taking place between R^- and I^- is given by

$$Q^+I^-_{(O)} + R^- \overset{K}{\rightleftharpoons} Q^+R^-_{(O)} + I^- \tag{16}$$

and thus

$$K = \frac{(K_{IP}K_D)_{QR}}{(K_{IP}K_D)_{QI}} = \frac{(Q^+R^-)_O(Q^+)(I^-)}{(Q^+)(R^-)(Q^+I^-)_O} = \frac{(Q^+R^-)_O(I^-)}{(Q^+I^-)_O(R^-)} \tag{17}$$

The values of $(Q^+I^-)_O$ and (I^-) are known by counting of the radioactivity. Stoichiometry requires that $(Q^+R^-)_O = (I^-)$ and $(R^-) = C_{R(\text{initial})} - (Q^+R^-)_O$ where C_R is the initial concentration of R^-. Thus since all the quantities are known, K can be evaluated. The values so calculated using water and decanol for a number of organic anions and anionic amino acids in place of R^- are listed in Table 16. From these values, the competitive extraction indicated by the ratio K_j/K_i can be derived. Also using the respective liquid membrane electrodes the selectivity constants K_{ij}^{pot} can be measured. These values are given in Tables 17 and 18. The correlation indicates that the extraction parameters quantitatively determine the selectivity characteristics of the liquid membrane electrodes. Similar considerations have been presented by Back in his studies of ion selectivities of anion-selective liquid membranes formed from solutions of tetrabutyl or tetrapropyl ammonium perchlorate[176] and of *n*-tetrahexyl ammonium picrate[177] in methylene chloride. In this work, the extraction ratio

$$E_{ij(QX)} = p\,\frac{E_{i(QX)}}{E_{j(QX)}} \tag{18}$$

where p is a proportionality constant, was used. This $E_{ij(QX)}$ is equivalent in significance to K_j/K_i given in Tables 17 and 18. The picrate electrode was tested in pure solutions containing Cl^-, Br^-, NO_3^-, and ClO_4^- ions whose

TABLE 16

Competition Extraction Constants Evaluated for Decanol and Water[a]

Anion	log K	Anion	log K
Formate	− 1.41	Glycine	− 2.04
Acetate	− 1.51	Serine	− 1.99
Propionate	− 1.42	Alanine	− 1.95
Benzoate	− 0.05	Valine	− 1.50
Salicylate	+ 0.73	Methionine	− 1.43
p-Toluene sulfonate	+ 0.27	Leucine	− 0.95
		Phenylalanine	− 0.43
		Tryptophan	− 0.41

[a]Measured at 25°C.

TABLE 17

Comparison of the Competitive Extraction Constants of Interfering Anions and Selectivity Constants of Various Organic Anion-Responsive Electrodes

Interfering anion	Membrane electrodes											
	Formate		Acetate		Propionate		Benzoate		*p*-Toluene sulfonate		Salicylate	
	$\log(K_j/K_i)$	$\log K_{ij}^{pot}$	$\log(K_j/K_i)$	$\log K_{ij}^{pot}$	$\log(K_j/K_i)$	$\log K_{ij}^{pot}$	$\log(K_j/K_i)$	$\log K_{ij}^{pot}$	$\log(K_j/K_i)$	$\log K_{ij}^{pot}$	$\log(K_j/K_i)$	$\log K_{ij}^{pot}$
Formate			+ 0.1	+ 0.04	0	− 0.2						
Acetate	− 0.1	− 0.3			− 0.09	− 0.31	− 1.46	− 2			− 2.24	− 2.4
Propionate	0	0.05	0.09	0.05								
Benzoate	1.36	− 1.0	1.46	1.4	1.37	1.2			0.32	− 0.24		
p-Toluene sulfonate			1.79	1.5								
Salicylate							0.78	0.75				

TABLE 18

Comparison of Competitive Extraction Constants of Interfering Anions and the Selectivity Constants for Various Amino Acid-Responsive Membrane Electrodes

Interfering anion	Membrane electrodes									
	Tryptophan		Phenylalanine		Leucine		Methionine		Valine	
	$\log(K_j/K_i)$	$\log K_{ij}^{pot}$	$\log(K_j/K_i)$	$\log K_{ij}^{pot}$	$\log(K_j/K_i)$	$\log K_{ij}^{pot}$	$\log(K_j/K_i)$	$\log K_{ij}^{pot}$	$\log(K_j/K_i)$	$\log K_{ij}^{pot}$
Glycine	− 1.52	− 1.8	− 1.61	− 1.4	− 1.11	− 1.2	− 0.61	− 0.8	− 0.54	− 0.9
Alanine	− 1.54	− 1.6	− 1.63	− 1.3	− 1.13	− 1.2	− 0.63	− 0.8	− 0.56	− 0.8
Valine	− 0.98	− 0.8	− 1.07	− 0.8	− 0.57	− 0.6	− 0.07	− 0.4		
Leucine	− 0.41	− 0.4	− 0.5	− 0.4			+ 0.50		+ 0.57	
Serine	− 1.47	− 1.9	− 1.56	− 1.4	− 1.06	− 1.5	− 0.56	− 1.2	− 0.49	− 0.9
Methionine	− 0.91	− 0.8	− 1.0	− 0.7	− 0.50	− 1.5			+ 0.07	− 0.1
Phenylalanine	+ 0.09	− 0.1			+ 0.5	+ 0.2	+ 1.0	+ 0.4	+ 1.07	+ 0.7
Tryptophan			− 0.09	− 0.1	+ 0.41	+ 0.2	+ 0.91	+ 0.1	+ 0.98	+ 0.7

potential responses in relation to Cl^- ion as the reference ion i have been compared in Table 19 with the $E_{ij(QX)}$ values taken from the work of Schill.[178]

TABLE 19

Responses of the *n*-Tetrahexyl Ammonium Picrate Liquid Membrane Electrode to Anions Compared with the Extraction Ratio[a]

Anion	$\log K_{ij}^{pot}$	$\log E_{ij(QX)}$	$\log p$
Br^-	0.70	1.40	0.47
I^-	1.48	3.12	0.48
NO_3^-	1.33	2.36	0.48
ClO_4^-	1.85	3.59	0.51

[a] Reference anion is Cl^-.

The linearity of the plot of $\log K_{ij}^{pot}$ vs. $\log E_{ij(QX)}$ testifies again to the correlation that exists between the extraction properties of the organic liquid in the membrane and the selectivity of the membrane electrode.

Baum[50] has used the data in Table 18 to show that the extent of interference caused by an amino acid (j) is roughly related to the free energy change accompanying the transfer of the amino acid (i) electrode from amino acid i to amino acid j. When $a_i = a_j$, the selectivity constant K_{ij}^{pot} is given by

$$\log K_{ij}^{pot} = (E_j - E_i)/S \tag{19}$$

where $S = 2.303RT/F$.

The free energy change ΔG is related to ΔE by Eq. 1 of Chapter 3. Thus ΔG (cal) = 23.05 ΔE (mV); so Eq. (19) can be written as

$$\log K_{ij}^{pot} = \frac{\Delta G_{t,j} - \Delta G_{t,i}}{23.05S} = \frac{\Delta G_t}{23.05S} \tag{20}$$

A plot of $\log K_{ij}^{pot}$ vs. ΔG_t for various amino acids in the case of leucine and phenylalanine electrodes gave a scatter of points in each case. However, the general trend of data followed Eq. (20), giving a slope corresponding to the calculated value (i.e., $1/23.05S$).

A number of other electrodes selective to a number of anions using a variety of compounds have been developed. A plastic membrane impregnated with dithizone was found selective to a variety of cations and anions, including uni-, bi-, and trivalent ions.[179] These electrodes have found applications in a variety of potentiometric titrations involving acid–base, precipitation, redox, and complexometric reactions. They are also useful in titrations involving nonaqueous media.[179, 180]

An ingenious technique for forming membrane electrodes has been described.[181] The ion-sensitive surface of a "liquid state" electrode is formed by a very thin layer of the organic liquid adsorbed on the surface of porous graphite. The composition of the organic phase, which must be immiscible with water, can be so chosen that it responds only to the activity of a particular ion in the aqueous sample phase. The construction and performance of such an electrode with solutions of I_2 in CCl_4, C_6H_6, and mesitylene have been described. In this type of electrode graphite is directly connected to the potentiometer so that no inner reference solution or electrode is required. An improved construction of this electrode, also called Selectrode, is shown in Fig. 1c.

Some of the common detergents (e.g., sodium salts of dodecyl sulfate, tetrapropylene benzene sulfonate, dioctylsulfosuccinate) have been used as liquid membrane electrodes selective to the anions concerned.[182, 183] The electrode showed a Nernstian response provided the detergents were completely dissociated. They have been used in potentiometric titrations.[182] At concentrations where they form micelles, the electrode potential changed little with concentration.[183] Consequently, these electrodes have been used to measure the critical micellar concentration (CMC). However, the applicability of this method has been questioned.[184] This was because the solvent used (nitrobenzene) to dissolve the anion exchanger, cetyltrimethyl ammonium dodecylsulfate, affected the CMC.[185] Consequently, an extensive search was made for a solvent that did not affect the CMC of sodium dodecyl sulfate. It was found that *o*-dichlorobenzene was water immiscible and had no effect on the concentration of sodium dodecyl sulfate, although the solvent had limited solubility for the anion exchanger complex $[C_{12}H_{35}SO_4^- C_{16}H_{33}(CH_3)_3N^+]$. This was overcome by adding small amounts of water-soluble hydrogen-bonding solutes. The most satisfactory solvent for a 10^{-3} *M* ion-association complex was found to be 0.17 *M* hexachlorobenzene plus 0.017 *M* bromoacetanilide in *o*-dichlorobenzene.[185] In addition, electrodes selective to a variety of surfactant types have been prepared and the manufacture of electrodes specific to surfactants of single alkyl chain length has been discussed. A dodecyl sulfate-selective electrode has been evaluated for its use in surfactant solutions containing polymers (dextran, polyvinyl alcohol, and polyvinyl pyrrolidone) and a protein (bovine serum albumin).[186] Thus the interactions of surfactant with the polymers and with the protein have been followed.

Liquid membrane electrodes responding to sulfonate ions have been described.[187] These were formed from four triphenylmethane dyes, Crystal Violet, Methyl Violet, Malachite Green, and Fuchsine Basic, and from a high molecular weight quaternary ammonium ion. An aromatic sulfonate (benzene sulfonate, α-naphthalene sulfonate) salt of Crystal Violet was formed by mixing solutions of sodium sulfonate and Crystal Violet. The precipitate was washed and dried. Aromatic sulfonate salts of other dyes and of quaternary ammonium ion (dodecyloctylmethyl ammonium) were prepared in a similar way. These salts were dissolved in nitrobenzene or 1, 2-dichloroethane to form the liquid membrane. The benzene sulfonate and α-naphthalene sulfonate electrodes based on Crystal Violet showed a nearly Nernstian response down to 10^{-4} M sulfonate. In the pH range 2.5–12.0, the potential of the electrode was constant. The selectivity constants K_{ij}^{pot} of the two electrodes are given in Table 20.

An anion exchange resin and a liquid anion exchanger (tridecylamine hydrochloride) have been used as electrodes to study their responses to the bicarbonate ion, the responses being Nernstian in pure solutions of $NaHCO_3$.[188] In mixed solutions of $NaHCO_3$–$NaCl$ and $NaHCO_3$–Na_2SO_4, the liquid membrane electrode as opposed to the resin electrode was specific to the HCO_3^- ion. Even in the presence of an excess of other anions (acetate, HPO_4^{2-}), the electrode showed high specificity to the HCO_3^- ion.

Early in the development of liquid membranes, complexes of positively charged transition metal complexes with a bulky organic ligand containing an orthophenanthroline chelating group were used.[3] Salts of the Fe or

TABLE 20

Selectivity Constants K_{ij}^{pot} of Crystal Violet-Based Aromatic Sulfonate Membrane Electrodes

		K_{ij}^{pot} for liquid membrane electrode	
Interfering ion (j)	Concentration of $j(C_j)$	Benzene sulfonate	α-Naphthalene sulfonate
Cl^-	0.5	3.0×10^{-3}	4.0×10^{-4}
NO_3^-	5×10^{-3}	0.76	3.0×10^{-2}
Phenol-4-sulfonate	5×10^{-3}	1.6×10^{-2}	
Benzene-m-disulfonate	5×10^{-3}	5.0×10^{-3}	
Benzoate	5×10^{-3}	4.0×10^{-2}	
α-Naphthalene sulfonate	2.5×10^{-4}	16	
1, 3, 6-Naphthalene trisulfonate	5×10^{-3}	8.0×10^{-4}	
	1×10^{-1}		6.0×10^{-5}
Benzene sulfonate	5×10^{-3}		7.0×10^{-2}
1, 5-Naphthalene disulfonate	5×10^{-3}		7.0×10^{-4}
4-Hydroxyl-2-naphthalene sulfonate	5×10^{-3}		2.5×10^{-2}
2, 3-Dihydroxynaphthalene-6-sulfonate	5×10^{-3}		4.5×10^{-4}

$NiL(NO_3)_2$ type, where L has the structure (XI) shown, function as anion exchangers. The nitrate- and the tetrafluoroborate-selective liquid membrane electrodes based on $NiL(NO_3)_2$ and the perchlorate-selective electrode based on $FeL(NO_3)_2$ gave the selectivity constants listed in Table 21. The solvents used in the preparation of these electrodes according to the patent[189] include nitrobenzene, decanol, dioctylphenyl phosphonate, and *p*-nitrocymene.

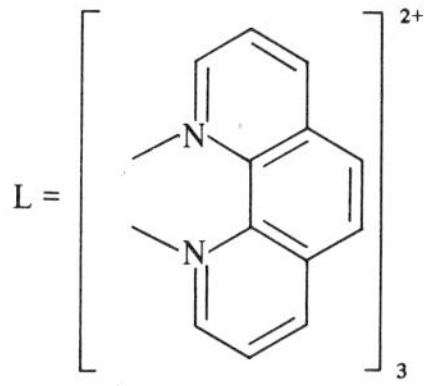

(XI)

TABLE 21

Selectivity Constants K_{ij}^{pot} of *o*-Phenanthroline-Based Liquid Anion-Selective Membrane Electrodes[a]

Interfering ion	Electrode selective to		
	NO_3^-	ClO_4^-	BF_4^-
NO_3^-	1	1.5×10^{-3}	0.1
ClO_4^-	10^3	1	
I^-	20	1.2×10^{-2}	20
ClO_3^-	2		
OH^-		1.0	1.0×10^{-3}
Br^-	0.9	5.6×10^{-4}	4.0×10^{-2}
S^{2-}	0.57		
NO_2^-	6.0×10^{-2}		
CN^-	2.0×10^{-2}		
HCO_3^-	2.0×10^{-2}		4.0×10^{-3}
Cl^-	6.0×10^{-3}	2.2×10^{-4}	1.0×10^{-3}
Acetate	6.0×10^{-3}	5.1×10^{-4}	4.0×10^{-3}
$S_2O_3^{2-}$	6.0×10^{-3}		
SO_3^{2-}	6.0×10^{-3}		
F^-	9.0×10^{-4}	2.5×10^{-4}	1.0×10^{-3}
SO_4^{2-}	6.0×10^{-4}	1.6×10^{-4}	1.0×10^{-3}
$H_2PO_4^-$	3.0×10^{-4}		
PO_4^{3-}	3.0×10^{-4}		
HPO_4^{2-}	8.0×10^{-5}		

[a]Data from Ross.[3]

1. Nitrate-Selective Membrane Electrode

Three of these electrodes are commercially available[190] from Beckman, Corning, and Orion Research.

Liquid membranes containing tris(1,10-phenanthroline)iron(II) [$Fe(phen)_3^{2+}$], tris(4,7-diphenyl-1,10-phenanthroline)iron(II) [Fe(d-phen)$_3^{2+}$], and tetraheptyl ammonium ion (tetrahep) in nitrobenzene, chloroform, and *n*-amyl alcohol have been constructed and evaluated.[191] The selectivity constants of these electrodes are given in Table 22.

TABLE 22

Selectivity Constants K_{ij}^{pot} of Phenanthroline- and Quaternary Ammonium Ion-Based NO_3-Selective Liquid Membrane Electrodes

	Fe(d-phen)$_3^{2+}$			$Fe(phen)_3^{2+}$	Tetrahep
Interfering ion (j)	1×10^{-3} *M* in nitrobenzene	2×10^{-3} *M* in chloroform	2×10^{-3} *M* in amyl alcohol	5×10^{-4} *M* in nitrobenzene	1×10^{-2} *M* in nitrobenzene
NO_3^-	1	1	1	1	1
Cl^-	40	21.7	4.0	3.0	27.5
Br^-	5.3	1.5	1.6	1.6	4.2
I^-	8.1×10^{-2}	1.8×10^{-2}	0.4	6.6×10^{-2}	9.1×10^{-2}
BF_4^-	4.2×10^{-2}	1.3×10^{-1}	1.2	2.1×10^{-1}	2.6×10^{-2}
SCN^-	2.9×10^{-2}	2.9×10^{-2}	1.9×10^{-1}	4.6×10^{-2}	3.3×10^{-2}
ClO_4^-	1.5×10^{-3}	8.3×10^{-3}	2.1×10^{-1}	5.2×10^{-3}	1.4×10^{-3}
PF_6^-	1.3×10^{-4}	1.3×10^{-3}	8.3×10^{-1}	6.3×10^{-4}	7.8×10^{-5}

It was found that with the exception of the amyl alcohol electrode the selectivity constants were relatively independent of membrane composition and followed a common sequence of decreasing selectivity: $PF_6^- > ClO_4^- > SCN^- \sim I^- \sim BF_4^- > NO_3^- > Br^- > Cl^-$. This sequence roughly parallels the order of increasing anion hydration energy, thereby indicating that the aqueous solvation energies play a predominant role in determining electrode selectivities for these ions.[191]

The Corning electrode uses tridodecylhexadecyl ammonium nitrate in *n*-octyl-2-nitrophenyl ether.[192] The Russian workers[193] have studied dimethylhexadecylbenzyl ammonium nitrate dissolved in decyl alcohol and tetraoctyl ammonium nitrate dissolved in octyl alcohol as exchanger materials for nitrate electrodes. They showed good permselectivity to NO_3^- ions in the presence of other anions such as I^-, Cl^-, NO_2^-, SO_4^-, CO_3^{2-}, or F^-.

The Orion nitrate electrode has been evaluated by Potterton and Shults.[194] A linear response with a slope of 56.6 mV/decade at 25°C was observed in the concentration range 10^{-1}–10^{-4} *M*. Nitrite (9×10^{-2}), Cl^-

(8×10^{-3}), EDTA (9×10^{-4}), and F^- (5×10^{-4}) interfered in the order indicated by the K_{ij}^{pot} values given in parentheses. Similarly, Srinivasan and Rechnitz[172] found values for K_{ij}^{pot} to be 1.71 for ClO_3^-, 0.15 for Br^-, and 16.2 for I^-.

The electrode has been used in the study of the complex formation of lanthanum ions with nitrate ions.[195] Two complexes with the stoichiometry of 1 : 1 and 1 : 3 (La : NO_3) are considered to exist in solution. It is used as an indicator electrode in the titration of nitrate with diphenylthallium(III) sulfate.[196] In practical analysis, the electrode has been used for the determination of nitrate in soil extracts,[197–203] in vegetable and other plant materials,[201, 203–207] in well and surface waters,[203, 208–210] in ion exchange liquid chromatography as a detector,[211] in microbial media,[212] in food,[213] limestone,[214] oleum,[215] and nitrites,[216, 217] and for determining traces of NO_2 and NO in gaseous mixtures.[218]

The electrode has been modified by using a wick of natural or synthetic porous polymer in place of the membrane. The wick is saturated with the ion exchanger.[219] This electrode has been used in the determination of nitrate in different waters.

2. Perchlorate-Selective Membrane Electrode

As indicated above, it is $Fe(phen)_3^{2+}$ (phen = phenanthroline) exhibiting the selectivities shown in Table 21 that has been used as a liquid membrane electrode selective to perchlorate ions.[3] In the concentration range 10^{-1}–$10^{-3.5}$ M, it responds selectively to ClO_4^- ion activity in the pH range 4–11.[220] The electrode has been noted to respond to permanganate, periodate, and dichromate ions.[221] In addition, it is sensitive also to perrhenate and thiocyanate activities. The response has been found to be Nernstian over several orders of magnitude with detection limits of about 10^{-6} M for perrhenate and 10^{-5} M for thiocyanate.[222] The titration curves for the addition of Hg(II) to SCN^-, Cl^-, Br^-, and I^- are U-shaped. The electrode seems to be highly sensitive to anionic mercury–halide complexes formed in the early stages of titration.

Metal chelates of *o*-phenanthroline, α,α'-dipyridyl, or bathophenanthroline have been used as the ion exchanger in liquid membranes of the perchlorate-selective electrode.[223] Electrodes with nitrobenzene or 1, 2-dichloroethane membrane containing tris(bathophenanthroline) ferrous perchlorate had the highest sensitivity. The electrode having the ferrous ion chelate of *o*-phenanthroline in nitrobenzene also showed excellent selectivity for the perchlorate ion. The selectivity constants of some of the perchlorate electrodes constructed by different investigators are shown in Table 23.

TABLE 23

Selectivity Constants K_{ij}^{pot} of Perchlorate (i)-Selective Membrane Electrodes

Interfering ion (j)	Orion	Dodecyloctylmethyl benzyl ammonium perchlorate		$Fe(phen)_3(ClO_4)_2$ in nitrobenzene	Brilliant Green perchlorate in chlorobenzene (10^{-1}–10^{-3} M)	*N*-Ethylbenzothiazole-2, 2′-azaviolene perchlorate		
		In nitrobenzene	In dichlorobenzene			In dichloro-benzene (3×10^{-4} M)	In dichloro-diethyl ether	Solid state
I^-	2.9×10^{-2}[a]	2.0×10^{-2}	1.7×10^{-2}	5.9×10^{-3}	8.0×10^{-2}–0.85	2.4×10^{-2}	3.0×10^{-1}	75
BF_4^-						1.2×10^{-1}	1.2×10^{-1}	1.2×10^{-1}
OH^-						1.6×10^{-3}	1.7×10^{-4}	6.3×10^{-3}
NO_3^-	4.3×10^{-3}	2.9×10^{-3}	1.0×10^{-2}	1.0×10^{-5}	1.3×10^{-2}–0.21	2.0×10^{-3}	1.5×10^{-3}	1.1×10^{-3}
ClO_3^-						1.8×10^{-3}	2.1×10^{-3}	8.7×10^{-4}
SO_4^{2-}						1.7×10^{-5}	4.4×10^{-5}	2.3×10^{-4}
Br^-	1.1×10^{-3}				6.6×10^{-3}–0.16	2.8×10^{-4}	3.9×10^{-4}	6.5×10^{-4}
F^-	2.9×10^{-4}				2.3×10^{-3}–0.12	1.8×10^{-4}	7.0×10^{-5}	4.2×10^{-4}
CH_3COO^-	1.7×10^{-3}				6.0×10^{-3}–0.21	4.1×10^{-5}	6.0×10^{-5}	3.2×10^{-4}
Cl^-					3.7×10^{-3}–0.17	2.5×10^{-5}	1.1×10^{-4}	7.8×10^{-4}
HCO_3^-	8.8×10^{-4}				2.0×10^{-2}–0.29			
Ref.	220	223	223	223	224	225	225	226

[a] A value of 1.3×10^{-2} was obtained by Srinivasan and Rechnitz.[172]

Ishibashi and Kohara[223] also investigated the effect of the central metal ion of *o*-phenanthroline chelate on the selectivity of the electrode for the perchlorate ion. The selectivity constants K_{ij}^{pot}, where $i = ClO_4^-$ and $j = I^-$, for the various central metal ions in the ligand were found to be 5.9×10^{-3} (Fe^{2+}), 1.3×10^{-2} (Ni^{2+}), 3.2×10^{-2} (Cu^{2+}), and 1.1×10^{-1} (Cd^{2+}). The least selective was the Cd ligand.

Fogg *et al.*[224] used Brilliant Green perchlorate in chlorobenzene saturating a commercial natural rubber sheeting to form a membrane electrode selective to ClO_4^- ions. It gave a steady potential in ClO_4^- solutions in the concentration range 10^{-3}–10^{-1} *M*. The interference of other ions is shown in Table 23 along with the selectivity constants of other perchlorate electrodes.

The organic radical ion salt derived from *N*-ethylbenzothiazole-2, 2′-azaviolene (NEBA-ClO_4) from which a solid state Selectrode is constructed (see Chapter 7) has been used to form liquid membranes in two solvents, 1, 2-dichlorobenzene (DCB) and β, β'-dichlorodiethyl ether (DDE).[225] Both these liquid electrodes gave a Nernstian slope of 59 mV/decade in the concentration range 1–10^{-4} *M*. The selectivities of these electrodes are given in Table 23 along with those of the solid state electrodes[226] for comparison. The worst interferences are from iodide and tetrafluoroborate ions. Low interference was shown by H^+ and OH^- ions, so the electrode may be used in the pH range 1–12.

The perchlorate electrode has been used in the low-temperature precipitation titration of perchlorate and tetrafluoroborate with tetraphenylarsonium chloride.[227] Use of a low temperature (2°C) sharpened the potentiometric titration curves. It has been used[220] for determining the solubility product of $KClO_4$ ($S_p = 9.6 \times 10^{-3}$ at 25°C, pH = 5.2), hexamine cobaltic perchlorate ($S_p = 3.4 \times 10^{-6}$ at 25°C, pH = 5.0), tetrapyridine cupric perchlorate ($S_p = 3.0 \times 10^{-5}$ at 25°C, pH = 6.4), and tris-1,10-phenanthroline ferrous perchlorate ($S_p = 2.2 \times 10^{-8}$ at 25°C, pH = 5.5).

From a systematic study of long-chain alkyl ammonium salts in ethyl bromide, tetra-*n*-heptyl ammonium perchlorate has been chosen as a promising liquid membrane perchlorate electrode.[228] It was used in an electrochemical cell of the type

$$H_2(1\ \text{atm}) \left| \begin{matrix} HClO_4 \\ m_1,\ \text{fixed} \end{matrix} \right\| \begin{matrix} \text{Liquid} \\ \text{membrane} \end{matrix} \left\| \begin{matrix} HClO_4 \\ m_2,\ \text{variable} \end{matrix} \right| H_2(1\ \text{atm})$$

to measure the emf of the membrane cell as a function of m_2. Mean molal activity coefficients of $HClO_4$ solutions at 25°C have been determined.[229] Some of the values of γ as a function of m were 0.989 (10^{-4} *m*), 0.965

(10^{-3} *m*), 0.905 (10^{-2} *m*), 0.793 (10^{-1} *m*), 0.730 (1.0 *m*), 0.970 (2.0 *m*), 1.38 (5.0 *m*), 7.45 (7.0 *m*), and 30.5 (10.0 *m*).

3. Phosphate-Selective Membrane Electrode

Two liquid systems, one made of a primary amine hydrochloride (Rohm & Haas, XLA3) and the other of quaternary amine chloride (General Mills, Aliquat 336) have been used (pH range 7.0–7.5) in the Orion barrel with a membrane to measure the divalent phosphate ion activities in dilute solutions.[230] Some heteropoly acids such as phosphomolybdic and phosphotungstic acids dissolved in *n*-pentanol have been tested as liquid ion exchange membranes selective to phosphate ions.[231] However, they were of little significance since their selectivity over other anions was very poor.

4. Tetrafluoroborate-Selective Membrane Electrode

Some of the phenanthroline-based electrodes (see Table 21) have been found to be selective to BF_4^- ions.[3] Recently Brilliant Green tetrafluoroborate dissolved in chlorobenzene and adsorbed on natural rubber sheeting has been used as a liquid membrane electrode.[232] It showed a Nernstian response to the BF_4^- ion in the concentration range 10^{-3}–10^{-1} *M* with a slope of 58.5 mV/decade. Its selectivity constants K_{ij}^{pot} in the concentration range 10^{-1}–10^{-4} *M* were as follows: BF_4^- = 1, ClO_4^- = 1.6–1.0, I^- = 1.2×10^{-2}–0.24, HCO_3^- = 4.0×10^{-3}–0.26, Br^- = 2.0×10^{-3}–0.26, NO_3^- = 2.0×10^{-3}–0.19, CH_3COO^- = 2.0×10^{-3}–0.15, Cl^- = 1.10^{-3}–0.19, $H_2PO_4^-$ = 1.0×10^{-4}–4.0×10^{-2}, F^- = 1.0×10^{-4}–2.0×10^{-2}, and borate = 4.0×10^{-5}–1.0×10^{-2}.

The *o*-phenanthroline-based electrode has been used in the potentiometric determination of boron after its conversion to tetrafluoroborate.[233] Similarly, it has been used to determine boron in crystalline solids such as aluminum oxide, boron carbide,[234] and in silicon.[235] Boron in silicon is converted to fluoroborate by treatment with HF and NH_4F in the presence of H_2O_2.

5. Thiocyanate-Selective Membrane Electrode

A quaternary ammonium salt (octadecyldimethylbenzyl ammonium thiocyanate, ODDBA-SCN) dissolved in a suitable solvent saturating a filter paper plug was used as the electrode.[236] The electrode response to the SCN^- ion in the presence of interfering ions as a function of the solvent used in the preparation of the electrode is shown in Table 24. The nitrobenzene electrode shows the best response characteristics.

TABLE 24

Selectivity Constants K_{ij}^{pot} of Thiocyanate-Selective Liquid Membrane Electrodes[a]

Interfering ion (j)	Octadecyldimethylbenzyl ammonium thiocyanate dissolved in		
	Chloroform (0.05 M)	Dichloroethane (0.01 M)	Nitrobenzene (0.01 M)
ClO_4^-	10.7	14.1	3.5
I^-	8.8×10^{-1}	6.0×10^{-1}	2.4×10^{-1}
ClO_3^-	3.8×10^{-1}	4.7×10^{-1}	5.0×10^{-1}
NO_3^-	4.6×10^{-1}	8.0×10^{-2}	3.0×10^{-2}
Br^-	1.4×10^{-1}	2.0×10^{-2}	6.0×10^{-3}
Cl^-	5.0×10^{-2}	6.0×10^{-3}	5.0×10^{-4}
CO_3^{2-}	1.0×10^{-3}	4.0×10^{-4}	6.0×10^{-4}
SO_4^{2-}	9.0×10^{-4}	4.0×10^{-4}	6.0×10^{-5}
$S_2O_3^{2-}$	2.0×10^{-4}	2.0×10^{-4}	3.0×10^{-5}

[a]Data from Stworzewicz *et al.*[236]

REFERENCES

1. J. Koryta, *Anal. Chim. Acta* **61**, 380 (1972).
2. M. Mioscu, E. Hopirtean, and C. Liteanu, *Rev. Roum. Chim.* **18**, 1637 (1973).
3. J. W. Ross, Jr., *in* "Ion Selective Electrodes" (R. A. Durst, ed.), Chapter 2. Nat. Bur. Std. Spec. Publ. 314, Washington, D.C., 1969.
4. K. Sollner, *in* "Diffusion Processes" (G. N. Sherwood, A. V. Chadwick, W. M. Muir, and F. L. Swinton, eds.), p. 655. Gordon and Breach, New York, 1971.
5. J. Sandblom and F. Orme, *in* "Membranes" (G. Eisenman, ed.), Vol. 1, p. 125. Dekker, New York, 1972.
6. G. M. Shean and K. Sollner, *Ann. N. Y. Acad. Sci.* **137**, 759 (1966).
7. K. Sollner, *Ann. N. Y. Acad. Sci.* **148**, 154 (1968).
8. K. Sollner and G. M. Shean, *J. Amer. Chem. Soc.* **86**, 1901 (1964); *Protoplasma* **63**, 174 (1967).
9. H. Brockman, M. Springorum, G. Traxler, and I. Hofer, *Naturwissenschaften* **50**, 689 (1955).
10. M. M. Shemyakin, N. A. Aldanova, E. I. Vinogradova, and M. Yu. Feigina, *Tetrahedron Lett.* No. 28, 1921 (1963).
11. Yu. A. Ovchinnikov, *in* "Mitochondria" (Biogenesis and Bioenergetics), Biomembranes (Molecular arrangements and transport mechanisms), p. 279. American Elsevier, New York, 1972.
12. V. I. Ivanov *et al.*, *Biochem. Biophys. Res. Commun.* **34**, 803 (1969).
13. M. M. Shemyakin *et al.*, *J. Membrane Biol.* **1**, 402 (1969).
14. P. A. Plattner, K. Vogler, R. O. Stander, P. Quitt, and W. Keller-Schierlam, *Helv. Chim. Acta* **46**, 927 (1963).
15. P. Quitt, R. O. Stander, and K. Vogler, *Helv. Chim. Acta* **46**, 1715 (1963).

16. R. L. Hamill, C. E. Higgens, H. E. Boaz, and M. Gorman, *Tetrahedron Lett.* No. 49, 4255 (1969).
17. V. T. Ivanov *et al.*, *Biochem. Biophys. Res. Commun.* **42**, 654 (1971).
18. Th. Wieland *et al.*, *Angew. Chem.* **80**, 209 (1968).
19. P. E. Hunter and L. S. Schwartz, *in* "Antibiotics" (D. Gottlieb and P. D. Shaw, eds.), Vol. 1, pp. 631, 642. Springer-Verlag, Berlin and New York, 1967.
20. E. G. Gale, E. Cundliff, P. E. Reynolds, M. H. Richmond, and M. J. Waring, "Molecular Basis of Antibiotic Action," p. 135. Wiley, New York, 1972.
21. M. C. Goodall, *Biochim. Biophys. Acta* **219**, 471 (1970).
22. W. Simon and W. E. Morf, *in* "Membranes" (G. Eisenman, ed.), Vol. 2, p. 329. Dekker, New York, 1973.
23. H. J. Moschler, H. G. Weler, and R. Schwyzer, *Helv. Chim. Acta* **54**, 1437 (1971).
24. H. K. Wipf, L. A. R. Pioda, Z. Stefanac, and W. Simon, *Helv. Chim. Acta* **51**, 377 (1968).
25. M. M. Shemyakin *et al.*, *Biochem. Biophys. Res. Commun.* **29**, 834 (1967).
26. Yu. A. Ovchinnikov, V. I. Ivanov, and I. I. Mikhaleva, *Tetrahedron Lett.* No. 2, 159 (1971).
27. V. I. Ivanov *et al.*, *Biochem. Biophys. Res. Commun.* **42**, 654 (1971).
28. W. K. Lutz, H. K. Wipf, and W. Simon, *Helv. Chim. Acta* **53**, 1741 (1970).
29. P. U. Fruh, J. T. Clerc, and W. Simon, *Helv. Chim. Acta* **54**, 1445 (1971).
30. J. H. Prestegard and S. I. Chan, *Biochemistry* **8**, 3921 (1969); *J. Amer. Chem. Soc.* **92**, 4440 (1970).
31. W. E. Morf and W. Simon, *Helv. Chim. Acta* **54**, 2683 (1971).
32. H. Diebler, M. Eigen, G. Ilgenfritz, G. Maass, and R. Winkler, *Pure Appl. Chem.* **20**, 93 (1969).
33. H. K. Frensdorff, *J. Amer. Chem. Soc.* **93**, 600 (1971).
34. R. M. Izatt, D. P. Nelson, J. H. Rytting, B. L. Haymore, and J. J. Christiansen, *J. Amer. Chem. Soc.* **93**, 1619 (1971).
35. J. J. Christiansen, J. O. Hill, and R. M. Izatt, *Science* **174**, 459 (1971).
36. N. Parthasarathy *et al.*, *Chimia* **27**, 368 (1973).
37. J. B. Harrell, A. D. Jones, and G. R. Choppin, *Anal. Chem.* **41**, 1459 (1969).
38. G. Scibona, L. Mantella, and P. R. Danesi, *Anal. Chem.* **42**, 844 (1970).
39. J. Ruzicka and J. C. Tjell, *Anal. Chim. Acta* **49**, 346 (1970).
40. J. Ruzicka and J. C. Tjell, *Anal. Chim. Acta* **51**, 1 (1970).
41. E. A. Materova, V. V. Muchovikov, and M. G. Grigorjeva, *Anal. Lett.* **8**, 167 (1975).
42. A. Ilani, *J. Gen. Physiol.* **46**, 839 (1963).
43. A. Ilani, *Biochim. Biophys. Acta* **94**, 415 (1965).
44. A. Ilani, *Biochim. Biophys. Acta* **94**, 405 (1965).
45. C. Liteanu and E. Hopirtean, *Stud. Univ. Babes-Bolyai Ser. Chem.* **16**, 83 (1971).
46. C. Liteanu, E. Hopirtean, and M. Mioscu, *Rev. Roum. Chim.* **18**, 1643 (1973).
47. E. Hopirtean, M. Mioscu, and C. Liteanu, *Stud. Univ. Babes-Bolyai Ser. Chem.* **18**, 133 (1973).
48. T. Higuchi, C. R. Illian, and J. L. Tossounian, *Anal. Chem.* **42**, 1674 (1970).
49. Corning #476200 Acetylcholine Ion Selective Electrode. Scientific Instruments, Corning Glass Works, Medfield, Massachusetts.
50. G. Baum, *J. Phys. Chem.* **76**, 1872 (1972).
51. G. Baum, *Anal. Lett.* **3**, 105 (1970).
52. G. Baum and F. B. Ward, *Anal. Biochem.* **42**, 487 (1971).
53. E. J. Cohn and J. T. Edsall, "Proteins, Amino Acids and Peptides." Van Nostrand-Reinhold, Princeton, New Jersey, 1943.
54. J. Sandblom, G. Eisenman, and J. L. Walker, Jr., *J. Phys. Chem.* **71**, 3862 (1967).

55. G. Baum, M. Lynn, and F. B. Ward, *Anal. Chim. Acta* **65**, 385 (1973).
56. G. Baum, *Anal. Biochem.* **39**, 65 (1971).
57. G. Baum and F. B. Ward, *Anal. Chem.* **43**, 947 (1971).
58. G. Baum, F. B. Ward, and S. Yaverbaum, *Clin. Chim. Acta* **36**, 405 (1972).
59. Z. Stefanac and W. Simon, *Chimia* **20**, 436 (1966); *Microchem. J.* **12**, 125 (1967).
60. W. Simon, L. A. R. Pioda, and W. K. Wipf, *Mosbach Colloq. Ges. Biol. Chem. 20th* p. 362 (1969).
61. L. A. R. Pioda and W. Simon, *Chimia* **23**, 72 (1969).
62. L. A. R. Pioda, V. Stankova, and W. Simon, *Anal. Lett.* **2**, 665 (1969).
63. W. Simon, H. R. Wuhrmann, M. Vasak, L. A. R. Pioda, R. Dohner, and Z. Stefanac, *Angew. Chem. Int. Ed.* **9**, 445 (1970).
64. Instruction Manual, Potassium Ion Electrode, Model 92–19, 1969. Orion Research, Inc., Cambridge, Massachusetts.
65. Liquid Membrane Potassium Electrode, Type No. IS 560-K. Philips Electronic Instruments Inc., Mount Vernon, New York.
66. M. S. Frant and J. W. Ross, Jr., *Science* **167**, 987 (1970).
67. S. Lal and G. D. Christian, *Anal. Lett.* **3**, 11 (1970).
68. E. Eyal and G. A. Rechnitz, *Anal. Chem.* **43**, 1090 (1971).
69. S. M. Hammond and P. A. Lambert, *J. Electroanal. Chem. Interfacial Electrochem.* **53**, 155 (1974).
70. R. P. Scholer and W. Simon, *Chimia* **24**, 372 (1970).
71. T. Anfalt and D. Jagner, *Anal. Chim. Acta* **66**, 152 (1973).
72. I. Sekerka and J. F. Lechner, *Anal. Lett.* **7**, 463 (1974).
73. M. M. Wise, M. J. Kurey, and G. Baum, *Clin. Chem.* **16**, 103 (1970).
74. L. A. R. Pioda, W. Simon, H. R. Bosshard, and H. Ch. Curtius, *Clin. Chim. Acta.* **29**, 289 (1970).
75. J. N. Butler and R. Huston, *Anal. Chem.* **42**, 676 (1970).
76. G. A. Rechnitz and M. S. Mohan, *Science* **168**, 1460 (1970).
77. M. S. Mohan and G. A. Rechnitz, *J. Amer. Chem. Soc.* **92**, 5839 (1970).
78. G. A. Rechnitz and E. Eyal, *Anal. Chem.* **44**, 370 (1972).
79. H. K. Wipf and W. Simon, *Biochem. Biophys. Res. Commun.* **34**, 707 (1969).
80. H. K. Wipf, A. Oliver, and W. Simon, *Helv. Chim. Acta* **53**, 1605 (1970).
81. H. K. Wipf, W. Pache, P. Jordan, H. Zahner, W. Keller-Schierlein, and W. Simon, *Biochem. Biophys. Res. Commun.* **36**, 387 (1969).
82. W. K. Lutz, H. K. Wipf, and W. Simon, *Helv. Chim. Acta* **53**, 1741 (1970).
83. J. H. Boles and R. P. Buck, *Anal. Chem.* **45**, 2057 (1973).
84. W. E. Morf, G. Kahr, and W. Simon, *Anal. Lett.* **7**, 9 (1974).
85. W. E. Morf, D. Ammann, and W. Simon, *Chimia* **28**, 65 (1974).
86. J. L. Walker, Jr., *Anal. Chem.* **43**, 89A (1971).
87. F. Vyskocil and N. Kriz, *Pflugers Arch. Eur. J. Physiol.* **337**, 265 (1972).
88. Liquid Membrane Ammonium Electrode, Type No. IS 560-NH_4. Philips Electronic Instruments Inc., Mount Vernon, New York.
89. J. Mertens, P. Van der Winkel, and D. L. Massart, *Anal. Lett.*, **6**, 81 (1973).
90. R. J. Levins, *Anal. Chem.* **43**, 1045 (1971).
91. R. J. Levins, *Anal. Chem.* **44**, 1544 (1972).
92. J. W. Ross, Jr., *Science* **156**, 3780 (1967).
93. M. E. Thompson and J. W. Ross, Jr., *Science* **154**, 1643 (1966).
94. B. Fleet, T. H. Ryan, and M. J. D. Brand, *Anal. Chem.* **46**, 12 (1974).
95. J. Bagg and R. Vinen, *Anal. Chem.* **44**, 1773 (1972).
96. J. Bagg, O. Nicholson, and R. Vinen, *J. Phys. Chem.* **75**, 2138 (1971).
97. B. Fleet and G. A. Rechnitz, *Anal. Chem.* **42**, 690 (1970).

98. M. Whitfield and J. V. Leyendekkers, *Anal. Chem.* **42**, 444 (1970).
99. H. B. Collier, *Anal Chem.* **42**, 1443 (1970).
100. M. Whitfield and J. V. Leyendekkers, *Anal. Chim. Acta* **45**, 383; **46**, 63 (1969).
101. M. Whitfield, J. V. Leyendekkers, and J. D. Kerr, *Anal. Chim. Acta* **45**, 399 (1969).
102. S. L. Tackett, *Anal. Chem.* **41**, 1703 (1969).
103. E. A. Moya and K. L. Chang, *Anal. Chem.* **42**, 1669 (1970).
104. T. P. Hadjiioannou and D. S. Papastathopoulos, *Talanta* **17**, 399 (1970).
105. J. A. King and A. K. Mukherji, *Naturwissenschaften* **53**, 702 (1966).
106. A. K. Mukherji, *Anal. Chim. Acta* **40**, 354 (1968).
107. W. H. Fertl and F. W. Jessen, *Clays Clay Mineral.* **17**, 1 (1969).
108. E. O. McLean, G. H. Snyder, and R. E. Franklin, *Soil Sci. Soc. Amer. Proc.* **33**, 388 (1969).
109. E. A. Woolson, J. A. Axley, and P. C. Kearney, *Soil Sci.* **109**, 270 (1970).
110. D. Jagner and K. Aren, *Anal. Chim. Acta* **57**, 185 (1971).
111. M. E. Thompson, *Science* **153**, 867 (1966).
112. R. W. Cummins, *Detergent Age*, **5** (March), 22 (1968).
113. D. R. Kester and R. W. Pytkowicz, *Limnol. Oceanogr.* **13**, 670 (1968); **14**, 686 (1969).
114. A. Hulanicki and M. Trojanowicz, *Anal. Chim. Acta* **68**, 155 (1974).
115. E. W. Moore, *Ann. N. Y. Acad. Sci.* **148**, 93 (1968); *in* "Ion Selective Electrodes" (R. A. Durst, ed.), Chapter 7. Nat. Bur. Std. Spec. Publ. 314, Washington, D. C., 1969.
116. M. J. Arras, *Postgrad. Med.* **45** (3), 57 (1969).
117. T. K. Li and J. T. Piechochi, *Clin. Chem.* **17**, 411 (1971).
118. W. H. Robertson, *Clin. Chim. Acta* **24**, 149 (1969).
119. G. A. Rechnitz and Z. F. Lin, *Anal. Chem.* **39**, 1406 (1967); **40**, 696 (1968).
120. T. R. Williams, B. Boettner, and S. Wakeham, *J. Chem. Educ.* **47**, 464 (1970).
121. F. S. Nakayama and B. A. Rasnick, *Anal. Chem.* **39**, 1022 (1967).
122. K. K. Tanji, *Environ. Sci. Technol.* **3**, 656 (1969).
123. F. S. Nakayama, *Soil Sci.* **106**, 429 (1968).
124. N. A. A. Mumallah and W. J. Popicl, *Anal. Chem.* **46**, 2055 (1974).
125. J. I. Watters, S. Kalliney, and R. C. Machen, *J. Inorg. Nucl. Chem.* **31**, 3823 (1969).
126. J. N. Butler, *Biophys. J.* **8**, 1426 (1968); *in* "Ion Selective Electrodes" (R. A. Durst, ed.), Chapter 5. Nat. Bur. Std. Spec. Publ. 314, Washington, D. C., 1969.
127. R. Kramer and H. Lagoin, *Naturwissenschaften* **56**, 36 (1969).
128. J. Bagg, *Aust. J. Chem.* **22**, 2467 (1969).
129. R. Huston and J. N. Butler, *Anal. Chem.* **41**, 200 (1969).
130. A. Shatkay, *Electrochim. Acta* **15**, 1759 (1970).
131. D. E. Arnold, M. J. Stansell, and H. H. Malkin, *Amer. J. Clin. Pathol.* **49**, 52 (1969).
132. W. G. Robertson and M. Peacock, *Clin. Chim. Acta* **20**, 315 (1968).
133. I. Oreskes, C. Hirsch, K. S. Douglas, and S. Kupfer, *Clin. Chim. Acta* **21**, 303 (1968).
134. D. S. Bernstein, M. A. Aliaponlions, R. S. Hattner, A. Wachman, and B. Rose, *Endocrinology*, **85**, 589 (1969).
135. J. Ruzicka and J. C. Tjell, *Anal. Chim. Acta* **47**, 475 (1969).
136. R. S. Hattner, J. W. Johnson, D. S. Bernstein, A. Wachman, and J. Brackman, *Clin. Chim. Acta* **28**, 67 (1970).
137. E. W. Moore, *J. Clin. Invest.* **49**, 318 (1970).
138. A. Raman, *Biochem. Med.* **3**, 369 (1970).
139. I. C. Radale, B. Hoffken, D. K. Parkinson, J. Sheepers, and A. Lackham, *Clin. Chem.* **17**, 1002 (1971).
140. E. W. Moore and A. L. Blum, *J. Clin. Invest.* **47**, 70a (1968).
141. E. W. Moore and G. M. Makhlouf, *Gastroenterology*, **55**, 465 (1968).
142. R. Kramer and H. Lagoni, *Naturwissenschaften* **56**, 36 (1969).
143. P. J. Muldoon and B. J. Liska, *J. Dairy Sci.* **52**, 460 (1969).

144. R. D. Allen, J. Hobley, and R. Carriere, *J. Ass. Offic. Agr. Chem.* **51**, 1177 (1968).
145. J. A. Blay and J. H. Ryland, *Anal. Lett.* **4**, 653 (1971).
146. J. S. Jacobs, R. S. Hattner, and D. S. Bernstein, *Clin. Chim. Acta* **31**, 467 (1971).
147. S. D. Hansen and L. Theodorsen, *Clin. Chim. Acta* **31**, 119 (1971).
148. H. D. Schwartz, B. C. McConville, and E. F. Christopherson, *Clin. Chim. Acta* **31**, 97 (1971).
149. J. A. Lott, *Crit. Rev. Anal. Chem.* **3**, 41 (1972).
150. C. Gavach, *C. R. Acad. Sci. Paris* **273C**, 489 (1971).
151. J. V. Leyendekkers and M. Whitfield, *Anal. Chem.* **43**, 322 (1971).
152. J. V. Leyendekkers and M. Whitfield, *J. Phys. Chem.* **75**, 957 (1971).
153. W. E. Morf and W. Simon, *Helv. Chim. Acta* **54**, 2683 (1971).
154. W. E. Morf, D. Ammann, E. Pretsch, and W. Simon, *Pure Appl. Chem.* **36**, 421 (1973).
155. D. Ammann, E. Pretsch, and W. Simon, *Tetrahedron Lett.* 2473 (1972).
156. D. Ammann, E. Pretsch, and W. Simon, *Helv. Chim. Acta* **56**, 1780 (1973).
157. W. Simon and W. E. Morf, *in* "Ion Selective Electrodes" (E. Pungor, ed.), p. 127. Akadémiai Kiado, Budapest, 1973.
158. D. Ammann, E. Pretsch, and W. Simon, *Anal. Lett.* **5**, 843 (1972).
159. G. A. Rechnitz and Z. F. Lin, *Anal. Lett.* **1**, 23 (1967).
160. P. Gabor-Klatsmanyi, K. Toth, and E. Pungor, *in* "Ion Selective Electrodes" (E. Pungor, ed.), p. 183. Akademiai Kiado, Budapest, 1973.
161. A. K. Covington and J. M. Thain, *Anal. Chim. Acta* **55**, 453 (1971).
162. S. Lal and G. D. Christian, *Anal. Chim. Acta* **52**, 41 (1970).
163. J. W. Ross, Jr. and M. S. Frant, *Anal. Chem.* **41**, 967 (1969).
164. A. G. Fogg, J. L. Kumar, and D. T. Burns, *Anal. Lett.* **7**, 629 (1974).
165. T. L. Malone and G. D. Christian, *Anal. Lett.* **7**, 33 (1974).
166. A. G. Fogg, A. A. Al-Sibaai, and C. Burgers, *Anal. Lett.* **8**, 129 (1975).
167. A. G. Fogg, M. Duzinkewycz, and A. S. Pathan, *Anal. Lett.* **6**, 1101 (1973).
168. R. N. Adams, *Anal. Chem.* **30**, 1576 (1958).
169. C. J. Coetzee and H. Freiser, *Anal. Chem.* **40**, 2071 (1968).
170. C. J. Coetzee and H. Freiser, *Anal. Chem.* **41**, 1128 (1969).
171. W. H. Haynes and J. H. Wagenknecht, *Anal. Lett.* **4**, 491 (1971).
171a. H. B. Hermans and G. A. Rechnitz, *Science* **184**, 1074 (1974); *Anal. Chim. Acta* **76**, 155 (1975).
171b. H. B. Hermans and G. A. Rechnitz, *Anal. Lett.* **8**, 147 (1975).
172. K. Srinivasan and G. A. Rechnitz, *Anal. Chem.* **41**, 1203 (1969).
173. A. Rouchouse and J. M. M. Porthault, *Anal. Chim. Acta*, **74**, 155 (1975).
174. M. Matsui and H. Freiser, *Anal. Lett.* **3**, 161 (1970).
175. H. J. James, G. P. Carmack, and H. Freiser, *Anal Chem.* **44**, 853 (1972).
176. S. Back, *Anal. Chem.* **44**, 1696 (1972).
177. S. Back, *Anal. Lett.* **4**, 793 (1971).
178. G. Schill, *Svensk. Kemi. Tidskrift* **80**, 10, (1966).
179. S. Lal and G. D. Christian, *Anal. Chem.* **43**, 410 (1971).
180. S. Lal, *Z. Anal. Chem.* **255**, 210 (1971).
181. J. Ruzicka and K. Rald, *Anal. Chim. Acta* **53**, 1 (1971).
182. P. Gavach and P. Seta, *Anal. Chim. Acta* **50**, 407 (1970).
183. P. Gavach and C. Bertrand, *Anal. Chim. Acta* **55**, 385 (1971).
184. B. J. Birch and D. E. Clarke, *Anal. Chim. Acta* **61**, 159 (1972).
185. B. J. Birch and D. E. Clarke, *Anal. Chim. Acta* **67**, 387 (1973).
186. B. J. Birch, D. E. Clarke, R. S. Lee, and J. Oakes, *Anal. Chim. Acta* **70**, 417 (1974).
187. N. Ishibashi, H. Kohara, and K. Horinouchi, *Talanta* **20**, 867 (1973).
188. A. L. Grekovich, E. A. Materova, and N. V. Garbuzova, *Zh. Anal. Khim.* **28**, 1206 (1973).

189. J. W. Ross, Jr., U. S. Patent, 3,483,112 (1969).
190. G. J. Moody and J. D. R. Thomas, "Selective Ion Sensitive Electrodes." Merrow Publ., Watford, Hertfordshire, England, 1971.
191. R. E. Reinsfelder and F. A. Schultz, *Anal. Chim. Acta* **65**, 425 (1973).
192. A. K. Covington, *Crit. Rev. Anal. Chem.* **3**, 355 (1974).
193. A. V. Gordievskii, Ya. A. Syrchenkov, V. V. Sergievskii, and N. I. Savvin, *Elektrokhimiya* **8**, 520 (1972).
194. S. Potterton and W. D. Shults, *Anal. Lett.* **1**, 11 (1967).
195. J. Knoeck, *Anal. Chem.* **41**, 2069 (1969).
196. J. DiGregorio and M. D. Morris, *Anal. Lett.* **1**, 811 (1968); *Anal. Chem.* **42**, 94 (1970).
197. J. M. Bremmer, L. G. Bundy, and A. S. Agarwal, *Anal. Lett.* **1**, 837 (1968).
198. R. J. K. Myers and E. A. Paul, *Can. J. Soil Sci.* **48**, 369 (1968).
199. M. K. Mahendrappa, *Soil Sci.* **108**, 132 (1969).
200. A. Øien and A. R. Selmer-Olsen, *Analyst* **94**, 888 (1969).
201. A. R. Mack and R. B. Sanderson, *Can. J. Soil Sci.* **51**, 95 (1971).
202. W. C. Dahnke, *Commun. Soil Sci. Plant Anal.* **2**, 73 (1971).
203. P. J. Milham, A. S. Awad, R. E. Paull, and T. H. Bull, *Analyst* **95**, 751 (1970).
204. J. Paul and R. M. Carlson, *J. Agr. Food Chem.* **16**, 766 (1968).
205. A. S. Baker and R. Smith, *J. Agr. Food Chem.* **17**, 1284 (1969).
206. P. J. Milham, *Analyst* **95**, 758 (1970).
207. A. S. Baker, N. H. Peck, and G. E. MacDonald, *Agron. J.* **63**, 126 (1971).
208. N. G. Bunton and N. T. Crosby, *Water Treat. Exam.* **18**, 338 (1969).
209. D. R. Keeny, B. H. Byrnes, and J. J. Genson, *Analyst* **95**, 383 (1970).
210. D. Langmuir and R. L. Jacobson, *Environ. Sci. Technol.* **4**, 834 (1970).
211. F. A. Schultz and D. E. Mathis, *Anal. Chem.* **46**, 2253 (1974).
212. S. E. Manahan, *Appl. Microbiol.* **18**, 479 (1969).
213. C. C. Westcott, *Food Technol.* **25**, 709 (1971).
214. P. M. Chalk and D. R. Keeney, *Nature (London)* **229**, 42 (1971).
215. J. M. C. Ridden, R. R. Barefoot, and J. G. Roy, *Anal. Chem.* **43**, 1109 (1971).
216. D. G. Gehring, W. A. Dippel, and R. S. Boucher, *Anal. Chem.* **42**, 1686 (1970).
217. G. P. Morie, C. J. Ledford, and C. A. Glover, *Anal. Chim. Acta* **60**, 397 (1972).
218. R. DiMartini, *Anal. Chem.* **42**, 1102 (1970).
219. A. Hulanicki, R. Lewandowski, and M. Maj, *Anal. Chim. Acta* **69**, 409 (1974).
220. T. M. Hseu and G. A. Rechnitz, *Anal. Lett.* **1**, 629 (1968).
221. R. J. Baczuk and R. J. Dubois, *Anal. Chem.* **40**, 685 (1968).
222. R. F. Hirsch and J. D. Portock, *Anal. Lett.* **2**, 295 (1969).
223. N. Ishibashi and H. Kohara, *Anal. Lett.* **4**, 785 (1971).
224. A. G. Fogg, A. S. Pathan, and D. T. Burns, *Anal. Chim. Acta* **73**, 220 (1974).
225. M. Sharp, *Anal. Chim. Acta* **65**, 405 (1973).
226. M. Sharp, *Anal. Chim. Acta* **62**, 385 (1972).
227. M. J. Smith and S. E. Manahan, *Anal. Chim. Acta* **48**, 315 (1969).
228. E. Dubini-Paglia, T. Mussini, and R. Galli, *Z. Naturforsch.* **26A**, 154 (1971).
229. T. Mussini, R. Galli, and E. Dubini-Paglia, *J. C. S. Faraday I* **68**, 1322 (1972).
230. I. Nagelberg, L. I. Braddock, and G. J. Barbero, *Science* **166**, 1403 (1969).
231. G. G. Guilbault and P. J. Brignac, Jr., *Anal. Chim. Acta* **56**, 139 (1971).
232. A. G. Fogg, A. S. Pathan, and D. T. Burns, *Anal. Lett.* **7**, 545 (1974).
233. R. M. Carlson and J. L. Paul, *Anal. Chem.* **40**, 1292 (1968); *Soil Sci.* **108**, 266 (1969).
234. H. E. Wilde, *Anal. Chem.* **45**, 1526 (1973).
235. P. Lanza and P. L. Buldini, *Anal. Chim. Acta* **75**, 149 (1975).
236. T. Stworzewicz, J. Czapkiewicz, and M. Leszko, *in* "Ion Selective Electrodes" (E. Pungor, ed.), p. 259. Akademiai Kiado, Budapest, 1973.

PART III

Glass and Other Related Membrane Electrodes

Chapter 9

GLASS MEMBRANE ELECTRODES

The most familiar glass membrane electrode is the pH type. Its potentiometric behavior can be described, despite some ambiguities in the ultimate definition of pH, by the Nernst equation for a wide range of hydrogen ion activities. Certain errors in the response of the glass electrode (e.g., alkaline error—deviation from the Nernst equation at high pH) gave impetus to the search for glass electrodes that would be responsive to other ions. The theoretical and experimental work done in this area has been summarized in a number of excellent articles[1–6] and monographs.[7–10] A summary of the work that has appeared in recent years and other material that is in consonance with the format of other chapters in the book is presented in this chapter.

A. STRUCTURE AND RESISTANCE OF GLASSES

Glasses are generally described as supercooled liquids or solids. Many oxides, selenides, sulfides, halides, and a few mixed carbonates and nitrates form glasses.[11] Little work on nonoxide glasses exists. The principal oxides used in the formation of glasses are SiO_2, B_2O_3, P_2O_5, As_2O_3, and GeO_2. Although others such as alkali oxides, alkaline earth oxides, high-valence oxides, and alumina, do not by themselves form glasses, they are used with the aforementioned oxides to form useful glasses. Silica forms a major constituent of the commercial glasses that have been used in the construction of glass electrodes.

The structure of glasses is governed mostly by short-range forces. A random network theory due to Zachariasen[12] is considered to hold for silicate glasses. According to this theory, four oxygen atoms form a tetrahedron, at the center of which is a silicon atom. In the assembly of silicon and oxygen atoms, there is only short-range order. The arrangement of the tetrahedra is not regular, however, as it would be in a crystal. The oxygen atoms connecting the silicon atoms are called bridging oxygens (I). When there is an univalent cation M^+, a nonbridging negatively charged oxygen is associated with it (II). There are two nonbridging oxygens in the presence of an alkaline earth cation. The cation fits into the interstices in the network adjacent to the nonbridging oxygen ion. Silicon can be replaced in the network by aluminum, but the excess negative charge of the AlO_4^- group must be neutralized, for example, by an alkali metal cation.

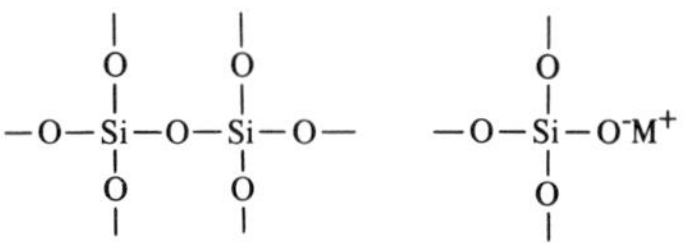

$$\begin{array}{ccc} & | \quad\; | & | \\ & O \quad\; O & O \\ & | \quad\; | & | \\ & -O-Si-O-Si-O- & -O-Si-O^-M^+ \\ & | \quad\; | & | \\ & O \quad\; O & O \\ & | \quad\; | & | \\ & (I) & (II) \end{array}$$

It was shown by Zachariasen[12] that, in general, a glass will be formed from an oxide only when the cation has a valency of 3 or more and if the ratio of cation to anion radii is small. Cations form coordinate links with three, four, or six oxygens and the oxygens are linked to no more than two cations. These conditions facilitate formation of a network structure. Accordingly the glass-forming oxides listed by Zachariasen[12] are B_2O_3, SiO_2, GeO_2, P_2O_5, As_2O_5, P_2O_3, As_2O_3, Sb_2O_3, V_2O_5, Sb_2O_5, and Ta_2O_5.

The glass compositions that may be used in the construction of glass membrane electrodes are limited by the electrical resistance of the glass membrane. It must be less than that of the measuring instrument and in general should not be larger than about 10^{12} Ω cm. Simple alkali silicate glasses are relatively highly conducting and the conductance increases with the alkali content. Substitution of divalent oxides decreases the conductivity markedly.[11] Also progressive substitution of B_2O_3 for SiO_2 in an alkali silicate glass lowers the conductance. Unlike these soda lime silica glasses in which conduction takes place by ions, glasses containing large amounts of V_2O_5 or Fe_2O_3 act as electronic conductors.

B. GLASSES FOR THE MEASUREMENT OF ION ACTIVITIES

The first glass to be prepared commercially for use in the manufacture of electrodes was Corning 015 of normal composition Na_2O (22%), CaO (6%), and SiO_2 (72%). This was arrived at during the period 1928–1929 through the work of Hughes,[13] Elder,[14] and MacInnes and Dole.[15] A similar glass was prepared by Sokolov and Passynsky.[16] These glass electrodes exhibit a Nernstian response to H^+ ion up to pH values of 11–12. The deviation from the Nernst equation depends on the kind and concentration of alkali metal cation present in the solution. Consequently, they proposed glasses of similar composition in which Na was replaced by Li or K. It was found that Li glass showed negligible deviation from a Nernstian response in 0.1 *M* KOH and a small deviation in 0.1 *M* NaOH.

Cary and Baxter[17] found that SrO, BaO, or PbO could be used in place of CaO. A BaO glass with the composition Li_2O (14.3%), BaO (7%), and SiO_2 (68.7%) was considered the best. Perley[18] studied the properties and stabilities of electrodes made from over 500 pH-sensitive glasses. The results of these studies showed that electrode glasses have at least three constituents—SiO_2, R_2O, and MO or M_2O_3 where R is an alkali metal and M is a bi- or trivalent metal. R_2O may be a mixture of two alkali metal oxides, and MO is often CaO, BaO, or SrO. Lanthanum oxide is considered by Perley to be the most effective of the M_2O_3 constituents. The limits of composition are SiO_2, 60–75 mole %; R_2O, 17–32 mole %; and MO or M_2O_3, 3–16 mole %. With respect to alkaline errors, lithium oxide glasses are superior to sodium oxide glasses. However, Li glasses have a higher resistance than soda glasses of similar composition. Glasses with different MO or M_2O_3 constituents show that large cations (Ba and Sr) are more effective than smaller ones (Ca, Mg, Be) in extending an ideal pH response in highly alkaline solutions. But replacement of Ca by Ba or Sr in glasses raises their electrical resistance. According to Perley, a suitable composition is SiO_2 (65 mole %), Li_2O (28 mole %), Cs_2O (3 mole %), and La_2O_3 (4 mole %).

The alkaline error of the glass electrode is caused mainly by its selectivity to the alkali metal ions. This led Lengyel and Blum[19] to investigate glasses based mainly on aluminum silicates and boron silicates which showed selectivity to alkali metal ions over a wide pH range. Eisenman and co-workers[20] initiated a thorough investigation of the electrode properties of sodium aluminum silicate glasses over a wide range of glass compositions. They demonstrated for the first time that the selectivity for different cations varied systematically with the glass composition. Eisenman's[21, 22] recommendations for the preparation of glass membrane electrodes selective to various cations are outlined in the following paragraphs.

The best glass for sensing the Li^+ ion in the presence of H^+ and Na^+ ions is LAS 15-25 (Li_2O = 15 mole %; Al_2O_3 = 25 mole %; SiO_2 = 60 mole %). The composition of NAS 11-18 (Na_2O = 11 mole %, Al_2O_3 = 18 mole %; SiO_2 = 71 mole %) gives the best Na electrode; its selectivity $K^{pot}_{Na\text{-}K}$ is 10^{-3} at high pH and 3.3×10^{-3} at neutral pH. The potassium-selective electrode is formed from NAS 27-4. Other alternatives recommended are KAS or RAS 20-5. If K is not an important contaminant, any K-selective electrode may be used for the determination of the Rb^+ ion. NAS 27.2-7.8 is recommended to be reasonably selective to Rb^+ or Cs^+ ions.

Glass electrodes selective to Ag^+ ions in relation to H^+ ions have the composition NAS 28.8-19.1, although for many applications NAS 11-18 is considered appropriate. For measurements of Tl^+ ions KAS 20-5 or KAS 20-4 glass is recommended. Since Tl^+ is similar in effect to that of the Rb^+ ion, compositions recommended for Rb^+ ions can be used for Tl^+.

All the present glass electrodes respond to the NH_4^+ ion. A typical glass selective to ammonium is NAS 27-3 + 3 mole % ZnO. If Na^+ and H^+ ions are the principal contaminants and the K^+ ion concentration is low, any K-selective electrode may be used to measure the NH_4^+ ion.

The aluminosilicate glass electrodes respond reproducibly to divalent cations. However, their selectivity to any particular divalent cation is not good. Most of these electrodes display several alkaline earth selectivity orders as a systematic function of glass composition.[23] Most of the alkali

TABLE 1

Composition (wt %) and Selectivity Constants (K_{ij}) of Some Divalent Cation-Sensitive Glasses[a]

Component	Tendeloo and Voorspuij[24]	Truesdell and Pommer[25] (phosphate)	Thompson[26] (916-P)	Truesdell[27] (NG-2)	Truesdell[27] (NG-6)
		Composition of glasses			
SiO_2	70	3.1	50	76.4	71.6
Al_2O_3	10	5.6	25	12.7	13.5
Fe_2O_3		16.0		0.56	0.70
FeO				0.58	4.3
MgO				0.4	2.5
CaO	6	0.2		0.3	2.8
Na_2O		6.1	6	4.1	1.4
K_2O		0.09		4.57	2.45
H_2O		0.19		0.67	0.20
P_2O_5		66.7			
Li_2O	14				
Cs_2O			6		
BaO			13		
		Values of K_{ij}			
K_{Ca-Na}	~15	0.56	13	24	11
K_{Ca-K}	~50	0.83	50	87	43
K_{Ca-Li}	~5	—			
K_{Ca-Mg}		0.19		1.0	0.77

[a] Data from Truesdell and Christ.[23]

aluminosilicate glasses strongly prefer the K^+ ion relative to the Ca^{2+} ion. Values for K_{Ca-K} range from 10 to 1000. However, some of these electrodes that prefer the Ca^{2+} ion are strongly sensitive to the Na^+ ion. The values of K_{Ca-Na} range from 50 to 500. Other glasses used in the construction of electrodes selective to divalent ions are listed in Table 1.

The electrodes of Tendeloo and Voorspruij[24] gave nearly theoretical slopes for pure solutions of Ca^{2+}, Ba^{2+}, and Li^+ ions. Solutions of H^+ and K^+, however, gave slopes that were less than the theoretical value. The values of K_{ij} given in Table 1 indicate that the selectivity of these electrodes to Ca^{2+} ion is not good.

Unlike the lithium–calcium–aluminosilicate glasses that were used in the form of bulbs, the natural glasses NG-2, NG-6 and the synthetic glass 916-P were used as flat membranes ground into thin sections 0.1–0.2 mm thick. Although these electrodes gave Nernstian responses in pure solutions, the selectivities (see Table 1) are not favorable to divalent cations. The electrodes have greater responses to monovalent cations than to divalent cations. On the other hand, the phosphate glass has significant selectivity to alkaline earth ions, although it is not very durable. The phosphate glass electrode showed, among the divalent cations, the selectivity order $Ba^{2+} > Sr^{2+} > Ca^{2+} > 2Na^+ > Mg^{2+}$.

C. PROPERTIES OF GLASS ELECTRODES

Experiments conducted by Haugaard[28] and others[15, 29] indicate that a freshly blown 015 glass membrane, on conditioning in an aqueous solution, undergoes ion exchange in which Na^+ ions are removed from the glass surface by an equal number of H^+ ions from the aqueous solution. A swollen layer of hydrous silica, which is considered necessary for the proper functioning of the pH electrode,[1] exists on the surface of the conditioned glass membrane. Changes in the dimension of the glass surface can be followed with an interferometer. The studies show that in alkaline and acidic solutions in which departures from a linear emf–pH relationship are found, the thickness of the swollen layer is decreased. This change in thickness is considered to be caused by an attack of the silicon–oxygen network in the case of high-pH solutions and by dehydration in the case of strong acid solutions.[1]

The electrical resistance of the glass membrane electrode increases considerably when the electrode is allowed to dry,[30] and the pH response falls below the Nernst theoretical value. Since hydration and dehydration of the glass network are reversible, unless subject to exhaustive drying, no harm is done to the electrode by drying. The conditioning of the electrode for a few hours in water usually restores its response. If the electrode response is erratic or "dead," it can generally be revived by dipping the electrode in a dilute solution of HF. Electrodes with imperfect responses can be corrected by treating them with superheated water under pressure.[1] This indicates that the hygroscopicity of the glass is an important factor in controlling the response of the pH electrode. Nonhygroscopic glasses such as quartz produce little or no electrode response. Either too much or too little water in the glass is not desirable. The optimum water uptake is in the range 50–100 mg/cm^3. The water uptake of some glasses, as it relates to electrode function, is shown in Table 2. The data of Hubbard[31] for a number of electrodes constructed from powdered glass are given in Table 3.

It is interesting that glasses which are far less hygroscopic than Corning 015 give a Nernstian slope. This suggests that little correlation exists between the electrical resistance of glass membranes and electrode function. Nevertheless, electrode resistance is of importance in designing the related instrumentation.

An interesting study of the effect of sintering on porous glass membranes has been made by Altug and Hair,[32] who determined the selectivity of the sintered membrane to Na^+, K^+, and Ca^{2+} ions. These results are given in Table 4. Sintering has the greatest effect on the selectivity of the membrane to K^+ over Ca^{2+}. In another parallel study[33] with a potassium-selective glass membrane electrode that was immersed for six weeks in $NaHCO_3$ solution, a value of 10 for $K^{pot}_{Na\text{-}K}$ and a negligible response to Ca^{2+} were observed. On leaving the electrode for 17 months in the leaching solution, the response to divalent ions increased and the $K^{pot}_{Na\text{-}K}$ value decreased to 5.6. Porous glass had a value of $K^{pot}_{Ca\text{-}K} = 15$, whereas

TABLE 2

Hygroscopicity and pH Response of Some Glasses According to Rechnitz[3]

Glass composition	Water uptake (mg/cm^3)	Electrode response (mV/pH)
20% Na_2O–10% CaO–70% SiO_2	≈ 60	59
20% Na_2O–5% CaO–75% SiO_2	75	59
14% Na_2O–86% SiO_2	63	59
20% Na_2O–80% SiO_2	110	54
10% Na_2O–90% SiO_2	40	47
25% Na_2O–75% SiO_2	135	40
4% Na_2O–96% SiO_2	30	35
30% Na_2O–70% SiO_2	160	23
2% Na_2O–98% SiO_2	22	15
40% Na_2O–60% SiO_2	320	12
100% SiO_2 (molten)	< 10	≈ 0

TABLE 3

Hygroscopicity and pH Response of Some Glasses According to Hubbard[31]

Glass	Water uptake (mg/cm^3)	Electrode response (mV/pH)
Corning 015	358	59
Dish	88	58
American Ceramic Society No. 1	40	57
Window	39	56
Electric hygrometer	39	56
Blue bottle	30	54
BSC 517	5.5	43
F 620	4.8	39
Ba C572	2.2	33
Pyrex	1.8	18

TABLE 4

Ion Selectivity of Sintered Porous Glass Membranes

Sintering membrane	Temperature	Time (hr)	Selectivity		
			K_{Na-K}	K_{Ca-K}	K_{Ca-Na}
1	750	0.5	1.8	15	6
2	750	—	2.0	28	8
3	770	—	2.1	110	25
4	770	1	3.6	690	55
5	775	1.33	4.5	1780	63
6	820	—	7.1	8700	170
7	800	—	8	—	—

$K^{pot}_{Na-K} = 1.8$. On heating, the Ca ion selectivity of the membrane decreased (see also Table 4) and prior to complete sintering K^{pot}_{Na-K} had a value of 10. These facts, also shown in Table 4, indicate that the selectivity of K^+ over Na^+ was dependent on the critical structure of the porous layer (pore dimension 2–3 Å). If the layer was too porous or too rigid, there was little selectivity.

How long a glass electrode can be used (i.e., its durability) is determined by a number of factors such as composition of the glass and of the solution in which it is used, the thickness of the pH-sensitive membrane surface, and the temperature of the solution in which it is used.

In general, it has been noted that the chemical durability of the glass electrode is decreased sharply with an increase in temperature.[1] The increase in the rate of attack has been estimated to be about twofold for an increase of 10°C and about tenfold for an increase of 25°C. Under normal circumstances the electrode will give good service for one to two years.

The attack of glass by aqueous solution takes place by the sorption of water and its deep penetration into the layers of bulk glass. The corrosive attack by the alkaline solution formed by the extraction of the alkali constituents of the body of the glass also takes place from the body side of the electrode. Thus the silicon–oxygen network is affected from both sides of the membrane. Eventually cracks develop, resulting in an unstable response.

Water and acids attack the glass electrode to release ionically bound basic constituents and replace them with hydrogen ions. The water-soluble reaction products are also released, leaving a swollen layer of hydrous silica that will resist further attack by water or dilute acid. Consequently, keeping the glass membrane electrode immersed in water extends the life of the electrode.[34] Conversely, alkalis attack the silica oxygen network and prevent formation of the protective layer.

An important property of the glass membrane electrode is its chemical stability. A number of investigators have attempted to study this problem by following the release of alkali from glass by titrations or by observation of changes in the pH level. An increase in the conductance of water in which the electrode is immersed has been employed to follow the resistance of glass to chemical attack.[34] Even interferometry has been applied to study the effects of chemical reagents on rectangular specimens of glass.[35] The results showed that there was normal swelling due to hydration of the surface layer and the chemical attack was evident at pH levels above 9.

The properties of the surface layer have been intensively investigated in recent years. Wikby[36–39] has identified the gel layer as a separate phase between two moving fronts. The boundary between the gel layer and bulk glass moves inward toward the bulk glass at a velocity determined by the rate at which hydration takes place. The boundary between the solution and gel layer moves in the same direction as the dissolution of the gel layer. When the rates of movement of these two fronts become equal, the gel layer will have a constant thickness that is time invariant, but which depends on the composition of the glass. For a glass of given composition the thickness also depends on the rates of movement of the two fronts referred to above. In view of this different values for thickness would be obtained despite use of the most accurate method for its measurement. For H^+ ion-selective electrodes, values ranging from 25 to 2000 Å have been derived from interferometry,[40] tritium uptake,[41] and ion-sputtering methods.[42] Karlberg[43] quotes the calculations of Wikby which showed that the low-temperature electrodes (Ingold LOT) hydrated for 44 and 340 hr had a thickness of 900 and 3000 Å, respectively. Values of 150 and 400 Å have been derived for general-purpose electrodes (Ingold 201) hydrated for 30 and 340 hr, respectively. Karlberg,[43] in his experiments with an Ingold

LOT electrode leached for 300 hr, has derived a value of 3300 Å for the gel layer.

The concentration profiles of silicon and the alkali metal constituent of glass have been established for glasses containing sodium and lithium using the fractional etching technique. Csakvari *et al.*,[44] using glass of the composition shown in Table 5, dissolved the surface films of glass samples gradually in HF solution and estimated the amounts of alkali and silicon removed. From these data, some characteristic information on the structure was obtained. The concentrations of alkali ions and silicic acid obtained by just leaching the rod-shaped samples of glass with water were never larger than 10^{-5} and 10^{-6} mole/liter, respectively. Leaching with water developed the surface glass, which was treated with a 0.1% HF solution that was changed every 3–7 min until the entire surface layer was dissolved. From the amount of silicon removed, an estimate of the thickness of the gel layer was made. The minimum thickness of the layer fraction that could be detected by this method was 2×10^{-6} cm. In Fig. 1,

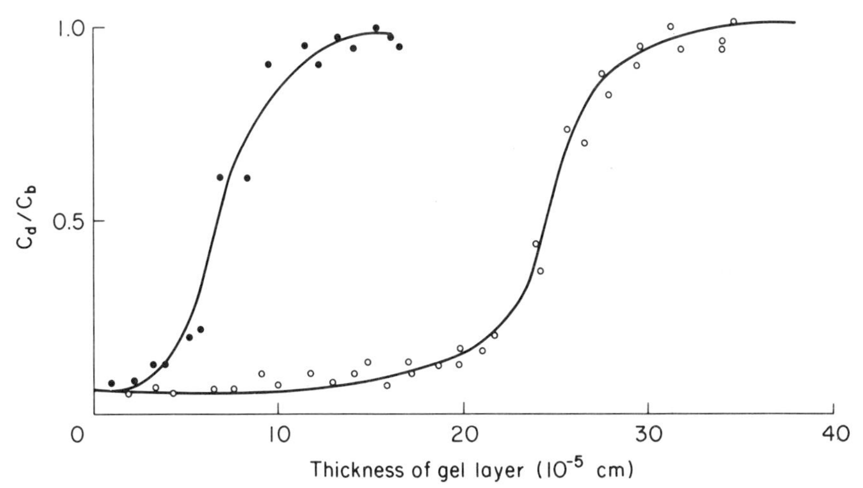

Fig. 1. Distribution of Na^+ ions in the surface layer of a MacInnes–Dole glass after 13 days' treatment in water: ● leached at 20°C; ○ leached at 40°C (after Csakvari *et al.*[44]).

TABLE 5

Composition of Na and Li Glasses

Glass	Ref.	Composition in mole percent						
		Na_2O	CaO	BaO	SiO_2	Li_2O	Cs_2O	La_2O_3
Dole glass	45	21.4	6.4	—	72.2	—		
Perley glass	18	—	—	—	65.0	28	3	4
Li–Ba glass	46	—	—	8	68	24	—	—

the concentration of Na (C_d) in the gel layer expressed as a fraction of the bulk Na concentration (C_b) of glass is plotted versus the thickness of the gel layer removed. The values of C_d/C_b increase from a very low value to the value that exists in the bulk glass. An increase in temperature hastens the formation of the surface layer. The results given in Fig. 2 show that the absence of hydrogen ions (commercial ethanol used for leaching) produces a very thin gel layer with probably little effect on its structure. The results obtained with Li glasses after they were leached in water at elevated temperatures (see Fig. 3) also show thin surface layers with "intact" structures. The low sodium error of electrodes made of Li glass is due to this retarded process of ion exchange between alkali ions in solution and hydrogen ions in glass.

The studies of Wikby on various glass membrane electrodes throw a great deal of light on their structures. He used a constant current pulse technique developed with Johansson[47] to follow the changes in the gel layer structure by following its resistance and time constant. The technique consisted in applying a 500 pA constant current (i) pulse directly to the glass electrode through a resistor and recording the voltage change as a function of time. The readings were taken over the range 100 μsec to several minutes. The voltage at any time t was subtracted from the steady state value at "infinite time." This difference in voltage was plotted against t on a semilogarithmic paper. A simple time constant gave a straight line. If several time constants were detected, they were resolved by evaluating the slowest component first and subtracting the values on its straight line from corresponding points on the earlier part of the graph. This procedure

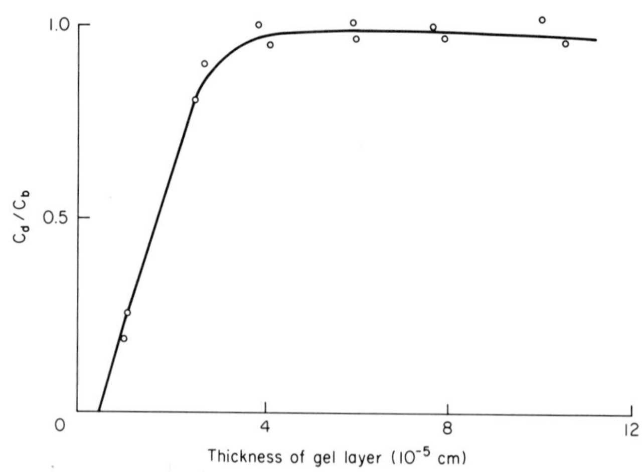

Fig. 2. Distribution of Na^+ ions in the surface layer of a MacInnes–Dole glass after 13 days' treatment in commercial ethanol at 40°C (after Csakvari *et al.*[44]).

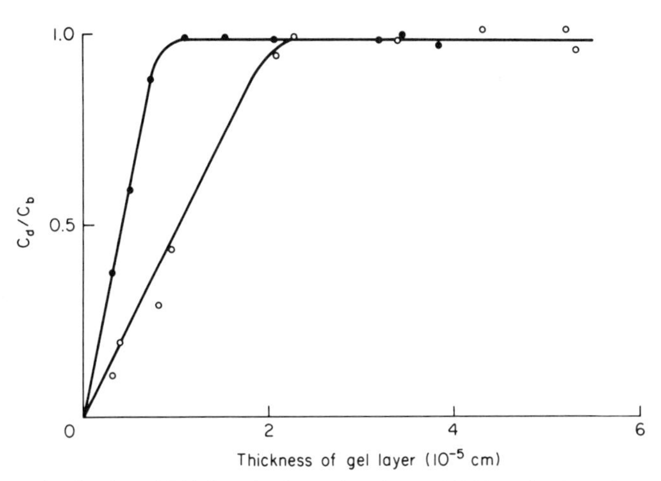

Fig. 3. Distribution of Li^+ ions in the surface layer of lithium–barium glass (○) after 13 days' treatment in water at 40°C, and of Perley glass (●) after 40 hr of treatment in water at 100°C (after Csakvari *et al.*[44]).

was repeated until all the time constants were resolved. From the value of the intercept on the ordinate corresponding to each time constant, a value for the resistance of that unit (i.e., V/i) was derived. The temperature dependences of both the time constant (τ) and the resistance (R) were also studied. From the plots of $\log(\tau)$ or $\log(R)$ vs. $1/T$, which were linear, values for the apparent activation energies were derived. Ingold electrode LOT 102 and Metrohm EA 107T were used in these studies after subjecting them to different periods of hydration in a phosphate buffer. The resolution of the voltage–time curves as described above gave four time constants and four resistances. The values obtained for these and the corresponding energies of activation are given in Table 6. These results show that there are two kinetic processes, one with a time constant in the millisecond range (fast) and the other (slower) with a time constant in the range of seconds. The fast process has a low activation energy (τ_1, τ_2, R_1, and R_2), whereas the slow process has a high activation energy (τ_3, τ_4, R_3, and R_4). The voltage–time curves of the LOT 102 electrode determined as a function of the period of hydration revealed that the fast process occurred with little change in resistance whereas the slow process occurred with considerable change in resistance. It was therefore considered that the fast process, involving τ_1, τ_2, R_1, and R_2, which was independent of hydration originated in the interior of the glass and the slow process, involving τ_3, τ_4, R_3, and R_4, originated from the hydrated surface layer. In order to confirm this, three electrodes of the same glass (Ingold LOT 102) were hydrated and their resistances and time constants were evaluated in terms of the thickness of the glass membrane of the electrode. The thicknesses of each electrode both at the bottom and the side were measured with the help of a microscope. The difference in the microscope readings when focused on the outer and the inner surfaces of the glass membrane multiplied by the refractive index gave the thickness. The membrane parameters are given in Table 7. The ratios pertaining to the resistances R_1 and R_2 are equal to the ratio of thicknesses measured around the circumference of the membrane bulb. The overall resistance R_{tot} is not proportional to the thickness and the ratios pertaining to this parameter change with temperature. As the temperature is raised the value of the ratio increases and at 45°C this ratio becomes 4.2. On the other hand, the ratios pertaining to the time constants are all unity at all temperatures. This was considered to prove conclusively that the quantities R_1, R_2, τ_1, and τ_2 pertain to the bulk glass and that R_3, R_4, τ_3, and τ_4 are parameters of the gel surface. Both the surface resistance R_s and glass body resistance R_g, which were followed as a function of current in the range 500 pA to 10 nA, showed that Ohm's law was obeyed. The change in composition of the aqueous solution had little effect on R_s.

Membrane resistance measurements have been extended to monitor the changes in the gel layer following treatment of the glass electrodes in acidic,[36] neutral,[37] and alkaline solutions.[39] The important conclusions of these studies are summarized in the following paragraphs.

TABLE 6

Resistance and Time Characteristics of Metrohm and Ingold Electrodes

Component	Resistance (MΩ)[b] EA 107T	ΔH from R vs. $1/T$ (kcal/mole)[a] EA 107T	ΔH from R vs. $1/T$ (kcal/mole)[a] LOT 102	Time constant	τ(m sec)[b] EA 107T	ΔH from τ vs $1/T$ (kcal/mole)[a] EA 107T	ΔH from τ vs $1/T$ (kcal/mole)[a] LOT 102
				τ_1	3.6	15.5	15.4
				τ_2	16.1	16.1	15.4
				τ_3	2.8×10^3	19.6	29.4
				τ_4	18.0×10^3	21.5	29.7
R_1	6.3	15.7	15.1				
R_2	12.4	16.6	15.3				
R_3	3.5	19.8	23.8				
R_4	2.7	27.4	23.8				
R_g	19.0	16.3	15.2				
R_s	6.4	24.7[c]	24.2				
R_{tot}	25.4	18.8[c]	17.8[c]				

[a]Values for ΔH derived for the temperature range 10–35°C.
[b]These values refer to 24.6°C.
[c]Plot of R vs. $1/T$ nonlinear.

TABLE 7

Resistances, Time Constants, and Thicknesses of Ingold Electrodes

	Temperature (°C)	Electrode			Ratio	
		Thin	Normal 1	Normal 2	(Normal 1)/Thin	(Normal 2)/Thin
Thickness (μm)						
Bottom		65	215	222	3.3	3.4
Side		89	383	367	4.3	4.1
Resistance (MΩ)						
R_1	15.2	7.1	29.5	28.2	4.2	4.0
	21.7	3.9	16.5	16.8	4.2	4.3
R_2	15.2	3.6	15.3	15.5	4.3	4.3
	26.7	2.0	8.4	8.1	4.2	4.2
R_{tot}	15.2	36.0	66.0	71.0	1.8	2.0
	21.7	15.5	33.0	36.0	2.1	2.3
Time constant (msec)						
τ_1	15.2	5.79	5.60	5.79	0.97	1.00
	21.7	3.13	3.03	3.09	0.97	0.99
τ_2	15.2	23.8	26.7	23.6	1.12	0.99
	21.7	14.7	14.7	14.9	1.00	1.01

The Ingold electrodes (LOT and 201) investigated showed that etching the hydrated electrode for 2 min in 5% aqueous HF eliminated the surface resistance because of the removal of the surface layers. Surface resistance, as pointed out already, is considered to arise at the phase boundary between the gel layer and the glass body when the electrode behaves ideally. In the range of acid error, an outer film is built in the gel layer which forces the resistance to go up. Hydrochloric acid in both water and isopropanol solvents allowed the surface resistance to rise. On the other hand, hydrofluoric acid eliminated the surface resistance, but in low concentrations had little effect on the surface resistance in water. In isopropanol the increase in surface resistance due to HF treatment was enormous on hydrated electrodes and moderate on etched electrodes. The acid error due to HCl was negligible for the hydrated 201 electrode, small for the etched LOT electrode, and large for the hydrated LOT electrode.

Similar conclusions with respect to the location of the active boundary layer were reached when surface resistance and destruction and dissolution of both Li and Si were studied simultaneously.[37] The concentration of Li ions in the gel layer was found to be about 10^{-2} times that in the bulk glass, but at the boundary between the gel layer and the bulk glass, it rose sharply to a value characteristic of the bulk glass. Surface resistance is considered to be located within this inner boundary. Measurements of both surface resistance and outflux of Li ions from the glass phase (leaching experiments) at different temperatures gave similar activation energies ($\sim$25 kcal/mole) both for conduction and outflux and so have been correlated by the relation

$$R_s = K \frac{1}{d(\mathrm{Li})/dt} + K_0$$

where K and K_0 had values of 6.8×10^{-3} (V mole/A-hr) and -0.01 (MΩ), respectively, for the LOT electrode and 6.2×10^{-3} (V mole/A-hr) and -0.015 (MΩ) for the 201 electrode.

In alkaline solutions the situation described above will be drastically altered because, as pointed out already, the alkaline attack increases rapidly above pH 9. Owing to the reduced concentration of hydrogen ions, the inner boundary between the gel layer and the bulk glass would move slowly. Accordingly, Douglas and El-Shamy[48] found that the rate of dissolution of Na^+ and K^+ ions from the glass decreased only slightly when the pH was varied from 2 to 9, and above 9 the rate decreased rapidly with increasing pH. The break at pH 9 was attributed to the paucity of hydrogen ions. Similarly, Bach and Baucke[42] found that the lithium glass hydrated for six days at pH 1 had a thicker gel layer (300 Å) than those hydrated for six days at pH 9. Also the alkali ions present in solution at very high pH will replace some of the hydrogen ions in the gel layer, which in turn will affect the resistance of the gel layer thickness. When the electrode behaves ideally, the surface resistance increases rapidly with storage time at higher pH values. This, as already surmised, is due to insufficient supply of hydrogen ions leading to retardation of movement of the gel layer–bulk glass boundary toward the dry glass. This effect is similar to the effect of transferring a hydrated electrode from distilled water into isopropanol.[38] In this case, the gel layer of the glass electrode (LOT) stopped increasing in thickness.

An increase of the alkali metal ion concentration (Na^+ or K^+) also increases the resistance of the gel layer. The dissolution rate of silicon from the silica network was higher in alkaline solutions than in neutral solutions. When there is an alkaline error the boundary between the gel layer and the glass bulk moves considerably slower than it does in neutral pH. On the other hand, the boundary between the gel layer and the solution will move more quickly because of dissolution. Consequently, the thickness of the gel layer will decrease with the time of storage in alkaline solutions. From this account it is evident that the ion exchange properties of the gel layer control the response of the glass electrode. It is difficult to study ion exchange in the gel layer without using highly basic solutions which chemically affect the gel layer, so Karlberg[43] resorted to using basic nonaqueous solutions to inhibit the dissolution and hydration of the gel layer.

To exchange H^+ ions in the gel layer with alkali metal ions, the glass bulb of the electrode was immersed for specific periods of time in a basic isopropanol buffer solution containing the concerned alkali metal perchlorate (0.02 *M* diisopropylamine solution, 5% of which titrated with $HClO_4$ + 0.01 *M* Na or $LiClO_4$). The electrode was removed and rinsed with isopropanol repeatedly. Transferring the electrode into 0.02 *M* $HClO_4$ for known periods of time allowed the alkali metal to be removed into the acid solution, and the amount removed was then estimated. The extent of ion exchange by some of the electrodes is given in Table 8.

TABLE 8

Ion Exchange Yield Obtained from Some H^+ Ion-Selective Glass Electrodes

Glass electrode	Alkali metal ion	Ion exchange yield (moles)	Estimated area of bulb (cm^2)
Low-temperature electrodes			
Beckman	Na^+	1×10^{-8}	0.8
Ingold LOT (No. 1)	Na^+	4.7×10^{-8}	2.4
Ingold LOT (No. 2)	Na^+	7.6×10^{-8}	2.4
Ingold LOT (No. 1)	Li^+	1.3×10^{-8}	2.4
General-purpose electrodes			
Ingold 201	Na^+	1.5×10^{-8}	3.0
Metrohm (No. 1)	Na^+	3.8×10^{-8}	3.6
Metrohm (No. 2)	Na^+	8.4×10^{-8}	4.4
Metrohm (No. 2)	Li^+	3.2×10^{-8}	4.4
High-temperature electrodes			
Ingold HA	Na^+	0.7×10^{-8}	3.4

The amount exchanged is a function of the acidity of the solution used. Ingold LOT electrode No. 3 gave 3.92×10^{-8}, 3.76×10^{-8}, 3.37×10^{-8}, 2.62×10^{-8}, and 1.59×10^{-8} mole of Na^+ when the acidity, expressed as $\log(p/(100 - p))$ (where p is the percentage of base titrated with $HClO_4$), was -1.0, -0.5, 0, 0.5, and 1.0, respectively. Similar values obtained for Metrohm electrodes ranged from 2.9×10^{-8} to 0.97×10^{-8} mole of Na^+. It was also found that the ion exchange process was faster when the H^+ ions in the gel layer were exchanged for Na^+ ions than when the reverse process occurred.

From the data given in Table 8 the exchange capacity of the gel layer is of the order of 10^{-8} mole/cm^2 of glass bulb area. It has been quoted[43] that the amount of Li ions leached from an Ingold LOT No. 2 electrode during 300 hr of hydration is 2×10^{-6} mole. Consequently, the exchange that takes place in a few hours involves only a few percent of the sites in the gel layer. Taking the value of 2×10^{-6} mole of Li as corresponding to 2.4 cm^2 bulk area and other related data (Ingold LOT glass contains 7% by weight of Li, and the density of the gel layer is 2.5 g/cm^3), a value of 3300 Å for the thickness of the gel area may be derived on the assumption that there is little Si dissolution involved.

D. ELECTRICAL IMPEDANCE OF GLASS ELECTRODES

As already pointed out, the pH response of the glass electrode has no bearing on the electrical resistance of the electrode although both properties are governed by the composition of the glass and to some extent by its thickness. The dc resistances of commercial glass electrodes at room temperature are generally in the range 5–500 MΩ. Reduction in this resistance below 1 MΩ will affect the durability of the electrode.

The ac resistance of the glass electrode depends on the frequency used in its measurements. Corning 015 glass electrode had an ac resistance of 2.55 MΩ at a frequency of 1020 Hz, and 1.27 MΩ at 3380 Hz, whereas its dc resistance was 81 MΩ.[49] The conductance of glass is low, so it acts primarily as a dielectric having considerable polarization capacity (ability to set up a back-emf).

The early model of a glass membrane electrode chosen is a simple electrical analog (Fig. 4a or 4b). In Fig. 4, C_p is the capacitance due to polarization and C_g is that due to static charges on the glass surface. If the resistance of the electrode is too high, a time lag in the response attributable to RC arises. Impedance studies of glass membrane electrodes involving network analysis have been conducted both theoretically and experimentally by Buck[50–52] who proposed the equivalent circuit shown in Fig. 5

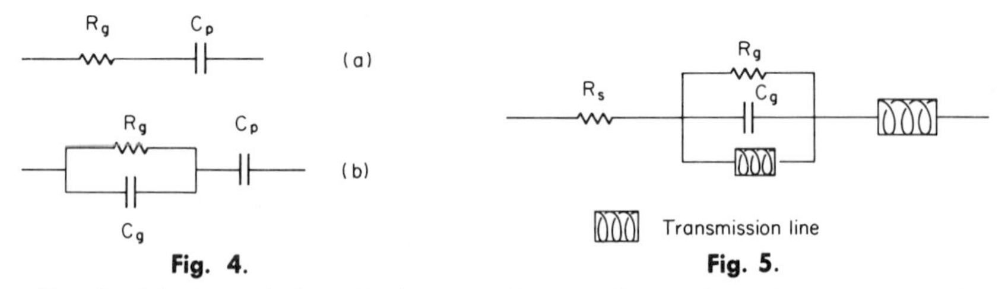

Fig. 4. **Fig. 5.**

Fig. 4. Simple equivalent circuit for a glass membrane electrode. (a) R_g, the glass membrane resistance is in series with the polarization capacity of the electrode C_p. (b) The glass membrane capacitance C_g is in parallel with R_g.

Fig. 5. Equivalent circuit for a glass membrane electrode due to Buck.[50] R_s is the series resistance due to the reference electrode and solutions. The surface film on the glass membrane electrode of resistance R_g and parallel capacitance C_g with diffusional Warburg impedance has been considered equivalent to a finite transmission line.

for a glass membrane with hydrolyzed surface film. The surface film in Fig. 5 is represented as a finite transmission line and the glass membrane as a parallel array of resistance, double layer capacitance, and Warburg diffusional impedance. This model was based on impedance measurements[52] of a variety of pH-sensitive and cation-sensitive glass electrodes. The impedance characteristics displayed by the various glass membranes followed the theoretical predictions of the model given in Fig. 5. The characteristics were the following: log Z_r (real part of impedance) and log Z_i (imaginary part of impedance) versus log frequency declined continually with an increase in frequency, with Z_i passing through a maximum. The intersection of the Z_i and Z_r curves at a frequency greater than that relating to $Z_{i(max)}$ indicated the presence of the transmission line at frequencies greater than that of $Z_{i(max)}$, Z_i declined as $1/\omega$, whereas Z_r declined as $1/\omega^{3/2}$ (where ω is the angular frequency given by $2\pi f$ and f is frequency in hertz). The shape of the plot of Z_i vs. Z_r corresponding to different frequencies was similar to a Cole–Cole plot.[53] A skewed semicircle with center below the Z_r axis, typical of all glass membranes, has been noted. The existence of a gel layer on the glass electrode is reflected in the appearance of a second distorted semicircle or a part of one at lower frequencies.

Values of the time constant τ, resistance, and capacitance have been calculated for the seven electrodes used in the study. The time constant τ was computed from the intersection point of the plots of log Z_i vs. log f and log Z_r vs. log f. The intersection point gives the value of the frequency at which $Z_r = Z_i$. For this condition $\tau = 1/\omega$. The resistance of the glass membrane, which was approximate, was obtained in the case of an ideal pH electrode from the constant value of the impedance at low frequencies. For the more permeable glasses, the resistance was estimated from the values of impedance in the vicinity of the inflection point of log Z vs. log f plots or by extrapolation of the low-frequency semicircle in the Cole–Cole plots to cut the real axis. Since $\tau = R_g C_g$, values for C_g may be obtained. Some of the values derived for the seven glass electrodes by Buck and Krull[52] are given in Table 9. Brand and Rechnitz also carried out impedance measurements using both liquid membrane[54] and glass membrane[55] electrodes. Their reexamination of the impedance properties of glass electrodes showed that at the highest frequencies used, the real part of the impedance Z_r of each electrode reached a finite value of about 10 kΩ and the imaginary part Z_i tended toward zero. The Cole–Cole plot of the monovalent cation electrodes as described above was a distorted arc of a circle having the center below the real axis and resembled the one determined earlier for liquid membrane electrodes.[54] This behavior was also shown by the pH- and the sodium-selective electrodes.[55] In addition,

TABLE 9

Values of τ, R_g, and C_g for Some Glass Membrane Electrodes as a Function of Temperature

Time constant τ (msec)	Glass resistance R_g (MΩ)	Capacitance (pF)	Temperature (°C)	Electrode area (two surfaces) (cm^2)	Average capacitance (pF units/cm^2)
Beckman Instruments E-2 Glass Electrode No. 41260					
320	830	385	15		
120	306	397	25	6.0	65
44	112	395	35		
Corning Glass Works Electrode No. 476020					
100	438	228	15		
37	173	214	25	3.5	63
14.5	67	216	35		
Leeds and Northrup Black Dot Electrode No. 117235					
210	592	358	15		
89	240	370	25	6.2	59
318	90	354	35		
Radiometer Type G202B Electrode					
88	567	156	15		
35	235	150	25	4.3	37
14.4	88	164	35		
Beckman Instruments General Purpose Glass Electrode No. 41263					
8.6	95	89	15		
3.6	42.5	85	25	1.81	47
1.45	17.5	83	35		
Beckman Instruments Sodium-Sensitive Electrode No. 39278					
9.95	53.7	186	15		
4.3	22	195	25	5.9	32
1.7	9.4	181	35		
Beckman Instruments Cation-Sensitive Electrode No. 39137					
4.5	20	222	15		
1.8	8.0	225	25	5.8	40
0.78	3.3	236	35		

at lower frequencies a second distorted circular arc was observed which was most pronounced for the pH electrode. This, as already described, indicated the presence of a surface film (gel layer) on the glass. Such a film was not present to an appreciable extent on the monovalent cation glass electrode. If the gel layer is not present, extrapolation to high frequencies of the arc to cut the real axis gives the value for R_s, the series resistance due to reference electrode and solutions. If Z_g is the impedance of the intact glass (without gel layer), then

$$Z_g = (Z_r - R_s) - jZ_i \tag{1}$$

where $j = \sqrt{-1}$. Similarly, if Z_f is the impedance of the film, then

$$Z_f = (Z_r - R_s') - jZ_i \tag{2}$$

where R_s' is the intercept on the real axis obtained by extrapolation to high frequencies of the second arc due to the film or gel layer.

Brand and Rechnitz[55] transformed the complex impedance of the film Z_f to the complex admittance plane and found that it resulted in a straight line admittance locus for both the pH and sodium glass electrodes. The following equations were derived for these straight lines:

$$Y_i = 1.02\,Y_r - 3.2 \times 10^{-3} \quad \text{for the pH electrode} \tag{3}$$

$$Y_i = 0.99\,Y_r - 8.9 \times 10^{-5} \quad \text{for the Na electrode} \tag{4}$$

where Y_r and Y_i (in units of $10^{-6}\ \Omega^{-1}$) are the real and imaginary admittances. The intercept (i.e. Y_r' when $Y_i = 0$) on the real admittance axis gave the reciprocal of the parallel resistance component of the film impedance. Hence the values derived for the film resistances of the pH and Na electrodes were 318 and 11,000 MΩ, respectively. The slope of the complex admittance plot is unity for the pH and Na electrodes, which is typical of a Warburg impedance. That this Warburg impedance is in parallel with the film resistance R_f has been shown by the linear dependence of $Y_r - Y_r'$ on the square root of the frequency for both pH and Na electrodes. The linear equations of best fit were

$$(Y_r - Y_r') = 0.0519\omega^{1/2} + 3.33 \times 10^{-4} \quad \text{for the pH electrode} \tag{5}$$

$$(Y_r - Y_r') = 0.0679\omega^{1/2} + 2.28 \times 10^{-3} \quad \text{for the Na electrode} \tag{6}$$

A similar analysis of the results obtained by the transformation of the complex impedance Z_g of the intact glass into the complex admittance plane showed that the admittance at low frequencies was directly proportional to frequency, a property typical of a capacitance. However, at frequencies greater than 20 kHz, the admittance of intact glass deviated from linearity. Further, the complex admittance plane plot of Y_r vs. Y_i gave evidence of the presence of a frequency-dependent resistance as well. The equivalent circuit for the glass membrane electrode proposed by Brand and Rechnitz[55] is shown in Fig. 6.

In the equivalent circuit proposed by Buck[51] the surface film (gel layer) was equated to a finite transmission line with a phase angle of 45° at high frequencies and a decrease in the phase angle as the frequency decreased.

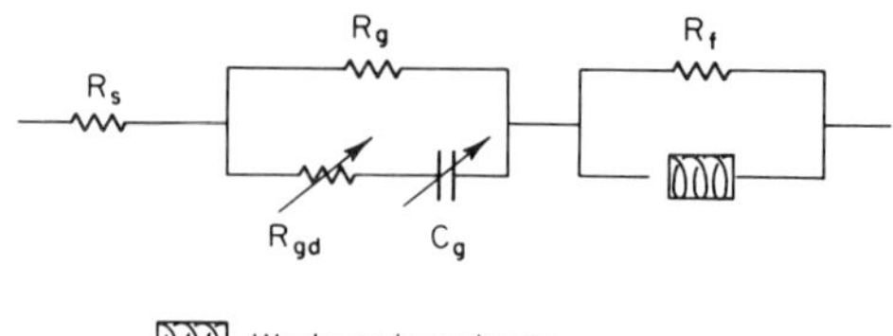

Fig. 6. Equivalent circuit for a glass membrane electrode due to Brand and Rechnitz.[55] R_s is the series resistance due to the reference electrode and solutions. The film resistance R_f is in parallel with the Warburg impedance. R_{gd} is the frequency-dependent resistance and C_g is the frequency-dependent membrane capacity.

Brand and Rechnitz,[56] using the appropriate equations, provided numerical solutions and constructed a complex impedance plane locus for a Warburg impedance (infinite transmission line) in parallel with a fixed resistor. The shapes of these were similar to those given by Buck but the maximum value of Z_i/R (where R is the low-frequency resistance of the film) was 0.206 (characteristic of a semi-infinite transmission line) as opposed to 0.417 (which is characteristic of a finite transmission line) obtained by Buck. Furthermore, the experimental data[55] obtained for the pH electrode agreed with those calculated for the circuit with Warburg impedance in parallel with a fixed resistor (right side of Fig. 6). However, with improved measurements of impedance in the low-frequency range Sandifer and Buck[57] showed that the low-frequency impedance response of glass electrodes (Beckman E2 and General Purpose) did not obey the simple finite Warburg model (see Fig. 5) but rather obeyed an empirical relation based on relaxation phenomena in dielectrics due to Cole and Cole.[53]

E. GLASS MEMBRANE POTENTIALS

In all of the theories of membrane potential reviewed in Chapter 3, it was assumed that the ion exchange sites were of one type, either weak acid or base type or strong acid or base type. It is possible at least in glasses that these sites could be of a mixed type containing strong and weak acid anionic sites. Altug and Hair,[58] in studying the ion exchange properties of porous glass by the pH titration method, found that the glass surface contained two types of ion exchange sites which differed in their acid strength. The pK_a of the stronger acid group was 5.1 while that of the weaker was about 7. The overall equilibrium selectivity of the glass was in the order $K^+ > Na^+ > Li^+$ with an exchange capacity of 0.07 mequiv/g of glass. In the theoretical development given in Chapter 3 concerning the membrane potential that occurs across glass membranes, this heterogeneity of the membrane which arises from the differences in either the nature of the sites or the degree of binding of the ions to the same sites, was not considered. This problem has been discussed at length by the Russian workers, a summary of which is given in the book edited by Eisenman.[59] The equation proposed for the glass electrode potential in which two different types of sites exist is of the form

$$E = \text{const} + \frac{RT}{2F} \ln(a_{H^+} + Ka_{M^+})$$

$$- \frac{RT}{2F} \ln\left(\frac{1}{a_{H^+} + \alpha_1 a_{M^+}} + \frac{\beta}{a_{H^+} + \alpha_2 a_{M^+}} \right) \tag{7}$$

where K is an ion exchange constant, α_1 and α_2 are constants characterizing the degree of bonding of H^+ ions to the two types of sites, and β is a constant characterizing the ratio of the products of the concentration of sites of each type and their acidic dissocation constants. The potential E which was measured as a function of pH for a variety of glasses [22 mole % Na_2O, x mole % Al_2O_3, $(78 - x)$ mole % SiO_2] in which the Al_2O_3 content of the glass was varied (x varied from 1 to 7%) and then plotted against pH, gave curves that showed step changes corresponding to the regions of acid error (shape of plot; a curve in the acid region), hydrogen function (straight line region), and alkali metal ion function (horizontal part). The straight line part became more pronounced as the Al_2O_3 content was reduced. These curves were quantitatively explained by the Russian workers on the basis of Eq. (7). However, for most glasses whose responses are dominated by one type of site, Eq. (118) of Chapter 3 is more useful. Equation (118) of Chapter 3, which has been verified by Eisenman *et al.*,[20] described the glass potential not only in H^+–cation mixtures but also in mixtures of two alkali metal cations at constant pH.[22, 60] In the case of Na^+–K^+ mixtures it was found that n [see Eq. (103) of Chapter 3] had a value of unity and that $K^{pot}_{Na-K} = \bar{P}_K/\bar{P}_{Na}$ where the $\bar{P}_i$'s are the permeabilities (see Lakshminarayanaiah[61]).

In order to check Eq. (118) of Chapter 3, Eisenman[22, 60] used a series of glass electrodes and measured potentials in pure 0.1 N NaCl and pure 0.1 N KCl solutions, referring them to a saturated KCl–calomel half-cell. By measuring uptakes of ^{24}Na and ^{42}K, values for the mobility ratio $\bar{u}_K/\bar{u}_{Na}$ were derived. The values derived for the two types of glass are shown in the accompanying tabulation, along with values of K derived potentiometrically (calculated) and from tracer diffusion studies. The agreement between them was considered adequate to substantiate the tenets of the theory from which Eq. (118) of Chapter 3 was derived.

Glass	K^{pot}_{Na-K} or $\bar{P}_K/\bar{P}_{Na}$	$\bar{D}_K/\bar{D}_{Na}$ or $\bar{u}_K/\bar{u}_{Na}$	K_{cal} [a]	K_{obs}
Type 1	10.3	0.18	57	90
Type 2	8.5	0.088	97	102

[a]Calculated from the relation $K^{pot}_{Na-K} = K(\bar{u}_K/\bar{u}_{Na})$ (see Chapter 3, page 76).

In Eq. (7), derived for the heterogeneous two-site glass, consideration has not been given to the glass phase ionic mobilities whose differences give rise to a diffusional potential. The potential E in Eq. (7), unlike Eq. (118) of Chapter 3, which contains the mobility term, corresponds to the algebraic sum of phase boundary potentials arising at various interfaces in the glass. For this reason, Buck[62] has made an attempt, based on solid state principles, to unify the ion exchange theories of the Russian workers[59] and Eisenman's n-type nonideal behavior.[21] The assumptions on which this solid state approach is based are that (1) a vacancy-defect equilibrium exists relating to the occupancy of the defect sites and the fixed sites by H^+ and M^+ ions, and that (2) the surface ion exchange equilibria exist at regions of two or more fixed sites. The equation derived for the glass membrane potential by algebraically adding the interfacial and diffusion potentials is complex; it contains a number of membrane parameters. However, the equation has been shown to predict the E vs. pH curves for a number of glasses of different alumina content. The complex equation reduces to a simpler form for homogeneous site glass membranes. This reduced equation is

$$E = \text{const} + \frac{RT}{F}\ln\left[a'_H + K_{H/M}(K_1 a'_H a'_M)^{1/2}\right]\left[1 + K_1 a'_M/a'_H\right]^{1/2} \quad (8)$$

where the a's are the activities of H^+ and M^+ ions in the sample at the (′) side, K_1 is the ion exchange equilibrium constant for metal ions replacing protons on silica sites, and $K_{H/M}$ is a mobility and defect generation ratio term given by

$$K_{H/M} = \frac{u_i^M K_{1i}^M \bar{\gamma}_{1M}}{u_i^H K_{1i}^H \bar{\gamma}_{1H}} \quad (9)$$

where u_i^M and u_i^H are the interstitial mobilities of ions M^+ and H^+, K_{1i}^M and K_{1i}^H are equilibrium constants for the generation of vacancies and interstitials for M and H in silica sites (type 1), and $\bar{\gamma}_{1M}$ and $\bar{\gamma}_{1H}$ are the activity coefficients of lattice site H^+ and M^+ ions in type 1 sites.

The K^{pot}_{H-M} of Eisenman's theory is related to $K_{H/M}$ by $K^{pot}_{H-M} = K_{H/M}K_1$. Thus $K_{H/M}$ is formally equivalent to the mobility ratio in the Eisenman model. Also, it has been shown[62] that the nonideality parameter

n in Eisenman's theory[21] is given by

$$n \approx \frac{59.14}{17.7} \log\left(1 + K_{\mathrm{H/M}}^{1/2}\right)\left(1 + \frac{1}{K_{\mathrm{H/M}}}\right)^{1/2} \tag{10}$$

For cation-selective glasses containing heterogeneous sites (containing Al_2O_3), the potential response was shown to be given by[62, 63]

$$\begin{aligned} E &= \text{const} + (RT/F)\ln\left[a'_{\mathrm{Na}} + K_{\mathrm{Na/M}}^{(1)}(a'_{\mathrm{Na}}a'_{\mathrm{M}})^{1/2}\right] \\ &\quad \times \left[1 + K_{\mathrm{Na/M}}^{(2)}(a'_{\mathrm{M}}/a'_{\mathrm{Na}})^{1/2}\right] \qquad (11a) \\ &= \text{const} + (RT/F)\ln\left(K_{\mathrm{Na/M}}^{(1)}K_{\mathrm{Na/M}}^{(2)}\right) + (RT/F) \\ &\quad \times \ln\left[a'_{\mathrm{M}} + K_{\mathrm{M/Na}}^{(1)}(a'_{\mathrm{Na}}a'_{\mathrm{M}})^{1/2}\right]\left[1 + K_{\mathrm{M/Na}}^{(2)}(a'_{\mathrm{Na}}/a'_{\mathrm{M}})^{1/2}\right] \qquad (11b) \end{aligned}$$

($K_{\mathrm{Na/M}}^{(2)}$ refers $K_{\mathrm{Na/M}}$ to sites of type 2 in alumina)

$$K_{\mathrm{Na/M}}^{(1)} = \frac{1}{K_{\mathrm{M/Na}}^{(1)}} = \frac{K_{2\mathrm{i}}^{\mathrm{M}}\bar{\gamma}_{2\mathrm{M}}}{K_{2\mathrm{i}}^{\mathrm{H}}\bar{\gamma}_{2\mathrm{H}}} K_2 \tag{12}$$

$$K_{\mathrm{Na/M}}^{(2)} = \frac{1}{K_{\mathrm{M/Na}}^{(2)}} = \frac{u_{\mathrm{i}}^{\mathrm{M}}}{u_{\mathrm{i}}^{\mathrm{Na}}} K_{\mathrm{Na/M}}^{(1)} \tag{13}$$

(where K_2 is the ion exchange constant applied to type 2 sites and $K_{2\mathrm{i}}$ is the equilibrium constant for the generation of vacancies and interstitials in sites of type 2) for a sodium-selective electrode and a_{M} is the interfering ion activity. In this case, the value of n is given by

$$n \approx \frac{59.14}{17.7} \log\left[\left(1 + \sqrt{\frac{u_{\mathrm{i}}^{\mathrm{M}}}{u_{\mathrm{i}}^{\mathrm{Na}}}}\right)\left(1 + \sqrt{\frac{u_{\mathrm{i}}^{\mathrm{Na}}}{u_{\mathrm{i}}^{\mathrm{M}}}}\right)\right] \tag{14}$$

Equation (8) was derived on the assumption that there were only a few mobile defect interstitial ions, whereas Eq. (11) was based on the assumption that cations relating to heterosites were completely ionized and mobile.

Using a number of electrodes, the potential response of each electrode was measured as a function of pH by Buck and co-workers.[62, 63] In the case of pH electrodes (Beckman General Purpose Glass No. 41263 and E-2 Glass No. 39004, among other commercially available pH electrodes), the experimental curves were fitted by Eq. (8) using the values for the various parameters given in Table 10. The responses were also measured at other temperatures, and the appropriate values of the various parameters to fit these E vs. pH curves have been given. Using Eqs. (11)–(13) the responses of a Beckman sodium-selective electrode were also computed with the help

TABLE 10

Metal Ion Response Parameters for Beckman Instruments Glasses[a]

Metal ion concentration (M)		$K_{H/M}$	K_1	$K_{H/M}^{pot}$
		General Purpose Glass Electrode No. 41263		
Na	3.0	20	7.5×10^{-14}	1.5×10^{-12}
	1.0	20	7.5×10^{-14}	1.5×10^{-12}
	0.3	20	5.0×10^{-14}	1×10^{-12}
	0.1	20	1.0×10^{-14}	2×10^{-13}
Li	3.0	20	1.0×10^{-13}	2×10^{-12}
	1.0	20	1×10^{-14}	2×10^{-13}
	0.3	18	6.5×10^{-15}	2×10^{-13}
	0.1	20	8.0×10^{-15}	2×10^{-13}
		E-2 Glass Electrode No. 39004		
Na	3.0	10	1×10^{-16}	1×10^{-15}
	0.1	10	1×10^{-16}	1×10^{-15}
Li	3.0	19	1×10^{-13}	2×10^{-12}
	1.0	15	1×10^{-13}	1.5×10^{-12}
	0.3	18	2.5×10^{-14}	4.5×10^{-13}
	0.1	18	2.5×10^{-14}	4.5×10^{-13}

[a]Measured at 25°C.

of the values of the parameters given in Table 11 and compared with experimental E vs. pH curves at constants Na^+ concentration.

The values of $K_{H\text{-}Na}^{pot}$ indicate that the electrode is more responsive to H^+ than to Na^+. The potassium and lithium interferences for the sodium-selective electrode have also been evaluated.

The values of K_{ij}^{pot} given in Tables 10 and 11 show that the experimental curves could not be fitted well by any one concentration-independent selectivity coefficient. This aspect of the dependence of the selectivity coefficient on the concentration or ratio of concentrations in mixtures for glass and other ion-selective electrodes has been discussed by Buck.[64]

TABLE 11

H^+ Ion Response Parameters of a Beckman No. 39278 Na Selective Glass Electrode[a]

Concentration of Na^+ (M)	u_i^{Na}/u_i^H	$K_{H/Na}^{(1)}$	$K_{H/Na}^{(2)}$	$K_{H/Na}^{pot}$
0.7	2.5	1.2×10^{-1}	2.3×10^{-3}	2.8×10^{-4}
0.07	1.4	7×10^{-2}	2.7×10^{-3}	1.9×10^{-4}
0.007	0.6	1×10^{-2}	2.7×10^{-2}	2.8×10^{-4}

[a]Measured at 25°C.

F. ELECTRODE RESPONSES IN AQUEOUS–NONAQUEOUS SOLVENTS

A very important characteristic of the glass electrode is its dynamic response which has been studied by a number of investigators. Disteche and Dubuisson[65] determined the time constant of a special electrode made of Corning 015 glass and found it to be of the order of 30 msec. Beck *et al.*[66, 67] studied the time dependence of glass electrodes whose potential variation was ascribed to changes in the asymmetry potential produced by the exchange of ions and water at the glass surface. Rechnitz and Hameka[68] related the experimental zero-current time responses of glass electrodes to a step change in the activity of the solution by the equation

$$E(t) = E^{\infty}(1 - e^{-t/\tau}) \tag{15}$$

where E^{∞} is the equilibrium potential reached following a step change in concentration, $E(t)$ is the potential at any time t after the step change, and τ is the time constant. The exponential form of this equation has been theoretically confirmed by Buck.[50] Johansson and Norberg[69] derived an exponential response time equation by applying the theory of electrode kinetics of charge transfer processes to ion exchange reactions at a glass surface.

The equation of electrode kinetics derived from the application of the principles of the theory of rate processes is[70]

$$\vec{i} = \frac{eKT}{h} a_{H^+} \exp\left[\frac{-\Delta G_0^{\#} - (1 - \beta)E_pF - \beta E_MF}{RT}\right] \tag{16}$$

where $\vec{i}$ is the current in the forward direction; $\Delta G_0^{\#}$ is the standard free energy of activity for the forward reaction; β, $0 < \beta < 1$, is a measure of the symmetry of the energy barrier; E_p is the potential of the preelectrolysis layer; E_M is the inner (Galvani) potential of the electrode site (both E_p and E_M are referred to the potential of the bulk of the membrane which is assumed zero); K is Boltzmann's constant; h is Planck's constant; e is the electronic change; R, T, and F have their usual meaning; and a_{H^+} is the activity of the hydrogen ion. The other parameters that usually appear in the general equation of electrode kinetics such as the transmission coefficient, stoichiometry number, the number of ions required to form one activated complex, and the valency of reactants are all equated to unity.

At equilibrium in a solution in which $a_{H^+} = a_1$ the current in the forward direction becomes equal to the current in the opposite direction, and both become equal to the exchange current i_0. Thus

$$i_0 = \frac{KTe}{h} a_1 \exp\left[\frac{-\Delta \overrightarrow{G}_0^{\#} - (1-\beta)E_pF - \beta E_M F}{RT} \right] \tag{17}$$

$$i_0 = \frac{KTe}{h} a_e \exp\left[\frac{-\Delta \overleftarrow{G}_0^{\#} - (1-\beta)E_pF + (1-\beta) E_M F}{RT} \right] \tag{18}$$

On the assumption that a_e and E_p are constants, the variation of the parameters with time t when the solution is changed rapidly from a_1 to a_2 can be written as

$$\overrightarrow{i}_t = \frac{KTe}{h} \exp\left[\frac{-\Delta \overrightarrow{G}_0^{\#}}{RT} \right] a_2 B \exp\left(\frac{-\beta E_t F}{RT} \right) \tag{19}$$

$$\overleftarrow{i}_t = \frac{KTe}{h} \exp\left[\frac{-\Delta \overleftarrow{G}_0^{\#}}{RT} \right] a_e B \exp\left[(1-\beta) \frac{E_t F}{RT} \right] \tag{20}$$

where

$$B = \exp\left[-(1-\beta) E_p \frac{F}{RT} \right] \tag{21}$$

Substituting the values for

$$\frac{KTe}{h} \exp\left[\frac{-\Delta \overrightleftarrows{G}_0}{RT} \right]$$

from Eqs. (17) and (18), Eqs. (19) and (20) become, on rearrangement,

$$\overrightarrow{i}_t = i_0 \left(\frac{a_2}{a_1} \right) \exp\left[-\frac{(E_t - E_1)F\beta}{RT} \right] \tag{22}$$

$$\overleftarrow{i}_t = i_0 \exp\left[\frac{(E_t - E_1)F}{RT} (1-\beta) \right] \tag{23}$$

The amount of charge Δq passing during the transient state is given by

$$\Delta q = -\int_0^t \left(\overrightarrow{i}_t - \overleftarrow{i}_t \right) dt \tag{24}$$

But

$$\Delta q = (E_t - E_1) C \tag{25}$$

It has been shown that the combination of Eqs. (22)–(25) with the assumption that $\beta = 0.5$ (symmetrical energy barrier), on integration, gives[69]

$$\frac{(E_t - E_1)F}{RT} = \ln \frac{a_2}{a_1} + 2 \ln\left[\frac{1 - \exp(-k_2 t + K')}{1 + \exp(-k_2 t + K')}\right] \tag{26}$$

where K' is the integration constant and k_2 is the rate constant. For the boundary condition when $t = 0$, $E_t = E_1$. Then Eq. (26) has been shown to reduce to

$$E_t - E_1 = \frac{RT}{F} \ln \frac{a_2}{a_1} [1 - \exp(-k_2 t)] \tag{27}$$

Also, the rate constant k_2 has been shown[69] to be given by

$$k_2 = \frac{eF}{RT} \frac{\overleftrightarrow{\kappa}}{C} \sqrt{a_2 a_e} \exp(-E_p F/RT) \tag{28}$$

where

$$\overleftrightarrow{\kappa} = \frac{KT}{h} \exp\left(\frac{-\overleftarrow{\Delta G_0^{\#}} - \overrightarrow{\Delta G_0^{\#}}}{2RT}\right)$$

is the specific rate at the standard potential.

Taking the logarithm of Eq. (28) gives

$$\log k_2 = \log A + \tfrac{1}{2} \log a_2 - \log C - 2.303 E_p F/RT \tag{29}$$

where

$$A = \frac{e\overleftrightarrow{\kappa}F}{RT} \sqrt{a_e}$$

The experimental data obtained for a number of glass electrodes in both aqueous and nonaqueous–aqueous solvents showed that a plot of $\log(E_t - E_\infty)$ versus time was linear according to Eq. (27) and the slope gave the value for the rate constant k_2. The logarithm of k_2 plotted against $\log a_2$ also gave a straight line in the acidic range according to Eq. (29). However, the values for the slopes obtained with different electrodes deviated from the theoretical value of 0.5 predicted by Eq. (29). The glass electrodes used (Metrohm EA 109 U, Beckman General Purpose) had time constants (time required to reach half of the maximum potential) of the order of seconds in isopropanol solution and of the order of milliseconds in water. The changes in the values of k_2 due to the addition of water to isopropanol are indicated in Table 12. In these studies the buffer system used was picric acid and tetraethyl ammonium picrate. A study of the behavior of the electrode (Metrohm 107 UX) in this buffer system showed that picric acid acted as a strong acid in isopropanol since the addition of picrate to picric acid whose concentration was held constant showed no change in the glass electrode potential.[71] It is also known that electrode response in poorly buffered media is slower than that in well-buffered media.[72] Consequently, the experimental finding that log k_2 varied linearly as log a_{H^+} in acid media, which is theoretically indicated, needs to be reinvestigated taking the buffer capacity into account.

TABLE 12

Values of the Rate Constant k_2, in Isopropanol at Different Concentrations of Water

Wt % water	3.0	1.5	0.43	0.29	0.28	0.19	0.10
log k_2	− 2.55	− 2.72	− 2.94	− 2.78	− 3.04	− 2.96	− 2.85

Karlberg[73, 74] has investigated the response characteristics of commercial glass electrodes. He used isopropanol solutions to avoid the possibility of chemical attack on glass. The electrode response to changes in the H^+ ion concentration made within the ideal response range was of the order of seconds.[74] On the other hand, changes around the two-ion response range brought about variations in the emf for several minutes. Dehydration of the glass surface resulted in a slow electrode response to changes in concentration, whereas etching shortened the response time. Optimal response was obtained when the ion exchange capacity of the gel layer attained its minimum values. General-purpose glass electrodes attained this value only after about 1 hr of hydration. Without loss of precision acid–base titration in isopropanol could be performed in less than 2 min, provided the glass electrode was properly etched. Etching generally reduced the gel layer thickness and brought about a faster response by depressing some of the ion exchange and diffusion-limited processes[73] and without affecting the electrode ion selectivity. To realize these characteristics careful etching is required. The sluggish response of glass electrodes continuously used in nonaqueous solvents can be eliminated by soaking them in a water solution. The presence of water in the gel layer helps the interdiffusion of ions and thus facilitates a rapid electrode response.

Time effects and time-dependent errors of hydrogen-responsive glass membrane electrodes in various aqueous standard buffer solutions, and in HCl and acetate buffer solutions in methanol–water and dimethylformamide–water mixtures have been investigated by Bottom and Covington.[75] The solvent used in the inner reference solution has also been varied. While some solvent mixtures profoundly altered the response characteristics of the electrode and the observed errors, other changes produced little or no effect on the time-dependent electrode characteristics.

Eisenman[22] has described the changes in response and selectivity of cation-selective glass electrodes when they are used in nonaqueous and nonaqueous–aqueous media. The electrode behavior in methanol was found to be similar to that in water with an increased selectivity to cation over hydrogen ions. The glass electrode has been found to respond satisfactorily to H^+ ions in hydrogen peroxide.[76] Nearly ideal responses have been obtained in acetonitrile,[77, 78] dimethylformamide,[79] and isopropanol, methyl ethyl ketone, and their mixture in equal volumes.[80] A glass electrode in which a buffered acetonitrile solution is used as the inner reference solution has been described.[81] Rechnitz and co-workers[82, 83] have made measurements with cation-selective electrodes in partially aqueous media. The electrodes behaved anomalously toward H^+ ions in ethanol–water mixtures. McClure and Reddy[84] measured the responses of the cationic glass electrode to alkali metal ions in propylene carbonate, acetonitrile, and dimethylformamide. They found that the glass electrode did not respond in acetonitrile after three months. The response time in solvents, however, was found to be between 5 and 10 sec.

Information on the alkaline error of the glass electrode in nonaqueous media has been provided by some investigators. The alkaline error in acetic acid has been studied by Wegmann *et al.*[85] Harlow[86] studied the influence of small amounts of K ion in a titrant (0.25 tetrabutyl ammonium hydroxide) in 80% pyridine and 20% isopropanol. K^+ ions depressed the electrode sensitivity to changes in acidity. The size of the effect varied from electrode to electrode and was dependent on the composition of glass and on the pretreatment. A systematic study of alkaline errors of glass electrodes in isopropanol has been made by Karlberg and Johanssen[87] who compared the electrode response in isopropanol to that of the hydrogen electrode. Glass electrodes that show low alkaline errors in water have been found to behave ideally in isopropanol. Bivalent ions were found to cause less deviations than monovalent ions. Hysteresis effects were observed in going from alkaline to acid solutions, although none were noted in the reverse direction. Therefore, to avoid hysteresis effects titrations should be made from acidic to basic values.

A new method for testing the performance of glass electrodes in the pH range 6.5–12.6 has been developed.[88] Cells without a liquid junction containing Ag–AgCl reference electrodes have been used. Test solutions were made from amine buffers and their hydrochlorides. With added sodium salts, the sodium error at a given pH and different values of pNa (e.g., 0, 1, or 2) were determined by transferring the glass electrode between solutions and comparing the potential difference obtained with that of a perfect glass electrode.

Glass electrode responses in heavy water (D_2O) have also been evaluated. The satisfactory response to deuterium ion in heavy water has been noted in some studies.[89, 90] Covington *et al.*[91] found the electrode response to pD in D_2O to be as good as it is to pH in ordinary water. However, Lowe and Smith[92] noted that when the electrode was transferred between acid solutions in ordinary water and deuterium oxide, the potential drifted systematically by 1–2 mV over several hours. After each transfer from HCl to DCl, the emf E followed as a function of time obeyed the empirical equation

$$E = E^{\infty} + A\exp(-gt^{1/2}) \tag{30}$$

where E^{∞} is the steady potential observed, t is the time from the moment of transfer, and A and g are constants whose values were found to be -1.42 mV and 0.25 $\text{min}^{-1/2}$ for 13 transfers from HCl to DCl and $+1.06$ and 0.25 for nine transfers from DCl to HCl. In view of this, electrodes standardized in H_2O cannot be readily used in D_2O for the precise measurement of the activity of the deuterium ion. Taking the necessary precautions, the responses of the cation-selective glass electrodes (CSGE) in D_2O have been investigated[93, 94] using the experimental cells,

$$\text{CSGE}|\text{NaCl}(m \text{ in } H_2O)|\text{AgCl–Ag} \tag{31}$$

$$\text{CSGE}|\text{NaCl}(m' \text{ in } D_2O)|\text{AgCl–Ag} \tag{32}$$

Unlike the drifts observed with hydrogen glass electrodes, the CSGEs showed little drift on changing from one solution in H_2O to another in D_2O, and steady potentials were observed quickly. The emf observed in the selectivity determinations were fitted to the equation[95]

$$E = E^{\circ} - (RT/F)\ln\{(a_{H^+} + K_{HM}a_{M^+})(a_{Cl})\} \tag{33}$$

in which M is Na or K. The values of the selectivity constant K_{HM} calculated on the assumption that the activity coefficients of H^+ and Na^+ ions were identical are given in Table 13. In the case of these cation-selective glass electrodes, the contributions of H^+ and D^+ to the emf observed in cells (31) and (32) are negligible.

TABLE 13

Molar Selectivity Constants of Glass Electrode

Electrode	$K_{HNa}(H_2O)$	$K_{DNa}(D_2O)$	K_{HK}	$K_{K\text{-}Na}$
GEA 33 (Na selective)	3.3×10^{-2}	0.9×10^{-2}	0.014×10^{-2}	230
GKN 33 (K selective)	7.8×10^{-2}	2.1×10^{-2}	155×10^{-2}	0.05

The change in emf ΔE following the transfer of the glass electrode in cell (31) from a solution of molality m_1 to one of molality m_2 is given by

$$\Delta E = E_2 - E_1 = 2S\log(m_1\gamma_1/m_2\gamma_2) \tag{34}$$

where $S = (RT/F)\ln 10 = 0.05916$ at 25°C. If m_1 is kept constant, then

$2S \log(m_1\gamma_1)$ may be treated as the effective standard emf for the experimental run.[94, 96] Hence Eq. (34) becomes

$$\Delta E = E^\circ - 2S \log(m\gamma) \tag{35}$$

Values of ΔE corresponding to the electrode transfer from the most dilute solution in each experimental run are given in Table 14. The molality of the solution in D_2O (aquamolality) is represented by m' and the corresponding activity coefficient by γ'.

TABLE 14

Cation-Selective Glass Electrode Transfer Between NaCl Solutions in the Same Solvent

In H_2O cell (31)				In D_2O, cell (32)			
Exp. 1		Exp. 2		Exp. 3		Exp. 4	
Molality (m)	ΔE (mV)	Molality (m)	ΔE (mV)	Aquamolality (m')	ΔE (mV)	Aquamolality (m')	ΔE (mV)
				GEA 33 Electrode			
0.3416	0	0.2406	0	0.2589	0	0.1711	0
0.5474	− 22.41	0.2513	− 2.06	0.3897	− 19.22	0.2235	− 12.47
0.6045	− 27.03	0.4946	− 34.07	0.5865	− 38.64	0.4654	− 46.93
0.8772	− 45.32	0.6117	− 44.27	0.8345	− 55.69	0.6485	− 62.65
1.0641	− 54.73	0.9196	− 64.23	0.9333	− 61.17	0.8978	− 78.39
						1.0129	− 84.38
				GKN 33 Electrode			
0.1615	0	0.0964	0	0.2590	0	0.1383	0
0.2414	− 18.80	0.1305	− 14.54	0.4241	− 23.24	0.2611	− 29.67
0.2583	− 2.95	0.3438	− 59.97	0.6136	− 40.90	0.4677	− 57.10
0.4779	− 51.13	0.4507	− 72.84	0.8888	− 58.98	0.6563	− 73.35
0.6682	− 67.19	0.6141	− 87.67	1.1187	− 70.46	0.7936	− 82.58
		0.7873	− 99.83			1.0263	− 95.36

Substituting for the activity coefficient given by Eq. (108) of Chapter 2, Eq. (35) can be expressed as

$$E' = E^\circ - 2Sbm \tag{36}$$

where

$$E' = \Delta E + 2S \log m - 2S \frac{Am^{1/2}}{1 + Bam^{1/2}} \tag{37}$$

The theoretical constants A and B have values of 0.5109 and 0.3287×10^8 in water, and 0.5132 and 0.3288×10^8 in D_2O, respectively, a is the distance of closest approach (ion-size parameter), and b is an empirical constant.

The experimental results of Table 14 fitted into Eq. (36) by a least squares procedure gave the value of 4 Å for the ion-size parameter. The average values for b obtained were 0.037 m^{-1} for water solutions and 0.033 for solutions in D_2O. Also the plot of $E' - E°$ versus aquamolality was linear according to Eq. (36) in both aqueous and D_2O solutions.

The changes in emf, $\Delta E (= E_D - E_H)$, measured when the glass electrode was transferred from cell (31) to cell (32) [$\Delta E(H \rightarrow D)$] and then returned to cell (31) [$\Delta E(D \rightarrow H)$] are given in Table 15. The average values of $\Delta E°$ obtained from the equation

$$\Delta E° = \Delta E + \frac{2RT}{F} \ln \frac{m'\gamma'}{m\gamma} \tag{38}$$

using the values of b given above (to calculate γ and γ') are also included in Table 15. The average values of $\Delta E°$ are -7.25 mV for GEA 33 and -7.45 for GKN 33 in the aquamolal range 0.1–1.0. Using these values and measuring ΔE at any other concentration, a value for the ratio γ/γ' at that particular concentration can be derived. Since the values of γ for solutions in ordinary water are tabulated in standard texts, values for γ' can also be determined. The aquamolal activity coefficient γ' so derived for NaCl solutions in D_2O are given in Table 16.

TABLE 15

Change in Potential Following Electrode Transfer between NaCl Solutions in H_2O and D_2O

Electrode	m (H_2O)	m' (D_2O)	$\Delta E(H \rightarrow D)$ (mV)	$\Delta E(D \rightarrow H)$ (mV)	$\Delta E°$ (mV)
GEA 33	0.0936	0.1042	− 11.74	− 11.78	− 6.79
	0.0982	0.1011	− 8.68	− 8.70	− 7.45
	0.1321	0.1280	− 5.92	− 5.96	− 7.55
	0.9882	1.1304	− 13.29	− 13.31	− 7.23
GKN 33	0.0936	0.1042	− 12.44	− 12.41	− 7.44
	0.0982	0.1011	− 8.89	− 9.01	− 7.69
	0.1321	0.1280	− 5.97	− 6.05	− 7.60
	0.9882	1.1304	− 13.33	− 13.30	− 7.06

TABLE 16

Values of γ' for NaCl Solutions in D_2O

Aquamolality (m')	− log γ'	Aquamolality (m')	− log γ'
0.1	0.110	0.6	0.175
0.2	0.135	0.7	0.179
0.3	0.151	0.8	0.183
0.4	0.161	0.9	0.185
0.5	0.170	1.0	0.188

The change in the emf $\Delta E^\circ = E^\circ_D - E^\circ_H$ derived above is related to the free energy of transfer of an ionic solute from H_2O to D_2O ($\Delta G^\circ_t = \Delta G^\circ_D - \Delta G^\circ_H$) by the relation

$$\Delta G^\circ_t = -F\Delta E^\circ \tag{39}$$

For NaCl solutions, the cation-selective glass electrode gave[93, 95] a value of -7.35 mV for ΔE°. Therefore this corresponds, according to Eq. (39), to a value of 169 cal/mole for ΔG°_t.

The temperature dependence of the response of glass electrodes in D_2O has also been investigated. KenForce and Carr[97] prepared solutions of NaOH of equal molality in H_2O and D_2O at constant ionic strength and calculated the hydrogen and deuterium ion concentrations at a given temperature. They also measured the concentrations at the same temperature using a glass membrane electrode. The ΔpD taken at that temperature as the difference between Δ_{cal} and Δ_{exp} where $\Delta = \text{pD} - \text{pH}$ was determined as a function of temperature. The data conformed to the relation

$$\Delta\text{pD} = \frac{2.02 \times 10^2}{T} - 0.204$$

provided the electrode showed low Na error.

The effect of light on several glass membrane electrodes has been investigated.[98, 99] In the case of the pH electrode both sunlight and artificial light caused a decrease in the measured pH values[98] and the effect was attributed to an increase in the conduction of the glass membrane. Similarly, the Na-sensitive glass electrode response was affected by light, causing a decrease of 0.1 pNa unit for an increase in potential of 6 mV.[99]

The similarity that the glass membrane electrode bears to the fluoride-selective electrode has been discussed by Vesely.[100] The response of the fluoride electrode is determined by the same parameters as the glass electrode, i.e., its selectivity is determined by the equilibrium constant of ion exchange between ions i and j in the gel layer and their mobilities. The exchange reaction involved in the case of the fluoride electrode seems to be

$$F^-(s) + (OH^-)(sol) \rightleftharpoons (F^-)(sol) + OH^-(s)$$

The charge is carried by the F^- ion, whereas it is carried by the alkali metal ions in glass. Consequently, the fluoride electrode acts as an anionic analog of the glass electrode. While the gel layer in glass is well defined, that in the single crystal of LaF_3 is less developed and probably has a different structure. The glass membrane electrodes have been used in a number of studies, some of which are indicated below.

The pH electrode has been used by Douheret[101] in galvanic cells with transport to determine diffusion potentials at the HCl–saturated KCl interface. The mechanism by which glass electrodes operate has been studied by measuring surface charge as a function of pH for Na- and K-responsive glasses in the presence of chlorides of Li, Na, K, Cs, and tetraethyl ammonium.[102] No relation has been found between the affinity of a given cation to a glass surface and the tendency of the glass to respond to that cation. The electrode response is governed chiefly by the ion mobility in the gel layer. Electrode behavior in liquid ammonia at −38°C has also been studied.[103] The alkali metal cation electrode responds to protonated solvent (NH_4^+) and so could be used to measure the activity of the NH_4^+ ion provided the correction for alkaline error is applied. The selectivity scale of different cations to a glass electrode changes drastically in going from water to liquid ammonia as solvent. A method for the determination of acidity constants of weak acids in ammonia using a glass electrode has been described.[104] The selectivities of some commercially available cation-selective glass membrane electrodes have been established.[99] In order to amplify the response of the electrode to H^+ ions or other glass membrane electrodes selective to NH_4^+ ions, prior enrichment of the species by using ion exchange membranes has been proposed.[105] The ion exchange membrane may be fixed to the surface of the glass membrane electrode that is used to sense the species. A similar procedure using a hydrophobic membrane in place of the ion exchange membrane has been used to determine ammonia in boiler feed water.[106] A sodium-responsive glass electrode has been used in studies of ion association in solutions of sodium tetrametaphosphate and trimetaphosphate[107] whose formation constants have been found to be 133.3 and 25.1 liter/mole, respectively, at 25°C. It has been used to monitor Na^+ ions continuously in high-purity water. Such sensing devices for Na are being used in power stations to monitor continuously the Na^+ concentrations in steam and boiler feed.[108] The electrode has been used in measurements of activity in NaCl solutions[109] and in concentrated NaCl–KCl solution mixtures,[110] and in the estimation of the Na content of various aqueous solutions[111–113] and of high-purity water.[114]

Sweat electrolytes in situ, particularly for Na^+ and Cl^-, have been estimated using sodium-selective glass and chloride-selective membrane electrodes.[115] A sodium-selective glass electrode has also been used in a nonaqueous medium for the microdetermination of halogens by argentometry.[116]

Microelectrodes constructed from glass are of considerable importance in recording electrical potentials in biological systems. Interesting papers concerned with the construction, behavior, and use of these microelectrodes in different biological systems have appeared.[117–120] In the case of biological systems glass membrane microelectrodes specific to ions (mostly Na^+ and K^+) have been used intracellularly to estimate the activities of these ions.[121] Glass electrodes of different design have been used to estimate activities of ions in biological fluids in vitro[122] and in vivo.[123] Other interesting possibilities of using glass microelectrodes of different construction and design in biological systems have been discussed by Rechnitz.[124]

REFERENCES

1. R. G. Bates, *in* "Reference Electrodes" (D. J. G. Ives and G. J. Janz, eds.), p. 231. Academic Press, New York, 1961.
2. G. A. Rechnitz, *Anal. Chem.* **37**, 29A (1965).
3. G. A. Rechnitz, *Chem. Eng. News* **45** (25), 146 (1967).
4. G. Eisenman, *Ann. N. Y. Acad. Sci.* **148**, 5 (1968).
5. R. H. Doremus, *in* "Ion Exchange" (J. A. Marinsky, ed.), Vol. 2, p. 1. Dekker, New York, 1969.
6. K. Schwabe, *Advan. Anal. Chem. Instrum.* **10**, 495 (1974).
7. M. Dole, "The Glass Electrode." Wiley, New York, 1941.
8. G. Eisenman, R. G. Bates, G. Mattock, and S. M. Friedman, "The Glass Electrode." Wiley (Interscience), New York, 1966.
9. G. Eisenman (ed.), "Glass Electrodes for Hydrogen and Other Cations, Principles and Practice." Dekker, New York, 1967.
10. R. G. Bates, "Determination of pH, Theory and Practice," 2nd ed. Wiley, New York, 1973.
11. J. O. Isard, *in* "Glass Electrodes for Hydrogen and Other Cations, Principles and Practice" (G. Eisenman, ed.), p. 51. Dekker, New York, 1967.
12. W. H. Zachariasen, *J. Amer. Chem. Soc.* **54**, 3841 (1932).
13. W. S. Hughes, *J. Chem. Soc.* 491 (1928).
14. L. W. Elder, *J. Amer. Chem. Soc.* **51**, 3266 (1929).
15. D. A. MacInnes and M. Dole, *J. Amer. Chem. Soc.* **52**, 29 (1930).
16. S. I. Sokolov and A. H. Passynsky, *Z. Phys. Chem (Leipzig)* **A160**, 366 (1932).
17. H. H. Cary and W. P. Baxter, quoted in reference 11.
18. G. A. Perley, *Anal. Chem.* **21**, 391, 394, 559 (1949).
19. B. Lengyel and E. Blum, *Trans. Faraday Soc.* **30**, 461 (1934).
20. G. Eisenman, D. O. Rubin, and J. U. Casby, *Science* **126**, 831 (1957).
21. G. Eisenman, *Biophys. J.* **2**, part 2, 259 (1962).
22. G. Eisenman, *Advan. Anal. Chem. Instrum.* **4**, 213 (1965).
23. A. H. Truesdell and C. L. Christ, *in* "Glass Electrodes for Hydrogen and Other Cations, Principles and Practice" (G. Eisenman, ed.), p. 293. Dekker, New York, 1967.
24. H. J. C. Tendeloo and A. J. Z. Voorspuij, *Rec. Trav. Chim.* **61**, 531 (1942).
25. A. H. Truesdell and A. M. Pommer, *Science* **142** 1292 (1963).
26. M. E. Thompson, quoted in reference 23.
27. A. H. Truesdel, *Amer. Mineralogist* **51**, 110 (1965).
28. G. Haugaard, *Nature (London)* **120**, 66 (1937); *Glastech. Ber.* **17**, 104 (1939).
29. K. Horovitz, *Z. Phys.* **15**, 369 (1923).
30. D. A. MacInnes and D. Belcher, *J. Amer. Chem. Soc.* **53**, 3315 (1931).
31. D. Hubbard, *J. Res. Nat. Bur. Std.* **36**, 511 (1946).
32. I. Altug and M. L. Hair, *J. Phys. Chem.* **72**, 2976 (1968).
33. I. Altug and M. L. Hair, *J. Electrochem. Soc.* **117**, 78 (1970).
34. L. Kratz, *Glastech. Ber.* **20**, 305 (1942).

35. D. Hubbard, E. H. Hamilton, and A. N. Finn, *J. Res. Nat. Bur. Std.* **46**, 168 (1951).
36. A. Wikby, *J. Electroanal. Chem. Interfacial Electrochem.* **33**, 145 (1971).
37. A. Wikby, *J. Electroanal. Chem. Interfacial Electrochem.* **38**, 429 (1972).
38. A. Wikby, *J. Electroanal. Chem. Interfacial Electrochem.* **38**, 441 (1972).
39. A. Wikby, *J. Electroanal. Chem. Interfacial Electrochem.* **39**, 103 (1972).
40. D. Hubbard and G. F. Rynders, *J. Res. Nat. Bur. Std.* **41**, 163 (1948).
41. K. Schwabe and H. Dahms, *Isotopentech* **1**, 34 (1971).
42. H. Bach and F. G. K. Baucke, *Electrochim. Acta* **16**, 1311 (1971).
43. B. Karlberg, *J. Electroanal. Chem. Interfacial Electrochem.* **45**, 127 (1973).
44. B. Csakvari, Z. Bossay, and G. Bouquet, *Anal. Chim. Acta* **56**, 279 (1971).
45. M. Dole, *J. Chem. Phys.* **2**, 862 (1934).
46. B. Lengyel, B. Csakvari, F. Till, and Z. Boksay. *Magy. Kem. Lapja* **9**, 265 (1954).
47. A. Wikby and G. Johansson, *J. Electroanal. Chem. Interfacial Electrochem.* **23**, 23 (1969).
48. R. W. Douglas and T. M. M. El-Shamy, *J. Amer. Ceram. Soc.* **50**, 1 (1967).
49. D. A. MacInnes and D. Belcher, *Ind. Eng. Chem. Anal. Ed.* **5**, 199 (1931).
50. R. P. Buck, *J. Electroanal. Chem. Interfacial Electrochem* **18**, 363 (1968).
51. R. P. Buck, *J. Electroanal. Chem. Interfacial Electrochem.* **18**, 381 (1968).
52. R. P. Buck and I. Krull, *J. Electroanal. Chem. Interfacial Electrochem.* **18**, 387 (1968).
53. K. S. Cole, "Membranes, Ions and Impulses." Univ. of California Press, Berkeley, California, 1968.
54. M. J. D. Brand and G. A. Rechnitz, *Anal. Chem.* **41**, 1185 (1969).
55. M. J. D. Brand and G. A. Rechnitz, *Anal. Chem.* **41**, 1788 (1969).
56. M. J. D. Brand and G. A. Rechnitz, *Anal. Chem.* **42**, 304 (1970).
57. J. R. Sandifer and R. P. Buck, *J. Electroanal. Chem. Interfacial Electrochem.* **56**, 385 (1974).
58. I. Altug and M. L. Hair, *J. Phys. Chem.* **71**, 4260 (1967).
59. B. P. Nicolsky, M. M. Shultz, and A. A. Belijustin, *in* "Glass Electrodes for Hydrogen and Other Cations, Principles and Practice" (G. Eisenman, ed.), p. 174. Dekker, New York, 1967.
60. G. Eisenman, *in* "Glass Electrodes for Hydrogen and Other Cations, Principles and Practice" (G. Eisenman, ed.), p. 133. Dekker, New York, 1967.
61. N. Lakshminarayanaiah, "Transport Phenomena in Membranes," p. 103. Academic Press, New York, 1969.
62. R. P. Buck, *Anal. Chem.* **45**, 654 (1973).
63. R. P. Buck, J. H. Boles, R. D. Porter, and J. A. Margolis, *Anal. Chem.* **46**, 255 (1974).
64. R. P. Buck, *Anal. Chim. Acta* **73**, 321 (1974).
65. A. Disteche and M. Dubuisson, *Rev. Sci. Instrum.* **25**, 86 (1954).
66. W. H. Beck and W. F. K. Wynne-Jones, *J. Chim. Phys.* **49**, C 97 (1952).
67. W. H. Beck, J. Caudle, A. K. Covington, and W. F. K. Wynne-Jones, *Proc. Chem. Soc.* 110 (1963).
68. G. A. Rechnitz and H. F. Hameka, *Z. Anal. Chem.* **214**, 252 (1965).
69. G. Johansson and K. Norberg, *J. Electroanal. Chem. Interfacial Electrochem.* **18**, 239 (1968).
70. R. Parsons, *Trans Faraday Soc.* **47**, 1332 (1951).
71. A. Wikby and B. Karlberg, *J. Electroanal. Chem. Interfacial Electrochem.* **43**, 325 (1973).
72. A. Disteche, *Mem. Acad. Belg. Ser. 2* **32**, 1 (1960).
73. B. Karlberg, *J. Electroanal Chem. Interfacial Electrochem.* **42**, 115 (1973).
74. B. Karlberg, *Anal Chim. Acta* **66**, 93 (1973).

75. A. E. Bottom and A. K. Covington, *J. Electroanal. Chem. Interfacial Electrochem.* **24**, 251 (1970).
76. A. G. Mitchell and W. F. K. Wynne-Jones, *Trans. Faraday Soc.* **51**, 1690 (1955).
77. E. Romberg and K. Cruse. *Z. Elektrochem.* **63**, 404 (1959).
78. J. F. Coetzee and G. R. Padmanabhan, *J. Phys. Chem.* **66**, 1708 (1962).
79. M. Teze and R. Schaal, *Bull. Soc. Chim. Fr.* 1372, (1962).
80. K. Norberg, *Talanta* **13**, 745 (1966).
81. J. Badoz-Lambling, J. Desbarres, and J. Tacussel, *Bull. Soc. Chim. Fr.* 53 (1962).
82. G. A. Rechnitz and S. B. Zamoschnik, *Talanta* **11**, 974 (1964).
83. G. A. Rechnitz and G. Krugler, *Z. Anal. Chem.* **214**, 405 (1965).
84. J. E. McClure and T. B. Reddy, *Anal. Chem.* **40**, 2064 (1968).
85. D. Wegmann, J. P. Escarfail, and W. Simon, *Helv. Chim. Acta* **45**, 826 (1962).
86. G. A. Harlow, *Anal. Chem.* **34**, 148 (1962).
87. B. Karlberg and G. Johansson, *Talanta* **16**, 1545 (1969).
88. M. Filomena, G. F. C. Camoes, and A. K. Covington, *Anal. Chem.* **46**, 1547 (1974).
89. P. K. Glasoe and F. A. Long, *J. Phys. Chem.* **64**, 188 (1960).
90. P. R. Hammond, *Chem. Ind.* 311, (1962).
91. A. K. Covington, M. Paabo, R. A. Robinson, and R. G. Bates, *Anal. Chem.* **40**, 700 (1968).
92. B. M. Lowe and D. G. Smith, *Anal. Lett.* **6**, 903 (1973).
93. B. M. Lowe and D. G. Smith, *Chem. Commun.* 989 (1972).
94. B. M. Lowe and D. G. Smith, *J. Chem. Soc. Faraday. Trans. I* **69**, 1934 (1973).
95. B. M. Lowe and D. G. Smith, *J. Electroanal. Chem. Interfacial Electrochem.* **51**, 295 (1974).
96. A. K. Covington and J. E. Prue, *J. Chem. Soc.* 3696 (1955).
97. R. KenForce and J. D. Carr, *Anal. Chem.* **46**, 2049 (1974).
98. A. F. Wilward, *Analyst* **94**, 154 (1969).
99. S. Phang and B. J. Steel, *Anal. Chem.* **44**, 2230 (1972).
100. J. Vesely, *J. Electroanal. Chem. Interfacial Electrochem.* **41**, 134 (1973).
101. G. Douheret, *Bull. Soc. Chim. Fr.* 2093 (1970).
102. T. F. Tadros and J. Lyklema, *J. Electroanal. Chem. Interfacial Electrochem.* **22**, 9 (1969).
103. W. M. Baumann and W. Simon, *Helv. Chim. Acta* **52**, 2054 (1969).
104. W. M. Baumann and W. Simon, *Helv. Chim. Acta* **52**, 2060 (1969).
105. W. J. Blaedel and T. R. Kissel, *Anal. Chem.* **44**, 2109 (1972).
106. D. Midgley and K. Torrance, *Analyst* **97**, 626 (1972).
107. G. L. Gardner and G. H. Nancollas, *Anal. Chem.* **41**, 514 (1969).
108. A. A. Diggens, K. Parker, and H. M. Webber, *Analyst* **97**, 198 (1972).
109. A. Shatkay and A. Lerman, *Anal. Chem.* **41**, 514 (1969).
110. R. Huston and J. N. Butler, *Anal. Chem.* **41**, 1695 (1969).
111. J. T. Pearson and C. M. Elstob, *J. Pharm. Pharmacol.* **22**, 73 (1970).
112. T. Y. Toribara and L. Koval, *Talanta* **16**, 529 (1969).
113. D. M. Nutbourne, *Analyst* **95**, 609 (1970).
114. H. M. Webber and A. L. Wilson, *Analyst* **94**, 209 (1969).
115. L. Kopito and H. Shwachman, *Pediatrics* **43**, 794 (1969).
116. K. Hozumi and N. Akimoto, *Anal. Chem.* **42**, 1312 (1970).
117. D. B. Carter and I. A. Silver, *in* "Reference Electrodes" (D. J. G. Ives and G. J. Janz, eds.), p. 464. Academic Press, New York, 1961.
118. M. Lavallee, O. F. Schanne, and N. C. Hebert (eds), "Glass Microelectrodes." Wiley, New York, 1969.
119. R. N. Khuri, *in* "Glass Electrodes for Hydrogen and Other Cations, Principles and

Practice" (G. Eisenman, ed.), p. 478. Dekker, New York, 1967.
120. P. Sekelj and R. B. Goldbloom, *in* "Glass Electrodes for Hydrogen and Other Cations, Principles and Practice" (G. Eisenman, ed.), p. 520, Dekker, New York, 1967.
121. J. A. M. Hinke, *in* "Glass Electrodes for Hydrogen and Other Cations, Principles and Practice" (G. Eisenman, ed.), p. 464. Dekker, New York, 1967.
122. E. W. Moore, *in* "Glass Electrodes for Hydrogen and Other Cations, Principles and Practice" (G. Eisenman, ed.), p. 412. Dekker, New York, 1967.
123. S. M. Friedman, *in* "Glass Electrodes for Hydrogen and Other Cations, Principles and Practice" (G. Eisenman, ed.), p. 442. Dekker, New York, 1967.
124. G. A. Rechnitz, *Chem. Eng. News* **53** (4), 29 (1975).

Chapter 10

ELECTRODES FOR SENSING GASES

A. AMMONIA SENSOR

In Chapter 7 a description of the membrane electrode responding to ammonium ion and/or ammonia gas was given. The electrode was prepared by coupling a hydrophobic membrane that was permeable to ammonia to a monovalent cation electrode.[1-3] The membrane, which was impermeable to ions such as Na^+, K^+, and NH_4^+, separated the alkaline test solution from an internal solution of 0.1 *M* NH_4Cl in which a glass pH electrode and a Ag–AgCl reference electrode were immersed. Diffusion of ammonia across the membrane brought about a change in the pH of the filling solution; this pH change was monitored by the glass electrode.

B. CARBON DIOXIDE SENSOR

The principle upon which the ammonia-sensing electrode is based was utilized almost two decades ago by Stow and co-workers[4, 5] and others[6, 7] to construct electrodes for the measurement of partial pressures of CO_2 in blood and other fluids. An enlarged view of the arrangement is shown in Fig. 1. In effect the CO_2 sensor acts as a pH electrode and measures the pH change in a thin film of bicarbonate solution in close contact with the pH-sensitive glass electrode (Fig. 2).

The equilibria existing in the system can be expressed as

$$NaHCO_3 \rightleftharpoons Na^+ + HCO_3^- \quad \text{(complete dissociation)} \tag{1}$$

$$CO_2 + H_2O \overset{K}{\rightleftharpoons} H_2CO_3 \overset{K_1}{\rightleftharpoons} H^+ + HCO_3^- \tag{2}$$

$$HCO_3^- \overset{K_2}{\rightleftharpoons} H^+ + CO_3^{2-} \tag{3}$$

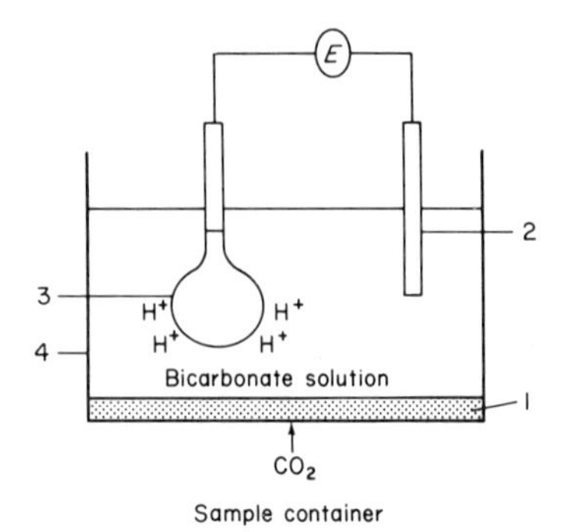

Fig. 1. Enlarged view of the cell assembly for the measurement of carbon dioxide: 1, membrane; 2, reference electrode; 3, glass electrode; and 4, container. E is the potential measured.

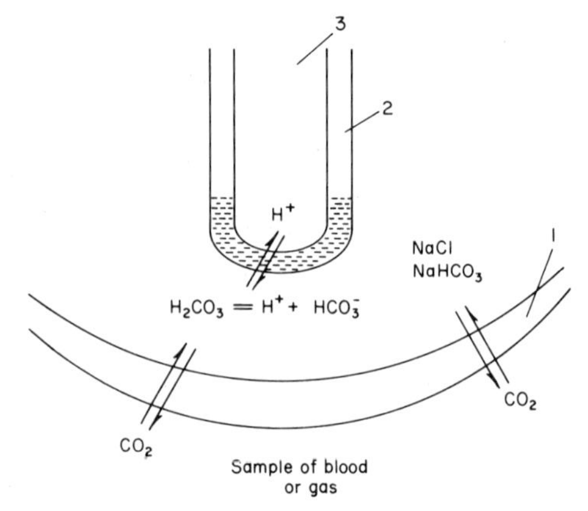

Fig. 2. Equilibria in a carbon dioxide electrode: 1, Teflon membrane; 2, glass; and 3, inside of glass electrode.

The first and second dissociation constants are given by

$$K_1 = \frac{(a_{\mathrm{HCO_3^-}})(a_{\mathrm{H^+}})}{a_{\mathrm{H_2CO_3}}} \tag{4}$$

$$K_2 = \frac{(a_{\mathrm{H^+}})(a_{\mathrm{CO_3^{2-}}})}{a_{\mathrm{HCO_3^-}}} \tag{5}$$

where the a's are the activities of the species indicated. Charge balance gives the relation

$$(a_{\mathrm{H^+}}) + (a_{\mathrm{Na^+}}) = (a_{\mathrm{OH^-}}) + (a_{\mathrm{HCO_3^-}}) + (2a_{\mathrm{CO_3^{2-}}}) \tag{6}$$

Substituting for $a_{\mathrm{OH^-}}$ (i.e., $a_{\mathrm{OH^-}} = K_w/a_{\mathrm{H^+}}$ where K_w is the ionic product of water) and $a_{CO_3^{2-}}$ from Eq. (5) into Eq. (6) gives, on rearrangement,

$$a_{HCO_3^-} = \frac{(a_{H^+}^2) + (a_{Na^+})(a_{H^+}) - (K_w)}{a_{H^+}[1 + (2K_2/a_{H^+})]} \tag{7}$$

Substituting eq. (7) into Eq. (4) gives, on rearrangement,

$$a_{H_2CO_3} = \frac{(a_{H^+}^2) + (a_{Na^+})(a_{H^+}) - (K_w)}{K_1[1 + (2K_2/a_{H^+})]} \tag{8}$$

$a_{H_2CO_3}$ in the bicarbonate film is proportional, by Henry's law, to the partial pressure of CO_2 (p_{CO_2}) in the sample outside the membrane and therefore can be replaced by αp_{CO_2} where α is the solubility coefficient of CO_2. Thus Eq. (8) can be written as[7]

$$\alpha p_{CO_2} = a_{H_2CO_3} = \frac{(a_{H^+}^2) + (a_{Na^+})(a_{H^+}) - (K_w)}{K_1[1 + (2K_2/a_{H^+})]} \tag{9}$$

The relationship between the change of pH (ΔpH) in the bicarbonate solution in a carbon dioxide electrode caused by a change of p_{CO_2} has been defined by Severinghaus and Bradley[7] as the sensitivity S of the electrode and is given by

$$S = \frac{\Delta \mathrm{pH}}{\Delta \log p_{CO_2}} \tag{10}$$

If no sodium bicarbonate is present in the trapped film of solution contacting the glass membrane, Eq. (9) approximates to

$$\alpha p_{CO_2} = (a_{H^+}^2)/K_1 \tag{11}$$

When p_{CO_2} changes from p'_{CO_2} to p''_{CO_2}, leading to changes in a_{H^+} from a'_{H^+} to a''_{H^+}, Eq. (11) can be written as

$$-\log \frac{p'_{CO_2}}{p''_{CO_2}} = -\log \frac{(a'_{H^+})^2}{(a''_{H^+})^2} \tag{12a}$$

or

$$\log p''_{CO_2} - \log p'_{CO_2} = 2(\mathrm{pH}' - \mathrm{pH}'') \tag{12b}$$

i.e.,

$$\Delta \log p_{CO_2} = 2\Delta \mathrm{pH} \tag{12c}$$

Comparing Eqs. (12c) and (10) yields a value of 0.5 for S. When sodium bicarbonate is added, the second term in Eq. (9) at concentration above

10^{-3} *M* becomes dominant and the equation reduces to

$$\alpha p_{CO_2} = \frac{(a_{Na^+})(a_{H^+})}{K_1} \tag{13}$$

Again taking two values, as stated above, yields

$$S = \frac{\Delta pH}{\Delta \log p_{CO_2}} = 1 \tag{14}$$

for the sensitivity. Thus a low concentration of bicarbonate ion doubles the sensitivity of the electrode.

Severinghaus and Bradley[7] found that the performance of the electrode agreed with theory in dilute solutions of bicarbonate but at higher concentrations, the sensitivity was less than the theoretically calculated value. Above a concentration of 0.02 *M*, the equilibration was prolonged, probably because of the presence of increased quantities of CO_3^{2-}.

Stow's carbon dioxide electrode[5] contained a very thin rubber membrane stretched over a glass electrode that was moistened with distilled water between the rubber and the glass. Severinghaus and Bradley[7] used a Teflon membrane and a very thin layer of cellophane (0.002 in. thick when wet) soaked in a solution of 0.01 *M* $NaHCO_3$ + 0.1 *M* NaCl. This cellophane held a better film of water between the glass and the Teflon membrane (see Fig. 3). Nylon mesh or Joseph paper could be used in place of Teflon.[8, 9]

Although the carbon dioxide electrodes described above are well suited for measurements of p_{CO_2} in blood and other biological fluids, they are too large to be used with biological tissues. Hertz and Siesjo[10] have described an electrode of suitable size and shape to be used with biological tissues. In this design, a calomel half-cell was used as a side extension to serve as the reference electrode. In place of the Ag–AgCl electrode, a calomel reference electrode is preferred because of its stability, and ease of preparation and handling.[11]

Calibration of this CO_2 electrode is simple. It responds to absolute values of p_{CO_2} and therefore gives identical values for blood, gas, or water having the same p_{CO_2} at the same temperature. This enables use of compressed gases for purposes of calibration. However, use of a gas with p_{CO_2} within the range of the samples is recommended for calibration.[9] Other problems such as the nature of the membrane and the existence of air bubbles, which affect the electrode response have been discussed by Severinghaus.[9]

The CO_2 electrodes are useful in studying normal or pathological respiration. They are used in measurements in vitro and to establish the acid–base status of blood,[12] which involves drawing samples of blood for

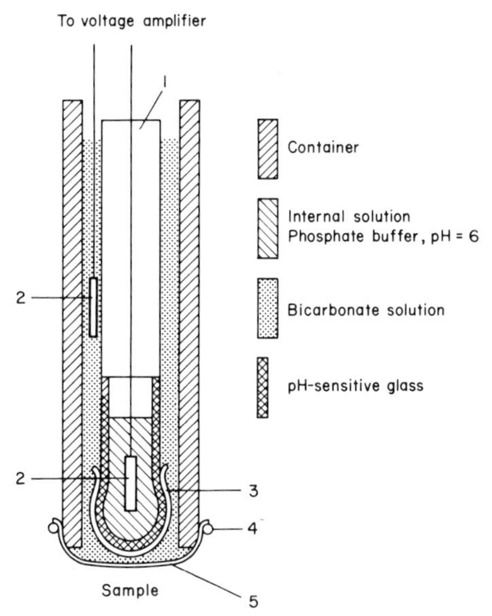

Fig. 3. An arrangement for sensing carbon dioxide (after Smith and Hahn[8]): 1, glass rod to hold the pH-sensitive glass; 2, Ag–AgCl reference electrodes; 3, cellophane or nylon mesh to hold the aqueous film; 4, O-rings; and 5, Teflon membrane.

estimation of its p_{CO_2} from the systemic arterial system. The smaller electrode developed by Hertz and Siesjo[10] has been used for measuring p_{CO_2} on the surface of the brain of anesthetized cats.[13] Even some microelectrodes have been devised and used in some physiological studies.[14, 15] The microelectrodes would be very useful for continuous in vivo monitoring of p_{CO_2} in fluids, particularly blood. However, the microelectrodes have been found to be less stable and respond very slowly.[16] Consequently, attempts have been made to develop a miniature electrode for in vivo monitoring of CO_2 pressures in fluids. A quinhydrone system generally used for measurement of H^+ ion activity was modified to measure p_{CO_2}.[16]

A combined platinum–calomel (or Ag–AgCl) electrode (see Fig. 4) immersed in 0.1 *N* KCl solution with 10^{-3} *M* quinhydrone and separated from the sample medium by a CO_2-permeable membrane was used. CO_2 diffusing through the membrane changed the activity of the H^+ ion in the 0.1 *N* KCl solution. The oxidation–reduction reaction and the corresponding electrode potential are given by

$$\underset{\text{Quinone}}{Q} + 2H^+ + 2e \rightleftharpoons \underset{\text{Hydroquinone}}{H_2Q} \tag{15}$$

$$E = E^\circ + \frac{RT}{2F} \ln \frac{a_Q a_{H^+}^2}{a_{H_2Q}} = E^\circ + \frac{RT}{2F} \ln \frac{a_Q}{a_{H_2Q}} + \frac{RT}{F} \ln a_{H^+} \tag{16}$$

Since a_Q and a_{H_2Q} are constant, the electrode potential is a function of a_{H^+} only. The change in a_{H^+} following entry of CO_2 through the membrane into the internal solution is reflected in the change in the value of E. Thus one can relate ΔE to $\Delta \log p_{CO_2}$ quantitatively after proper calibration with solutions of known p_{CO_2}. The stability, response time, and error in determining the p_{CO_2} of this electrode were comparable to those of the Stow electrode.

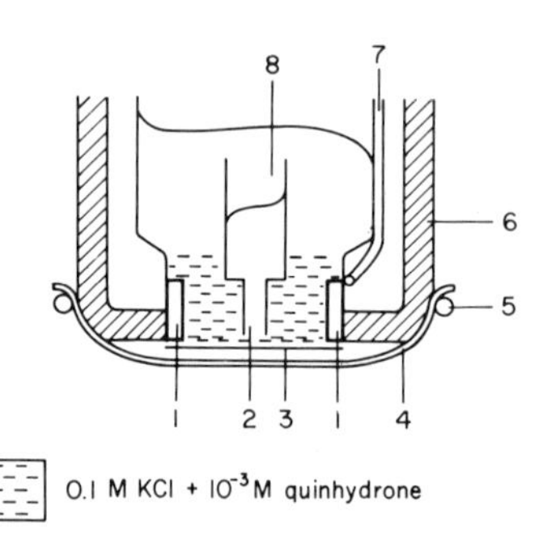

Fig. 4. A magnified view of the CO_2–quinhydrone electrode (after Van Kempen *et al.*[16]): 1, is a platinum ring connected to silver wire 7 through which an external connection for measurement can be made; 2 is the calomel–liquid junction; 1 and 2 are fixed in the lucite housing 6; 3 is lens paper to prevent bulging of the Teflon membrane 4, rubber O-ring 5, and calomel or Ag–AgCl electrode 8.

Another recent electrode (see Chapter 7) is the so-called air-gap electrode in which an air gap instead of a gas-permeable membrane separates the electrolyte layer from the sample solution.[17] The glass and the calomel reference electrodes were held in a Teflon body and the reference electrode communicated through a ceramic pin with the electrolyte layer at the surface of the glass membrane. This electrolyte layer was a solution (2×10^{-2} M) of $NaHCO_3$ saturated with a nonionic wetting agent. The sample solution (50 μl) was kept in a polyethylene cup to which 100 μl of 0.1 M lactic acid solution was added. This cup covered the glass membrane and thus created an air gap between the sample solution and the bicarbonate layer on the glass membrane. The solution in the cup was kept well stirred. The CO_2 evolved brought about the change in pH of the electrolyte layer which was sensed by the glass membrane. An improved version[18] of this air-gap electrode in which a Ag–AgCl reference electrode instead of the calomel electrode is used is shown in Fig. 5.

If a NH_4Cl solution is used as the electrolyte on the glass membrane, the air-gap electrode will sense ammonia gas. Similarly, other gases, e.g., SO_2 or oxides of nitrogen, can be sensed provided an appropriate electrolyte layer is used on the surface of the glass membrane.

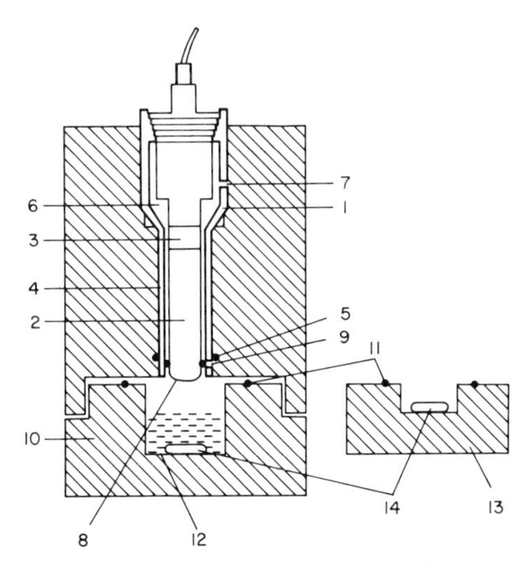

Fig. 5. An air-gap electrode (after Hansen and Ruzicka[18]): 1 is the cylindrical Teflon body holding the glass 2 and reference 3 electrodes in a common Perspex tube 4; 5 is the O-ring; and 6 is the electrolyte solution surrounding the Ag–AgCl reference electrode and contained in the Perspex tube, the solution being introduced through a small hole 7. The same solution covers the pH-sensitive surface 8 of the glass electrode. 9 is an O-ring that has been degreased and made hydrophilic in a strong solution of detergent and prevents leakage of electrolyte from the reference compartment. 10 is the sample chamber (macrochamber, and 13 is the microchamber) with a smooth coaxial hole and fitted to the main body of the equipment through the O-ring 11, which serves as a gasket. 12 is a polythene vial fitting closely into the cavity of the macrochamber. The microchamber can be used by itself as a sample holder. Both chambers are equipped with Teflon-coated stirring bars 14.

A number of advantages for the air-gap electrode have been claimed by its discoverers.[17] They are

(1) quick response since the diffusion of gas in air is faster than it is in solid, aqueous, or porous media;
(2) quick response since the layer of liquid is thinner than the membrane;
(3) no interference from surfactants, particulate matter, or organic solvents due to the absence of physical contact;
(4) easy renewal of the electrolyte layer, so the same electrode can be used for measurements of, e.g., ammonia, CO_2, or SO_2, since the species-dependent membrane effects are absent; and
(5) simpler construction.

On the other hand, proponents[19] of the polymer-based gas sensors list the following advantages:

(1) There is no need to renew the electrolyte layer on the glass

membrane after every measurement. The polymer membrane retains so much of the bulk internal electrolyte that it stabilizes the composition of the thin layer on the glass membrane surface by slow but steady diffusion.

(2) The probe can be used in continuous operation and no special equipment is required to hold the sample.

(3) The danger of the tip of the probe drying out and thereby impairing the response of the electrode is greatly minimized.

(4) The thickness of the electrolyte layer can be easily adjusted by pressing the glass membrane electrode on the polymer membrane.

(5) The electrolyte film on the glass membrane is protected by the polymer membrane and so the chances of air or other oxidizing gases attacking the film are reduced.

The air-gap electrode has been used to determine the CO_2 and ammonia contents of a series of samples,[17] to determine urea in whole blood, plasma, and serum,[18] and to determine the ammonia content of waste waters.[20]

C. SENSORS FOR OTHER GASES

The air-gap and the polymer membrane-based electrochemical systems described above may be modified to sense other gases. Necessary changes to be incorporated into the system will depend on the nature of the gas to be sensed. This in turn will dictate the nature of the polymer membrane and the internal solution forming the thin film on the glass membrane to be used. The characteristics of membranes and the possible chemical equilibria to be established are discussed by Ross *et al.*[21]

Two types of membranes have been considered in the construction of gas-sensing probes. To the first type belongs the heterogeneous membranes having a microporous structure; they are available in a wide variety of materials—cellulose acetate, Teflon, polyvinyl chloride, polyvinyl fluoride, polypropylene, and polyethylene. The second type is composed of homogeneous plastic films such as Teflon, silicone rubber, and mylar. In both types of membranes gas molecules diffuse through the membrane, but the mechanism of diffusion differs. In the heterogeneous type of membrane, the gas molecules diffuse by entering the pores in the membrane, whereas in the homogeneous type of membrane, the gas molecules cross the membrane by dissolving in the membrane phase.

The behavior of the gas-sensing system—its time response, sensitivity, limits of detection, etc.—is dependent on a number of variables such as geometry, membrane properties, and the internal electrolyte used. The concentration profile of the diffusing species at the sensing end of the electrochemical system is shown in Fig. 6, where C_1 is the sample con-

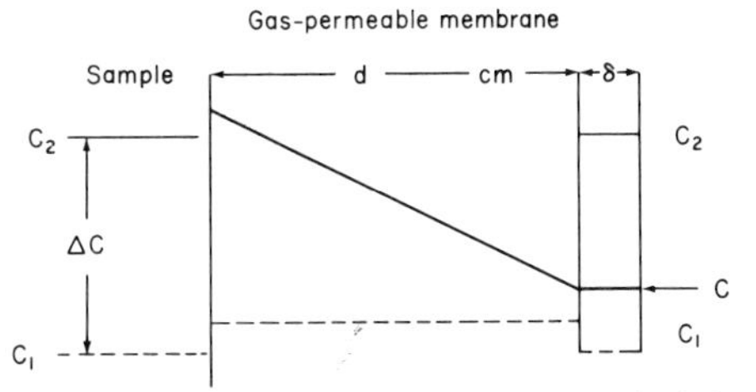

Fig. 6. Steady state concentration profile at the sensing end of the glass electrode (after Ross *et al.*[21]): δ (cm) represents the thickness of the internal electrolyte layer contacting the glass membrane of thickness d cm. C_1 is the sample concentration of the species to which the membrane electrode system is permeable and C_2 is the concentration of the species changed at time $t = 0$. ΔC is the change in concentration.

centration of the species to which the membrane (thickness d cm) of the electrode system is permeable. The concentration of the species in the internal electrolyte (film of thickness δ cm) is also C_1. The concentration in the membrane is given by

$$\overline{C} = \alpha C \tag{17}$$

where α is the partition coefficient of the species between the aqueous and the membrane phases (the membrane phase is represented by an overbar). When the concentration of the species in the sample solution is changed suddenly (i.e., at time $t = 0$) to C_2, the establishment of an immediate partition equilibrium at the membrane–sample solution interface leads to $\overline{C}_2$, the concentration of the species at the interface. In some cases, such as CO_2, this equilibrium may not be reached immediately. This will lead to a limitation on the time response of the electrode.

The concentration gradient of the species thus established leads to its flux across the membrane, which according to Fick's law is given by

$$J_g = -\frac{AD\Delta\overline{C}}{d} = -\frac{AD\Delta C\alpha}{d} \tag{18}$$

Because of this flux, the concentration of the species in the internal film layer will change and the total concentration will be equal to $A\delta C_t$. So the steady state flux, again J_g, across the film layer is given by

$$J_g = A\delta \frac{dC_t}{dt} \tag{19}$$

where C_t is the total concentration of the diffusing species in the internal electrolyte where the diffusing species can exist either as neutral, ionized, or complexed species, the mobilities of which are assumed to be equal. Thus

$$C_t = C + C_b \tag{20}$$

where C is the concentration of neutral species and C_b is the concentration of the species existing in other forms.

Differentiation of Eq. (20) with respect to C gives the relation

$$dC_t = \left[1 + \frac{dC_b}{dC} \right] dC \tag{21}$$

Substituting Eq. (21) into Eq. (19) and then equating the result to Eq. (18) gives, on rearrangement,

$$\frac{[1 + (dC_b/dC)]\, dC}{C_2 - C} = -\frac{D\alpha}{d\delta}\, dt \tag{22}$$

Introducing the parameter x, defined by

$$x = \left| \frac{C_2 - C}{C_2} \right|$$

(i.e., the fraction of C_2 approaching equilibrium) into Eq. (22) yields

$$\left[1 + \frac{dC_b}{dC} \right] d \ln x = \frac{D\alpha}{d\delta}\, dt \tag{23}$$

If it is assumed that dC_b/dC is very small or remains constant when $C_2 - C$ (i.e., ΔC) is small, Eq. (23) can be integrated. Thus

$$t = \frac{\delta d}{D\alpha} \left[1 + \frac{dC_b}{dC} \right] \ln \frac{\Delta C}{C_2} \tag{24}$$

According to Eq. (24), when $C_2 \gg C_1$, $\Delta C \approx C_2$, t (response time) becomes independent of the change in concentration. On the other hand, when $C_1 \gg C_2$, $\Delta C \approx C_1$, the response time will be determined by the ratio C_1/C_2. Ross *et al.*[21] have computed the theoretical curves that are shown in Fig. 7. These show that the time taken to reach 99% equilibrium ($t_{0.01}$) is approximately 13 times greater in going from 10^{-1} to 10^{-5} M than it is for the reverse process.

Equation (24) thus predicts the effects of the different parameters such as geometry $d\delta$, membrane characteristics $D\alpha$, electrolyte composition dC_b/dC, and the experimental conditions $\Delta C/C_2$ on the time responses of the electrode. Of these the most important are the membrane characteristics and the internal electrolyte solution. Some of the characteristic values of D and α for some membranes given by Ross *et al.*[21] for different gases are shown in Table 1.

From Eq. (24), it is seen that to obtain a satisfactory time response for the electrode the value of $D\alpha$ must be as large as possible. The data given in Table 1 show that the microporous membrane is superior in this respect

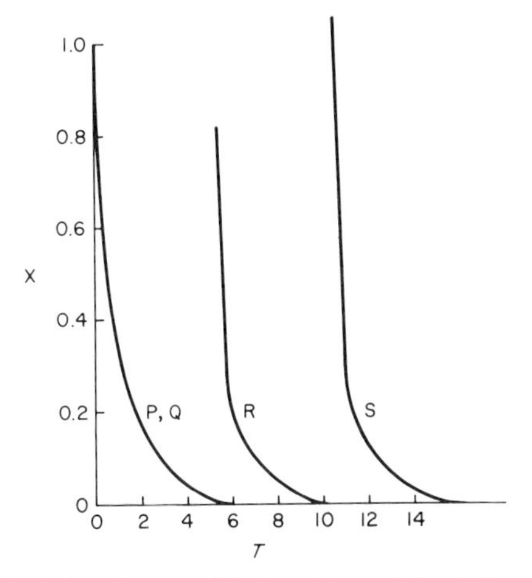

Fig. 7. Fractional deviation from equilibrium (x) vs. T for different values of C_1 and C_2 (after Ross *et al.*[21]): Curve P: $C_1 = 10^{-5}$ M, $C_2 = 10^{-1}$ M, $T_{0.01} = 4.6$; curve Q: $C_1 = 10^{-3}$ M, $C_2 = 10^{-1}$ M, $T_{0.01} = 4.6$; curve R: $C_1 = 10^{-1}$ M, $C_2 = 10^{-3}$ M, $T_{0.01} = 9.2$; curve S: $C_1 = 10^{-1}$ M, $C_2 = 10^{-5}$, $T_{0.01} = 13.8$; $dC_b/dC = 0$; $T = (D\alpha/\delta d)t$.

to other homogeneous membranes, and that silicone rubber membranes are preferable to polyethylene membranes.

The characteristics of the electrolyte chosen for the internal solution and its equilibrium with the diffusing species S have been discussed by Ross *et al.*[21] For a general case, the equilibrium reaction is

$$s\mathrm{S} + a\mathrm{A} \rightleftharpoons p\mathrm{P} + s\mathrm{S_b} \tag{25}$$

where P is the ion sensed by the glass membrane electrode and its concentration is given by

$$(\mathrm{P}) = \frac{K(\mathrm{S})^{s/p}(\mathrm{A})^{a/p}}{(\mathrm{S_b})^{s/p}} \tag{26}$$

or

$$(\mathrm{P}) = \left[\frac{K(\mathrm{A})^a}{(\mathrm{S_b})^s}\right]^{1/p}(\mathrm{S})^{s/p} \tag{27}$$

If the concentrations of the species A and $\mathrm{S_b}$ are high and they do not cross the membrane, then (A) and ($\mathrm{S_b}$) will be constant and (P) will be proportional to $(\mathrm{S})^{s/p}$, showing a response according to the Nernst equation with a slope of $2.303(RT/F)(s/p)$ per decade change in S. The sensitivity of the electrode will be determined by the stoichiometric ratio s/p. The conditions that maintain the constancy of A and $\mathrm{S_b}$ are that K, the equilibrium constant, should not be too large or too small. In any case,

TABLE 1

Values for diffusion (D) and Partition (α) Parameters for Some Membranes and Gases[a]

Gas	Heterogeneous membrane (microporous)			Homogeneous membrane: Silicone rubber			Homogeneous membrane: Low-density polyethylene		
	D (cm^2/sec)	α	$D\alpha$ (cm^2/sec)	D (cm^2/sec)	α	$D\alpha$ (cm^2/sec)	D (cm^2/sec)	α	$D\alpha$ (cm^2/sec)
O_2	1.8×10^{-1}	32	5.8	1.6×10^{-5}	9.9	1.6×10^{-4}	9×10^{-7}	2.2×10^{-1}	2×10^{-7}
CO_2	1.3×10^{-1}	1.2	1.6×10^{-1}	1.1×10^{-5}	2.6	2.9×10^{-5}	8×10^{-7}	4.8×10^{-4}	3.8×10^{-7}
H_2S			7.7×10^{-2}			3.4×10^{-5}			
Cl_2			5.4×10^{-2}						
SO_2			3.7×10^{-3}			3.8×10^{-6}			
NO_2			2.2×10^{-3}			1.0×10^{-6}			
HF			9.9×10^{-4}						
NH_3			5.3×10^{-4}			9.8×10^{-8}			
CH_3NH_2			3.3×10^{-4}						
HOAc			2.5×10^{-6}						
H_2O			1.3×10^{-7}			5.1×10^{-9}			
HCl			2.3×10^{-9}						

[a] According to Ross *et al.*[21]

TABLE 2

Gas Sensing Electrodes and Their Characteristics

Species	Internal electrolyte	Possible equilibria	Sensor electrode	Lower limit of detection	Slope	Interference	Applications
NH_3	0.01 *M* NH_4Cl	$NH_3 + H_2O \rightleftharpoons NH_4^+ + OH^-$	H^+–glass	10^{-6} *M*	− 60	Volatile amines	Soil, water, nitrogen in organic compounds
CO_2	0.01 *M* $NaHCO_3$	$CO_2 + H_2O \rightleftharpoons H^+ + HCO_3^-$	H^+–glass	10^{-5} *M*	+ 60	—	Blood, vats, fermentation
SO_2	0.01 *M* $NaHSO_3$	$SO_2 + H_2O \rightleftharpoons H^+ + HSO_3^-$	H^+–glass	10^{-6} *M*	+ 60	Cl_2, NO_2 must be absent	Stack gases, foods wines, S in fuels
NO_2	0.02 *M* $NaNO_2$	$2NO_2 + H_2O \rightleftharpoons NO_3^- + NO_2^- + 2H^+$	H^+–glass NO_3-selective	5×10^{-7} *M*	+ 60	SO_2 must be absent CO_2 interferes	Stack gases, ambient air, in foods
H_2S	Citrate buffer (pH 5)	$H_2S + H_2O \rightleftharpoons HS^- + H^+$	S^{2-}-selective (Ag_2S)	10^{-8} *M*	− 30	O_2	Pulps, anaerobic muds, fermentation
HCN	$KAg(CN)_2$	$Ag(CN)_2^- \rightleftharpoons Ag^+ + 2CN^-$	Ag^+-selective (Ag_2S)	10^{-7} *M*	− 120	H_2S	Plating baths, wastes
HF	1.0 *M* H^+	$HF \rightleftharpoons H^+ + F^-$	F^--selective (LaF_3)	10^{-3} *M*	− 60		Etching baths, steel pickling
HOAc	0.1 *M* NaOAc	$HOAc \rightleftharpoons H^+ + OAc^-$	H^+–glass	10^{-3} *M*	+ 60		
Cl_2	HSO_4 buffer	$Cl_2 + H_2O \rightleftharpoons 2H^+ + ClO^- + Cl^-$	H^+–glass Cl-selective	5×10^{-3} *M*	− 60		Bleaching
X_2 (X = Br, I)		$X_2 + H_2O \rightleftharpoons 2H^+ + XO^- + X^-$	Br-selective I-selective				

ideal behavior can be expected only over a limited range of (P). The considerations presented in the case of the CO_2 electrode have been applied by Ross *et al.*[21] to describe the behavior of the SO_2 electrode.

Bailey and Riley[19] used the commercial gas electrodes to measure ammonia (EIL Model 8002-200) and sulfur dioxide (Model 810-200) in aqueous solutions and also constructed another gas electrode similar to the one shown in Fig. 3 to measure nitrogen oxides in aqueous solution. The membrane used to trap the internal solution (0.4 M KNO_3 + 0.1 M $NaNO_2$ + 0.1 M KBr saturated with AgBr) was 0.025-mm-thick polypropylene. The standards were prepared from NH_4Cl (ammonia probe), potassium metabisulfite (sulfur dioxide probe), and $NaNO_2$ (nitrogen oxide probe). The buffers used for pH adjustments were 1 M NaOH, 2 M $HClO_4$, and 5.4 M H_2SO_4, respectively. A number of other gas-sensing electrodes constructed and evaluated by Ross *et al.*[21] are given in Table 2 along with their performance characteristics.

In the construction of these gas-sensing electrodes, the factors that require consideration besides those already mentioned are[19] (a) sample pH, (b) sample temperature, (c) sample osmotic pressure, (d) state of determinand, (e) potential interferents, (f) volatility of the determinand, (g) stirring of the sample, and (h) standardization. All these factors must be maintained at their optimum levels to get the best results. If the pH of the sample is not maintained at the level at which all of the determinand is available in the state of dissolved gas, the accuracy of the measurement will be impaired. This applies to factors (d) and (e). Proper precautions must be taken to release the determinand from the complexes, if any, and to eliminate the interferents from the sample. Temperatures of standards and samples must be similar. The osmotic pressure of the sample and that of the internal electrolyte must be made equal by addition of some inert electrolyte if necessary, otherwise water transport across the membrane will alter the concentration of the internal electrode and lead to drift of electrode potential. Stirring in closed systems not only prevents escape of volatile species but also eliminates the particles of the determinand from adhering to the sensor and to the surfaces of the sample container. Since the accuracy of measurements obtained with a sensor is limited by the accuracy of the standards, scrupulous care must be taken in the preparation of standard solutions.

D. OXYGEN ELECTRODE

Measurement of the partial pressure of oxygen (p_{O_2}) in gaseous, liquid, and semiliquid media is now daily routine work in most hospitals and physiological laboratories. The polarographic principles enunciated by Heyrovsky[22] and his co-workers are utilized in the analysis of many substances including oxygen. The polarographic method was introduced to biology by Prat.[23] Davies and Brink[24] covered the cathode with a membrane for measurements in blood to eliminate the poisoning of the electrode by protein. Similarly, Clark *et al.*[25] used a cellophane membrane to cover the platinum cathode to eliminate the undesirable effects of red blood cells and stirring. However, the results obtained with the cathode protected from a variety of materials proved unsatisfactory[26] until the situation was changed by Clark,[27] who introduced the principle of the closed system in which both the oxygen cathode and the reference anode were covered by a single membrane permeable to oxygen.

Several comprehensive review articles have been published concerning the measurements of pH and oxygen tension in various biological tissues and blood. The different cathodes used in various configurations to measure p_{O_2} levels in different biological preparations have been described by Davies[28] and others.[7–9, 29–34] A brief description of the characteristics of the Clark type of oxygen sensor is now given.

The basic arrangement of a simple electrolytic cell to illustrate the polarographic principle upon which the oxygen sensor is based is shown in Fig. 8. When the platinum electrode (cathode) is made a few tenths of a volt negative with respect to the anode in a solution containing dissolved oxygen as shown in Fig. 8, the oxygen is reduced at the cathode to give a current that can be measured by a galvanometer. When the cathode is made more negative by increasing the voltage, electrolysis occurs more

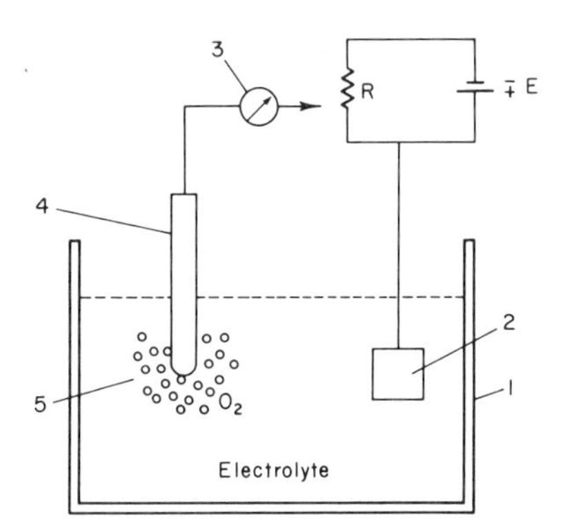

Fig. 8. Basic arrangement of a simple electrolytic cell to illustrate the polarographic principle upon which the oxygen sensor is based: 1, container; 2, anode (Ag–AgCl); 3, current-measuring device (galvanometer); 4, cathode (platinum); and 5, dissolved oxygen. *E* is the voltage source connected to the resistance *R*, by varying which the applied potential can also be varied.

rapidly and increases the current in the manner shown in Fig. 9. The rising phase at low currents is limited by the rates of diffusion of the gas to the electrode surface and its reduction. At more negative values of applied potential, the curve forms a plateau because the rate of reduction is faster than the rate at which oxygen molecules are supplied to the cathode by diffusion. Thus the electrolysis current is affected only to a minor extent by the electrode potential. Usually when diffusion and convection are limited, the current in the plateau region becomes independent of potential. At still higher voltages, current rises due to production of current-carrying reduction products.

The current at any time t due to linear diffusion to a plane electrode is given by[35]

$$i_t = nFAD \frac{C}{\sqrt{\pi D t}} \tag{28}$$

where n is the number of electrons per mole of O_2 electrolyzed, C is the initial concentration of oxygen, A is the area of platinum surface in square centimeters, and D is the diffusion coefficient of oxygen.

The current due to spherical diffusion is given by[35]

$$i_t = nFADC\left[\frac{1}{\sqrt{\pi D t}} + \frac{1}{r}\right] \tag{29a}$$

$$= nFDC4\pi\left[\frac{r^2}{\sqrt{\pi D t}} + r\right] \tag{29b}$$

where r is the radius of the electrode. According to Eqs. (28) and (29), the steady current is first proportional to the diffusion coefficient of oxygen in the medium. Second, according to Eq. (29) the current is the sum of a constant part and a transient portion. The transient portion [first term of Eq. (29)] will fall off with time; its contribution to the total current will

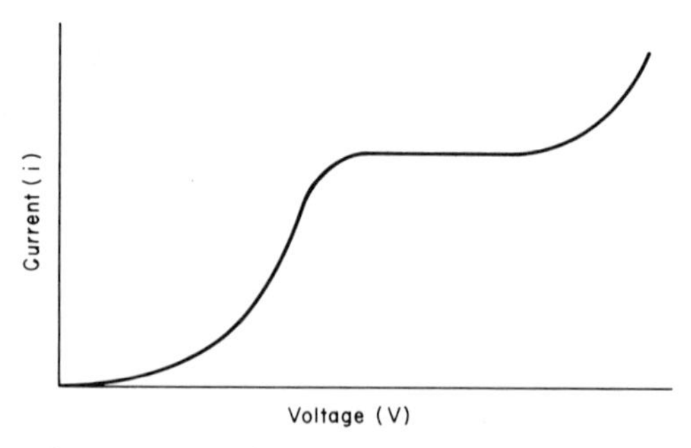

Fig. 9. Current–voltage curve for the reduction of oxygen at a constant concentration of oxygen.

depend on the size of the electrode. Thus for a sphere 1 mm in diameter the two terms become equal when $t \approx 31$ sec ($D = 2.6 \times 10^{-5}$ for oxygen in water[36]) and the first term is one-tenth of the second after 3100 sec. For a sphere 10 μ in diameter, the first term is one-eighteenth of the second after 1 sec. It is obvious that a steady state is rapidly attained if the electrode is small.

The shape of the curve in Fig. 9 (polarogram) indicates the nature of the substance in the sample that is being reduced; the height of the plateau gives its concentration. If the applied voltage in the plateau region is held constant, the heights of a series of plateaus will be proportional to a series of concentrations of the substance that is being reduced (in this case oxygen).

The nature of the anode used in the electrolysis cell is governed by the limitations of space. If space is not at a premium, a calomel half-cell can be used; if space becomes restricted, a Ag–AgCl electrode is very convenient.

The open system shown in Fig. 8, although adequate for simpler solutions, is not suitable for use with complex samples such as blood and biological tissues. In these studies, the Clark type of electrode (see Fig. 10)

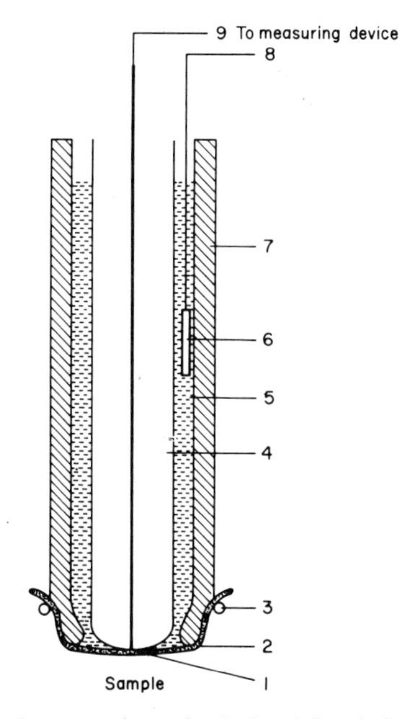

Fig. 10.. Clark type of oxygen electrode: 1, tip of the platinum cathode 9; 2, membrane (Teflon); 3, rubber O-ring; 4, insulating glass rod with platinum wire sealed in; 5, electrolyte solution; 6, Ag–AgCl reference anode 8; and 7, insulated container.

has proved very useful. Membranes such as polyethylene, polypropylene, Teflon, or mylar are used to separate the electrolyte (KCl or KCl with buffer) from the sample medium in which oxygen is to be measured. These membranes must be permeable to oxygen and impermeable to other ionized contaminants.

A very simplified steady state result for this membrane-covered electrode is to express current (i) as being proportional to p_{O_2} outside the membrane. This is depicted in Fig. 11 for a theoretically perfect electrode.[8] It is assumed that the electrode is polarized in the plateau region.

The oxygen electrodes are usually polarized between -0.6 and -0.8 V with respect to the anode or reference electrode; however, the actual voltage depends on the particular electrode and the manufacturer. Therefore, the linearity of the electrode response to p_{O_2} must be checked and the calibration of the electrode carried out with oxygen-free N_2 gas (for instrument zero) and room air [20.93% O_2 (v/v)] for the span. The time response of the electrode is generally much longer when it is changed from a high p_{O_2} to a low p_{O_2} than when the change is in the other direction. After the electrode has been exposed to a sample with high p_{O_2}, it may take several minutes to reach the true zero value, probably due to the reaction

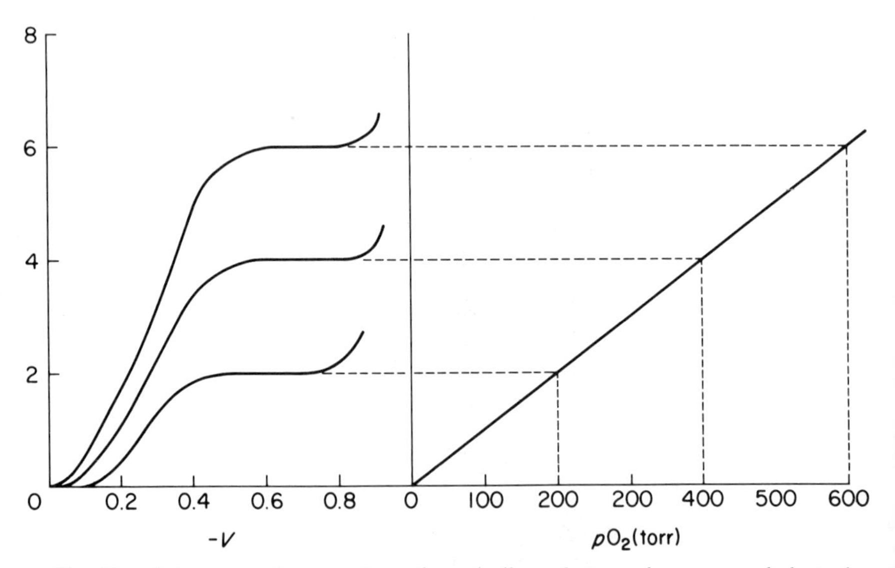

Fig. 11. Polarogram of oxygen for a theoretically perfect membrane-covered electrode. Electrode current i is in arbitrary units (after Smith and Hahn[8]). V is the cathode polarization voltage with respect to the anode. In the plateau region the current i is directly proportional to the partial pressure of oxygen p_{O_2}.

taking place at the cathode:

$$O_2 + 2H_2O + 2e \rightleftharpoons H_2O_2 + 2OH^-$$

$$H_2O_2 + 2e \rightleftharpoons 2OH^-$$

Since no H^+ ions are involved, these reactions should be insensitive to pH provided H_2O_2 conversion proceeds fast enough. Otherwise the medium becomes alkaline. If all the H_2O_2 is electrolyzed, the overall reaction requires four electrons per molecule of oxygen. The Clark type of oxygen electrode has been used in the chronoamperometric determination of dissolved oxygen in situ in natural and waste waters.[37]

If blood contains any anesthetics in the determination of oxygen, care is required in the estimation of the oxygen, since several halogenated hydrocarbons including the commonly used anesthetic halothane (containing bromine) are known to be polarographically reduced in standard Clark-type oxygen electrodes.[38] Severinghaus *et al.*[38] recommend that if the electrode must be used in the presence of halothane, the polarizing voltage should be reduced to about −0.5 V and the possibility of its interference with the response of the electrode should be tested by introduction of gas mixtures containing halothane at both zero and a known p_{O_2}. If there is interference, a slow upward drift of a zero p_{O_2} reading will occur.

REFERENCES

1. G. I. Goodfellow and H. M. Weber, *Analyst* **97**, 95 (1972).
2. D. Midgley and K. Torrance, *Analyst* **97**, 626 (1972).
3. J. G. Montalvo, Jr., *Anal. Chim. Acta* **65**, 189 (1973).
4. R. W. Stow and B. F. Randall, *Amer. J. Physiol.* **179**, 678 (1954).
5. R. W. Stow, R. F. Baer, and B. F. Randall, *Arch. Phys. Med. Rehabil.* **38**, 646 (1957).
6. K. H. Gertz and H. H. Loeschke, *Naturwissenschaften* **45**, 160 (1958).
7. J. W. Severinghaus and A. F. Bradley, *J. Appl. Physiol.* **13**, 515 (1958).
8. A. C. Smith and C. E. W. Hahn, *Brit. J. Anaesth.* **41**, 731 (1969).
9. J. W. Severinghaus, *Ann. N. Y. Acad. Sci.* **148**, 115 (1968).
10. C. H. Hertz and B. Siesjo, *Acta Physiol. Scand.* **47**, 115 (1959).
11. D. B. Cater and I. A. Silver, *in* "Reference Electrodes" (D. J. G. Ives and G. J. Janz, eds.), p. 464. Academic Press, New York, 1961.
12. O. Siggaard-Andersen, "The Acid-Base Status of Blood." Munksgaard, Copenhagen, 1972.
13. D. H. Ingvar, B. Siesjo, and C. H. Hertz, *Experientia* **15**, 306 (1959).
14. H. P. Constantine, M. R. Craw, and R. E. Forster, *Amer. J. Physiol.* **208**, 801 (1965).
15. B. Rybak and H. Penforms, *J. Physiol. (Paris)* **61**, 394 (1969).
16. L. H. J. Van Kempen, H. Deurenberg, and F. Kreuzer, *Resp. Physiol.* **14**, 366 (1972).
17. J. Ruzicka and E. H. Hansen, *Anal. Chim. Acta* **69**, 129 (1974).
18. E. H. Hansen and J. Ruzicka, *Anal. Chim. Acta* **72**, 353 (1974).
19. P. L. Bailey and M. Riley, *Analyst* **100**, 145 (1975).

20. J. Ruzicka, E. H. Hansen, P. Bisgaard, and E. Reymann, *Anal. Chim. Acta* **72**, 215 (1974).
21. J. W. Ross, J. H. Riseman, and J. A. Krueger, *Pure Appl. Chem.* **36**, 473 (1973).
22. J. Heyrovsky, *Chem. Listy* **16**, 256 (1922).
23. S. Prat, *Biochem. Z.* **175**, 268 (1926).
24. P. W. Davies and F. Brink, *Rev. Sci. Instrum.* **13**, 524 (1942).
25. L. C. Clark, R. Wolf, D. Granger, and Z. Taylor, *J. Appl. Physiol.* **6**, 189 (1953).
26. R. H. Shepard, *Fed. Proc.* **15**, 169 (1956).
27. L. C. Clark, *Amer. Soc. Art. Int. Org.* **2**, 41 (1956).
28. P. W. Davies, *in* "Physical Techniques in Biological Research" (W. L. Nastuk, ed.), Vol. IV, p. 137. Academic Press, New York, 1962.
29. D. B. Cater and I. A. Silver, *in* "Reference Electrodes" (D. J. G. Ives and G. J. Janz, eds.), p. 503. Academic Press, New York, 1961.
30. M. B. Laver and A. Seifen, *Anesthesiology* **26**, 73 (1965).
31. A. P. Adams, J. O. Morgan-Hughes, and M. K. Sykes, *Anesthesia* **22**, 575 (1967); **23**, 47 (1968).
32. A. P. Adams and J. O. Morgan-Hughes, *Brit. J. Anaesth.* **39**, 107 (1967).
33. I. Fatt, *Ann. N. Y. Acad. Sci.* **148**, 81 (1968).
34. L. C. Clark and G. Sachs, *Ann. N. Y. Acad. Sci.* **148**, 133 (1968).
35. J. Heyrovsky and J. Kuta, "Principles of Polarography," p. 76. Academic Press, New York, 1966.
36. C. R. Wilke, *Chem. Eng. Progr.* **45**, 218 (1949).
37. M. D. Lilley, J. B. Story, and R. W. Raible, *J. Electroanal. Chem. Interfacial Electrochem.* **23**, 425 (1969).
38. J. W. Severinghaus, R. B. Weiskopf, M. Nishimura, and A. F. Bradley, *J. Appl. Physiol.* **31**, 640 (1971).

Chapter 11

ENZYME ELECTRODES

The enzyme electrodes are similar to the gas-sensing membrane electrode systems described in Chapter 10. The significant difference lies in the immobilization of the enzyme at the sensing electrode surface. Probably the first description of an enzyme electrode system was given by Clark and Lyons[1] who determined glucose amperometrically by means of glucose oxidase held as a thin layer between two cuprophane membranes. The reaction involved the consumption of oxygen according to the equation

$$\text{Glucose} + O_2 + H_2O \underset{\text{oxidase}}{\overset{\text{glucose}}{\rightleftharpoons}} \text{gluconic acid} + H_2O_2 \tag{1}$$

Updike and Hicks[2] introduced enzyme electrode, and made a Clark type of oxygen electrode that contained dual cathodes with glucose oxidase immobilized in a polyacrylamide gel. The principle upon which the enzyme electrode is based is illustrated in Fig. 1 and the dual cathode enzyme electrode used by Updike and Hicks[2] is shown in Fig. 2.

In order to measure glucose concentrations with the glucose oxidase enzyme electrode, the electrode should be calibrated with standard glucose solutions like any other ion-selective membrane electrode. The calibration curves are affected by both the intragel enzyme concentration and the pore size. An increase in enzyme concentration increases the electrode sensitivity but decreases the range of linear response of the electrode. On the other hand, an increase of gel concentration, which decreases the pore size, decreases the sensitivity of the electrode but increases the linear concentration range.

The platinum tip diameter is less than 25 μ and the plastic membranes are 25 μ thick. The purpose in using a dual cathode arrangement (Fig. 2) is to eliminate the interference of O_2 with the enzyme electrode since the

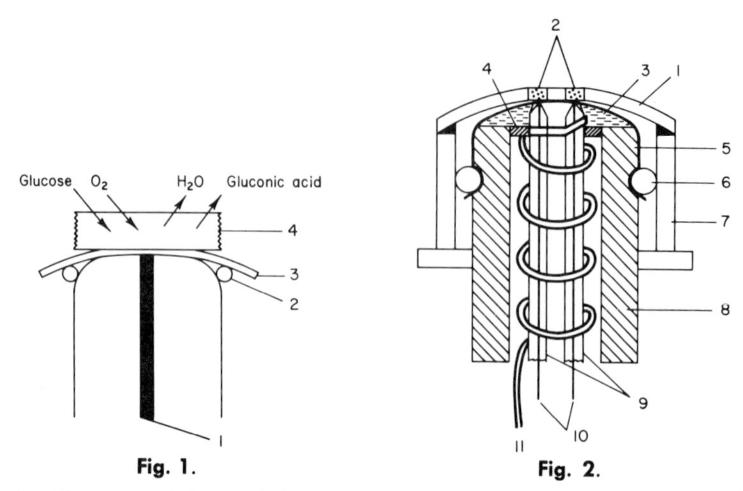

Fig. 1. Illustration of the principle of the enzyme electrode (after Updike and Hicks[2]): 1, platinum cathode; 2, silver anode; 3, plastic membrane; and 4, enzyme gel layer.

Fig. 2. Enzyme electrode with dual cathodes (after Updike and Hicks[2]): 1, nylon net impregnated with silicone rubber; 2, orifice for enzyme gel; 3, electrolyte solution; 4, epoxy seal; 5, plastic membrane; 6, rubber O-ring; 7, plastic cup; 8, electrode housing; 9, glass capillary tubes to hold the platinum cathodes 10; and 11, silver anode.

electrode is sensitive to both glucose and oxygen. Glucose oxidase exists at two sites (see Fig. 2); one site is made unresponsive to glucose by raising its temperature[3] above 7°C to destroy the activity of the enzyme. However, this part still responds to p_{O_2}. By recording the difference between the output of the two cathodes with the same Ag–AgCl as the reference electrode, the enzyme electrode response to changes in glucose concentrations can be obtained.

A. IMMOBILIZATION OF ENZYME

The success of the enzyme electrode depends on the immobilization of the enzyme in the gel layer. This problem seems to be very important in the use of enzymes in chemical analysis. If enzymes have to be used in ordinary chemical analysis, large quantities are required. This is an expensive proposition. Consequently, considerable time and effort have been spent in finding ways to use as small a quantity of pure enzyme as possible in chemical analysis. The enzyme electrode in which small quantities of the concerned enzyme are fixed in such a way that the enzyme is allowed to retain its activity for a long time has attracted considerable attention. In the last six years or so, a few reviews[4–8] have been published which emphasize the various aspects of enzyme electrode construction, function, and applications. A book[9] on immobilized enzymes reviews the various methods available for fixing enzymes to various matrices. Some of the techniques that are useful in the construction of electrodes are the following:

(a) Enzymes may be entrapped within a hydrophilic membrane as described above or cross-links may be established between molecules of the enzyme in the formation of membranes.

(b) The enzyme may be chemically bound to membranes or other surfaces.

(c) The enzyme may be copolymerized with other enzymes or proteins.

(d) The enzyme may be immobilized in a hydrocarbon-based liquid surfactant membrane to form microcapsules.[10]

(e) The enzyme, for example urease, may be incorporated into collagen by electrolysis to form a urease–collagen membrane on the platinum cathode.[11]

In using any of these procedures, it is important to avoid steps that would denature the enzyme partly or fully. In order to be certain of this, the end product must be monitored for its enzyme activity before it is used in any sensing device. The specific method to be used for immobilization of the enzyme would depend on the particular enzyme electrode system. The sensing device used in the enzyme electrode will depend on the products of the enzymatic reaction. Invariably, one of the solid or liquid ion-selective electrodes described in the preceding chapters would be used.

B. ENZYME ELECTRODE FOR THE ESTIMATION OF GLUCOSE

There are several methods available for the determination of glucose. Frequently, spectrophotometric[12, 13] and electrochemical[14–16] methods are used for the determination of glucose in biological fluids. Electrochemical methods are relatively simpler and amenable to automation. Most of them involve an enzyme-catalyzed system whose rate of reaction is measured. In the method used by Updike and Hicks[2] and others,[17] the oxygen sensor is used to measure the decrease of p_{O_2} in the glucose sample [see Eq. (1)] to monitor the changes of glucose concentration. Williams *et al.*[16] replaced oxygen as the hydrogen acceptor with benzoquinone to monitor glucose according to the following equations:

$$\text{Glucose} + \text{quinone} + H_2O \xrightarrow[\text{oxidase}]{\text{glucose}} \text{gluconic acid} + \text{hydroquinone} \tag{2}$$

$$\text{Hydroquinone} \xrightarrow{\text{platinum}} \text{quinone} + 2H^+ + 2e \tag{3}$$

The reaction potential is 0.4 V with respect to the standard calomel electrode. The enzyme electrode is composed of the sensor (Pt electrode), enzyme reaction layer (glucose oxidase entrapped in a porous or gel layer), and dialysis layer containing a cellophane membrane. The method of glucose measurement in blood requires the addition of buffer salts (0.04 *M* phosphate, pH 7.4 containing 0.026 *M* NaCl + 0.004 *M* KCl) and quinone to the blood. The enzyme membrane electrode is introduced and the application of 0.4 V gives the current, which is proportional to the concentration of glucose. Another experiment using a similar electrode without the enzyme is performed to get the blank or the background current which, subtracted from the value obtained with the enzyme electrode, gives a measure of the blood glucose. In a similar manner lactate has been determined using an enzyme electrode containing lactate dehydrogenase (cytochrome b_2). The reactions involved are

$$\text{Lactate}^- + 2\text{Fe(CN)}_6^{3-} \xrightarrow{\text{LDH}} \text{pyruvate}^- + 2\text{Fe(CN)}_6^{4-} + 2\text{H}^+ \qquad (4)$$

$$2\text{Fe(CN)}_6^{4-} \xrightarrow{\text{Pt}} 2\text{Fe(CN)}_6^{3-} + 2\text{e} \qquad (5)$$

The reaction potential is 0.4 V with respect to the standard calomel electrode (SCE).

Another electrochemical method (constant current voltametry) for the estimation of D-glucose is to use the immobilized enzyme (glucose oxidase) in a polyacrylamide gel–platinum gauge matrix as a catalyst.[7, 18] This enzyme–Pt matrix serves as one electrode and using another suitable electrode (Pt), constant dc current is passed to oxidize glucose at constant pH, and the potential is measured. The potential versus current data were linear, with a slope of -56 mV.

Oxidation of glucose [see Eq. (1)] gives H_2O_2. If this production of H_2O_2 can be measured amperometrically, glucose can be easily estimated. This possibility was explored by Guilbault and Lubrano[19] who prepared three types of enzyme electrodes. Type 1 contained glucose oxidase chemically bound to polyacrylamide, a thin layer of which was trapped between the electrode (Pt) and a layer of cellophane with the help of an O-ring. Type 2 contained glucose oxidase physically entrapped over the electrode using a matrix of polyacrylamide gel. Type 3 contained powdered glucose oxidase held between the electrode and cellophane membrane with a rubber O-ring (solubilized). The electrode was placed in a solution of glucose and a potential of 0.6 V (vs. SCE) was impressed across the electrode. H_2O_2 formed by the enzyme reaction diffused out of the enzyme layer toward the Pt sensor where it was oxidized according to the reaction.

$$H_2O_2 \rightarrow O_2 + 2H^+ + 2e \qquad (6)$$

Typical response curves of the glucose electrode showed that the current initially decreased for about 2 sec because of the diffusion of glucose through the cellophane and/or enzyme gel layers, and then started to increase with time reaching a steady state. Thus two methods of analysis based on the initial rate of increase of current and the final steady state current (about 1 min) can be used. Both these measurements, as a function of glucose concentration, give satisfactory calibration curves. Study of the effect of pH on the current indicated that the optimal pH was 6.6 for the rate method and 6.9 for the steady state current method. Of the three types of enzyme membrane electrodes used, the long-term stability of the electrodes decreased in the order chemically bound > physically bound > solubilized.

This method, which is used in the determination of blood glucose, compared favorably with the other commonly used methods with respect to accuracy, precision, and stability. However, it was found[20] that the sensitivity of this method is not as good as the method in which p_{O_2} in the solution is monitored. This was probably due to the unstability of H_2O_2 which could be consumed in other ways. Despite this, a double enzyme electrode in which two consecutive reactions occur has been constructed to measure glucose activity.[21] The two reactions are that given in Eq. (1) and

$$H_2O_2 + 2I^- + 2H^+ \xrightarrow{\text{catalyst}} I_2 + 2H_2O \tag{7}$$

Reaction (7) can be catalyzed by molybdate ions[14, 15] or by a peroxidase enzyme which has a higher efficiency at low concentrations of iodide. As H_2O_2 reacts in a stoichiometric manner with iodide, the change in iodide concentration can be measured by using an iodide-selective electrode. The decrease in iodide activity has been monitored both in flow streams[21, 22] and in stationary solutions.[21]

Two iodide-selective electrodes (reference and indicator electrodes) separated by a delay coil and a chamber in which the reaction takes place were used in the flow stream. The double enzyme electrode was used for measurements in stationary solutions.[21] The flow system of Nagy *et al.*[21] is shown in Fig. 3. A phosphate-buffered substrate solution (glucose + KI) introduced at 2 and an enzyme solution with an optimal ratio of peroxidase to glucose oxidase of 1 : 2 introduced at 1 were mixed at 4. The mixture passed through 5 for a certain time for the reaction to proceed. The decrease in the concentration of I^- was monitored at 6. The electrode response to glucose concentration was calibrated under the following conditions: 10^{-4} *M* KI, 1 *M* phosphate buffer, pH 6.0, enzymes (peroxidase 0.3 mg/ml, and glucose oxidase 6 mg/ml) temperature 26°C, and reaction time 235 sec. In the range of glucose concentration 10^{-3}–10^{-4} *M*, a change in potential of 40 mV was observed. Llenado and Rechnitz[22] used

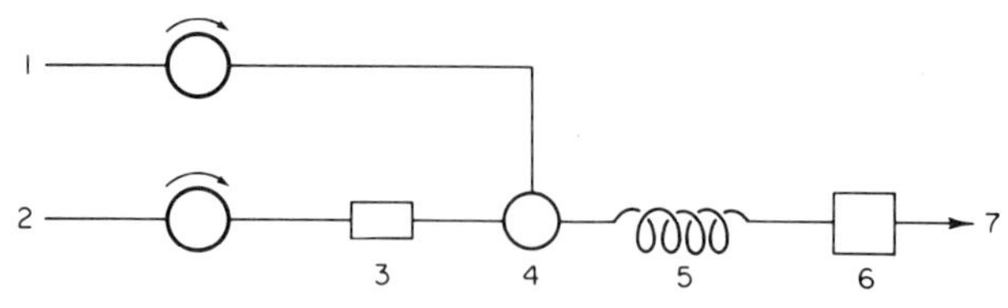

Fig. 3. Flow system for the determination of glucose: 1, glucose oxidase and peroxidase solution; 2, substrate solution; 3, reference iodide-selective electrode; 4, mixing chamber; 5, delay coil kept at constant temperature; 6, indicator iodide-selective electrode; and 7, outflow.

molybdenum ions instead of peroxidase to catalyze the reaction in their automatic glucose analysis system. They have studied the responses of the flow-through iodide-selective electrode using the fixed time kinetic approach. They have optimized the different conditions, such as influence of pH, type and concentration of buffer, time and temperature of incubation for the reaction to proceed, and amount of enzyme, that affect the electrode signal. The stream of glucose-containing sample was mixed with the other streams of air or oxygen, enzyme, iodide, and catalyst and the reactions were allowed to proceed for a fixed time at constant temperature. After the allowed time, reaction (1) was terminated by a strong acid (0.25 *M* $HClO_4$) which also drove reaction (7) to completion. The sample stream was debubbled and allowed to enter the iodide-selective flow-through electrode which monitored the iodide content of the sample.

In the case of measurements using stationary solutions, three different procedures were used.[21] In procedure 1, the enzyme layer consisted of a solution of enzymes at concentrations of 100 mg/ml of glucose oxidase and 50 mg/ml of peroxidase. In procedure 2, the enzyme mixture was physically entrapped in polyacrylamide which was polymerized by light on a nylon mesh. In procedure 3, the enzyme mixture was chemically bound to the polyacrylic acid-based matrix. In all cases the enzyme layer on the iodide-selective membrane electrode was covered with a tight cap made from dialysis paper. The electrodes were stored in phosphate buffer solutions at pH 6.

The double enzyme electrode used in the glucose solution containing KI allowed reactions (1) and (7) to proceed at the sensor surface. This coupled to the calomel reference electrode enabled the potential to be measured as a function of glucose concentration. Steady state potentials measured as a function of log(glucose) gave an S-shaped curve. The non-Nernstian response was very marked in the glucose concentration range 10^{-5}–10^{-7} *M* and the electrode prepared by procedure 1 gave the best results. The double enzyme electrode had no interferences from anions such as Cl^-, F^-, and PO_4^{3-} or cations such as Ca^{2+}, Mg^{2+}, Fe^{3+}, K^+, and Na^+.

Protein (egg albumin) and urea did not interfere but strong interference was obtained from uric acid, tyrosine, ascorbic acid, and Fe^{2+} which should therefore be removed from the samples before analysis.

C. ENZYME ELECTRODE FOR THE ESTIMATION OF UREA

An enzyme electrode for the determination of the amount of urea in a solution or biological fluid can be constructed by using an ammonium ion-selective glass membrane electrode (Beckman Nos. 39047 and 39137) whose outer surface is suitably treated to hold an enzyme. For this purpose Guilbault and co-workers[23, 24] evaluated the selectivity characteristics of the glass membrane electrodes and used them for the assay of deaminase enzyme systems. The No. 39047 electrode showed a selectivity order which decreased as $Ag^+ > H^+ > K^+ > NH_4^+ > Na^+ > Li^+ \gg Mg^{2+}, Ca^{2+}$ whereas the No. 39137 electrode showed the selectivity order $Ag^+ > K^+ > H^+ > NH_4^+ > Na^+ > Li^+ > Mg^{2+}, Ca^{2+}$. At pH 7, serious interferences in the assay of NH_4^+ ion came from Ag^+ and K^+ ions.

Katz and Rechnitz used electrode No. 39137 for the determination of urea[25] and urease[26] after complete conversion of urea to NH_4^+ ion by enzymatic hydrolysis. Thus

$$CO(NH_2)_2 + H_3O^+ + H_2O \xrightarrow{\text{urease}} 2NH_4^+ + HCO_3^- \tag{8}$$

The potential sensed by the electrode in the solution after hydrolysis is determined by the amount of ammonium ion produced and therefore is given by the concentration of urea or urease. The effects of various ions on the urease-catalyzed reaction of urea were followed by Katz and Cowans.[27] In the study of Guilbault *et al.*, [23] an attempt was made to assay both the enzyme and the substrate continuously. The enzymatic reactions followed were

$$\text{Urea} \xrightarrow{\text{urease}} NH_4^+$$

$$\text{Glutamine} \xrightarrow{\text{glutaminase}} NH_4^+$$

$$\text{Asparagine} \xrightarrow{\text{asparaginase}} NH_4^+$$

$$\text{Amino acid} \xrightarrow{\text{oxidase}} NH_4^+$$

$$\text{Amine} \xrightarrow{\text{oxidase}} NH_4^+$$

$$\text{Glutamic acid} \xrightarrow{\text{dehydrogenase}} NH_4^+$$

Before initiation of the enzymatic reaction, a zero potential should be measured and upon initiation a potential due to production of the NH_4^+

ion following the equation

$$E = E^\circ + 0.059 \log(\mathrm{NH_4^+}) \quad (9)$$

should be observed. Differentiation of Eq. (9) with respect to time gives

$$dE/dt = 0.059[1/(\mathrm{NH_4^+})][d(\mathrm{NH_4^+})/dt] \quad (10)$$

Provided $1/(NH_4^+)$ is constant, dE/dt will be proportional to the rate of change of (NH_4^+) with time. This has been shown to be so when the initial rate of the reaction was taken by drawing a tangent to E versus time curves. A plot of $\Delta E/\Delta t$ versus concentration of enzyme and/or substrate always gave a straight line in all the enzyme systems studied.

Again Montalvo and Guilbault[24] found that when the cation-selective electrode (No. 39137) was coated with an enzyme gel layer reinforced with a nylon net, the useful range of the electrode was extended down to an ammonium ion concentration of 5×10^{-5} M as opposed to the limit of 10^{-3} M obtained with the uncoated electrode. Also, the coating had no effect on the selectivity characteristics of the electrode.

Using a layer of acrylamide gel (60–350 μ thick) in which urease was immobilized on the surface of the glass membrane, an enzyme electrode was constructed. When this was placed in a solution containing urea, the substrate diffused into the gel layer of immobilized enzyme and underwent hydrolysis according to Eq. (8). The NH_4^+ ion produced was sensed by the ammonium ion-selective glass membrane electrode.[28] Improvements were introduced in the construction of the electrode by exploring other possibilities of immobilizing the enzyme urease.[29, 30] Three types of urease electrodes were prepared. Type 1 was prepared by coating the No. 39137 Beckman glass electrode with an enzyme gel (urease–acrylamide); type 2 had a cellophane membrane over the enzyme layer; and type 3 had two cellophane films (i.e., glass surface–cellophane, enzyme layer–cellophane). The three types of electrodes with a 350 μ netting and 175 mg/ml of gel gave the same response to urea. Cellophane layers prevented the leaching of urease into the surrounding solution and had no effect on the electrode response irrespective of the urea concentration. An improved urea electrode has been described by Montalvo.[31] Despite the good stability and dynamic response times,[5, 29, 30] the urease electrode showed interferences from Na^+ and K^+ ions. Placing a cation resin membrane covering in place of the outer cellophane membrane should help in reducing this interference from Na^+ and K^+ ions, but may present substrate diffusional problems. Therefore Guilbault and Hrabankova[32] used cation-exchange resin granules with the sample solution to remove the Na^+ and K^+ ions present in blood and urine. Two procedures were used. In one, a small amount of the resin (1–2 g 50 ml of sample) was added directly into the solution to be analyzed and the electrode response was measured after stirring the mixture. In the second procedure, 2.5–10 g of ion exchange resin was mixed with a small volume of the sample (5 ml of sample + 5 ml of water) and an aliquot of the mixture (5 ml) was added to 50 ml of the buffer solution in which the electrode response was measured. Procedure 1 was found to be fast and gave reproducible results. In simple solutions of urea, addition of 2 g of Dowex 50W-X2 to the solution and as much as 5×10^{-3} M Na or 10^{-4} M K did not interfere in the determination of 10^{-5} M urea. However, this procedure is useless for the determination of urea in biological fluids such as blood and urine because some of the components of these fluids will affect the enzyme electrode response. In order to eliminate these interferences, use of an uncoated cation-selective glass membrane electrode (Beckman No. 39137 or 39047) as the reference electrode has been recommended. In this case an accuracy of about 2–3% for the direct assay of urea in blood and urine has been claimed.[32]

Other approaches to eliminate this problem of interferences from Na^+ and K^+ ions have also been made. One method was to use the CO_2 electrode to sense the CO_2 released according to reaction (8). Guilbault and Shu[33] used this procedure by coupling a urease layer with a CO_2 electrode. However, the response of the electrode was sluggish compared to that of a pure CO_2 electrode. The pH of the solution significantly affected the electrode response. In most measurements a pH of 6.2 was used.

Another approach made by Guilbault and Nagy[34] was to prepare an entirely different membrane electrode that had a high selectivity for NH_4^+ ions over Na^+ and K^+ ions. This NH_4^+ ion-selective electrode was made by using the antibiotic nonactin (see Chapter 8) as the active ingredient embedded in a silicone rubber matrix. The membrane was made by mixing 500 mg of silicone rubber thoroughly with 300 mg of nonactin. The paste was pressed between two glass plates coated with paraffin and left for two days to form the solid membrane from which the electrode was constructed by attaching a piece of the membrane (diameter 5 mm; thickness 0.2 mm) to a glass tube. A silver tape in 0.1 N NH_4Cl solution was used as the internal reference electrode. A nylon net was placed on the outer surface of the NH_4^+ electrode and a solution of 175 mg urease in 0.9 ml monomer solution (0.58 g of N,N'-methylene bisacrylamide, 5 g acrylamide, 3 mg potassium persulfate, and 3 mg of riboflavin per 25 ml of solution) was dropped on for light polymerization for 60 min. The electrode was stored in 0.1 M Tris (hydroxymethyl)amino-methane chloride buffer solution, pH 7. As the selectivity of the ammonium electrode to NH_4^+ ion over K^+ and Na^+ ions was 6.5 and 750 respectively (poor response to Li^+ ions), the enzyme-covered ammonium electrode used in urea solutions showed a selectivity ratio of urea to K^+ that was higher than that of NH_4^+ to K^+. This is due to the stoichiometry of the decomposition of urea in which two NH_4^+ ions are produced from one urea molecule. In any case, the interference of K^+ with the enzyme electrode response to urea in pure solutions and in biological fluids was eliminated. A further improvement to this nonactin–silicone-based enzyme electrode was introduced by Guilbault *et al.*[35] who used an improved method for insolubilizing the urease by covalently coupling it to a cross-linked acrylic acid-based polymeric network. This enzyme electrode had good stability, little interference from Na^+ and K^+ ions, and lasted for over a month. Use of this electrode in the analysis of urea in artificial and clinical blood serum samples gave results that were in agreement with those obtained with a standard spectrophotometric method.

Still other approaches made to eliminate the interferences with the enzyme electrode response are based on the use of gas electrodes (ammonia-sensing and air-gap electrodes). Rogers and Pool[36] used the Orion Model 95-10 ammonia gas-sensing electrode to determine the amount of ammonia released from urea when it was hydrolyzed by the enzyme urease. A similar procedure was used to determine urea in raw sewage samples. The ammonia electrode has been used in an automated continuous flow system to assay urea in aqueous samples and in serum.[37] As opposed to this procedure of using the ammonia electrode, Anfalt *et al.*[38] fixed urease directly on the ammonia gas diffusion membrane of the ammonia probe by means of intermolecular cross-linking of protein molecules. For this purpose glutaraldehyde was used. Urease fixation into the membrane was done in two ways. In the first procedure, 0.1 ml of urease solution was dropped on the gas diffusion membrane and set aside for 12 hr at 4°C for the solvent to evaporate. Next glutaraldehyde solution was added dropwise and set aside for 1.5 hr at 4°C and rinsed. The second procedure was the same as the first except that 0.2 ml of urease solution was used and the period of drying was extended to four days. The membrane prepared by method 1 was thinner and had a response time of 1.5–2 min. The membrane prepared by method 2 had a response time of approximately 10 min. The calibration curves were established for known concentrations of urea in 0.1 *M* tris. Also theoretical calibration curves were determined with the help of a computer program HALTAFALL.[39] In these computations it was assumed that all urea within the membrane phase was converted to NH_4^+ and HCO_3^- ions and that the electrode responded to the ammonium concentration in the membrane. The program for each total urea concentration gave values for the free ammonia concentration in the membrane when relevant values for the dissociation constants of NH_4^+, HCO_3^-, and tris were used. Thus $-\log(NH_3)$ can be plotted against $-\log(\text{urea})$ to calculate theoretical slopes at different pH levels. These slopes, when multiplied by the Nernstian factor 59 mV/decade, gave the values for the slopes of the theoretical curves of E vs. $-\log(\text{urea})$ (see Table 1). The close agreement of data (Table 1) shows that the Orion ammonia probe could be adapted for analysis in the enzyme system.

TABLE 1

Comparison of Theoretical and Experimental Slopes of Urea Calibration Curves at Different pH Values for the Urease Electrode[a]

pH (0.1 M tris)	Slope of $-\log(NH_3)$ vs. $-\log(\text{urea})$	Theoretical slope	Experimental slope
7.0	1.2	72.0	69.5
7.4	1.12	66.0	67.0
8.0	1.06	62.5	61.5
9.0	1.02	60.2	57.2

[a]Ammonia probe.

Hansen and Ruzicka, who introduced the air-gap electrode,[40] refined[41] its construction and used it to assay urea in serum and whole blood. This electrode, which is described in Chapter 10, sensed the ammonia evolved from the hydrolysis of urea (urease + NaOH, pH 8–9). This involved two steps–conversion of urea with urease at pH 7 to NH_4^+ ion, and addition of strong base to convert all of the product to ammonia. Guilbault and Tarp[42] have made this into a direct one-step process by means of chemically immobilized urease that was held at the bottom of the sample chamber. In this method of analysis, none of the common ions present in blood (Na^+, K^+, or NH_4^+) interfered in the assay of urea. This method is considered the best[42] among all the other methods described above. It is considered to be fast, economical, reliable (accuracy 2%), specific, and easy to perform.

The procedure described above for the analysis of urea using immobilized urease has been reversed by Montalvo[43] to measure the activity of the enzyme urease. Beckman No. 39137 cation electrode was covered with 15 μ cellophane membrane equipped with a spacer and two polyethylene microtubings. One of the tubings was attached to a syringe to maintain a flow of urea solution into the space between the surfaces of the glass membrane electrode and the cellophane membrane. When this assembly and a reference electrode were introduced into a urease test solution, a response was obtained that depended on the urea concentration, its rate of flow, and urease activity. The potential measured varied logarithmically with enzyme activity in the range 0.05–1.06 Sumner units/mg.

D. ENZYME ELECTRODE FOR THE DETERMINATION OF URIC ACID

Uricase (urate oxidase, EC 1.7.3.3) oxidizes uric acid according to the reaction

$$\text{Uric acid} + 2H_2O + O_2 \xrightarrow{\text{uricase}} H_2O_2 + CO_2 + \text{Allantoin} \qquad (11)$$

Thus immobilized uricase (uricase in glutaraldehyde) was mounted on the surface of a disk platinum electrode and secured with a nylon cloth and O-rings and stored in a borate–ammonium sulfate buffer, pH 9.2. This electrode was used with a calomel electrode in the buffer solution at 0.6 V vs. SCE. When the current became steady, samples were introduced into the buffer solution and the initial rate of change in the dissolved oxygen current was recorded.[44] This current, which is a measure of the disappearance of dissolved oxygen, is proportional to the uric acid present in the sample fluid. In the assay of serum, 5 ml of 0.3 glycine buffer (pH 9.2) with 0.5 ml of serum was used. In the case of urine, 0.5 ml of urine was added to 10 ml of the buffer solution. Instead of measuring the initial current due to consumption of oxygen, the steady state current (2–3 min) could be used. The estimation of uric acid in serum and urine thus determined amperometrically agreed with the values determined by the spectrophotometric method.

E. ENZYME ELECTRODE FOR THE DETERMINATION OF AMINO ACIDS

Many methods for the analysis of amino acids have been reviewed by Guilbault.[45] Guilbault and Hrabankova[46] have described an electrode based on the enzyme, L-amino acid oxidase (L-AAO), which was placed as a thin layer over a Beckman monovalent cation electrode.[47] The L-amino acid diffused into the enzyme layer where the enzyme catalyzed the decomposition of the amino acid to NH_4^+ ion by the reaction

$$RCHNH_3^+COO^- + H_2O + O_2 \xrightarrow{L-AAO} RCOCOO^- + NH_4^+ + H_2O_2 \qquad (12)$$

The H_2O_2 formed reacted nonenzymatically with the α-ketoacid product

$$RCOCOO^- + H_2O_2 \longrightarrow RCOO^- + CO_2 + H_2O \qquad (13)$$

The NH_4^+ ion produced at the surface of the electrode was sensed by the Beckman cation electrode No. 39137. The oxygen dissolved in the solution was utilized by the reaction.

Immobilization of the enzyme was carried out in two ways. First, the enzyme was held by polymerization in an acrylamide gel[30] which was placed on a 350-μ-thick nylon net over the glass electrode bulb. Second, a solution of L-AAO (20–200 mg/ml) was dropped on the nylon netting placed over the glass membrane electrode (liquid enzyme layer) and covered with a cellophane membrane. Studies of the stability of these electrodes showed[46] that type 2 electrodes were more stable and gave better responses than the type 1 electrodes. This may possibly be due to the partial decomposition of the enzyme during the polymerization process. The liquid enzyme layer electrode, which is simpler to prepare, gave a steady response 1 hr after its preparation whereas the other required conditioning in a buffer solution (phosphate buffer with 10^{-2} M Cl^-) overnight before use. The stability of some of the electrodes is shown in Table 2.

Storage of the electrode in phosphate buffer at pH 5.5 containing 10^{-2} M Cl^- instead of tris buffer at pH 7.2 improved the stability of the electrode. It was also found to improve the reproducibility of the electrode responses.[46]

TABLE 2

Stability of Four Types of L-Amino Acid Electrodes

Electrode	Decrease in response of electrode to L-phenylalanine (mV/day) 2×10^{-4} M^a	2×10^{-2} M^a
1. Liquid layer: 20 mg L-AAO/ml	0.35	2.6
2. Acrylamide gel: 100 mg L-AAO/mg gel solution	2.5	4.2
3. Liquid layer: 100 mg L-AAO/ml	0.13	2.1
4. Same as 3 but stored in phosphate buffer	0.05	1.5

[a]Concentration of L-phenylalanine.

The enzyme catalase destroyed H_2O_2; therefore if catalase were added to the amino acid solution, the following overall reaction would take place:

$$2RCHNH_3^+COO^- + O_2 \rightarrow 2RCOCOO^- + 2NH_4^+ \quad (14)$$

Two molecules of amino acid require only one molecule of oxygen. Addition of catalase should shift the equilibrium of reaction (12) to the right and make that reaction complete. An improvement of the electrode response was obtained when small amounts of catalase solution (50 mg/ml of enzyme) were added to the L-AAO solution immediately before the preparation of the electrode. Also, the stability of the electrode improved. Responses of electrodes of type 2 have been recorded for different L-amino acids such as L-leucine, -methionine, -alanine, and -proline.[46] As already pointed out, interferences from Na^+, K^+, and H^+ ions were typical and so a proper method must be used to eliminate the interferences. To do this a silicone rubber-based antibiotic (nonactin) type of electrode that is highly selective to NH_4^+ ions has been used in place of the cation-selective glass membrane electrode.[48] Also a dual enzyme reaction layer—L-AAO and horseradish peroxidase (HRP)—chemically immobilized in polyacrylamide gel and coupled to an iodide-selective membrane electrode [see reactions (12) and (7)], has been used in the determination of L-phenylalanine.[48] Similarly, electrodes for D-amino acids by using a D-amino acid oxidase enzyme layer with cation-selective glass electrode No. 39137 have been constructed and evaluated.[49] Electrodes have been constructed for the D-amino acids, D-phenylalanine, -alanine, -valine, -methionine, -leucine, -nor leucine, and -isoleucine, which undergo reactions (12) and (13) in the presence of D-amino acid oxidase. But since D-proline undergoes the following deamination reaction:

$$\text{D-proline } (^+H_2N\text{-ring: } HC(COO^-)\text{—}CH_2\text{—}CH_2\text{—}CH_2) + O_2 + H_2O \longrightarrow CH_3NH_2 + H_2O_2 + COO^-\text{—}C{=}O\text{—}CH_2CH_3 \quad (15)$$

without producing any NH_4^+ ions, no useful electrode based on the D-AAO enzyme for D-proline could be formed.

Based on these techniques of enzyme electrode formation, electrodes for asparagine[49] and glutamine[50] using the enzymes asparaginase and glutaminase III, respectively, have been constructed. The reactions upon which the electrodes are based are

$$\begin{matrix} CO{-}NH_2 \\ | \\ CH_2 \\ | \\ CH{-}NH_2 \\ | \\ COOH \end{matrix} + H_3O^+ \xrightarrow{\text{asparaginase}} \begin{matrix} COOH \\ | \\ CH_2 \\ | \\ CH{-}NH_2 \\ | \\ COOH \end{matrix} + NH_4^+ \quad (16)$$

$$\text{Glutaminc} \xrightarrow{\text{glutaminase}} \text{glutamate} + NH_4^+ \quad (17)$$

In the case of the electrode for glutamine, three types of glutaminase catalyze reaction (17). Glutaminase I (enzyme of kidney) and glutaminase II (pyruvate activated) were considered unsuitable for purposes of electrode construction. Glutaminase III, prepared from bacteria and commercially available, was used in the construction of the electrode[50] whose response time was 1–2 min in the concentration range 10^{-1}–10^{-4} *M* glutamine.

Arginine is an amino acid that is hydrolyzed by the enzyme arginase to form urea and the amino acid ornithine:

$$\underset{\text{Arginine}}{HOOC{-}\underset{\underset{NH_2}{|}}{HC}{-}(CH_2)_3{-}NH{-}C\begin{matrix}\nearrow NH \\ \searrow NH_2\end{matrix}} + H_2O \xrightarrow{\text{arginase}} \underset{\text{Ornithine}}{HOOC{-}\underset{\underset{NH_2}{|}}{HC}{-}(CH_2)_3{-}NH_2} + \underset{\text{Urea}}{O{=}C\begin{matrix}\nearrow NH_2 \\ \searrow NH_2\end{matrix}} \quad (18)$$

The product urea can be hydrolyzed by using urease according to Eq. (8) to produce NH_4^+ which can be monitored by an NH_4^+ ion-selective electrode. This procedure was followed by Neubecker and Rechnitz[51] who used the nonactin antibiotic in diphenyl ether as a liquid membrane to sense the NH_4^+ ion. The dual enzymes, arginase (40 units/ml of buffer, 0.5 *M* tris at pH 8.0, properly activated with Mn^{2+} ions) and urease (40 units/ml of buffer) mixed in equal volumes, were added (0.1 ml) to 10 ml of standard arginine solution and incubated for about 3 hr at room temperature. The NH_4^+ ion produced was determined with the help of the NH_4^+ ion-selective electrode. Since CO_2 is also produced in the same enzyme system, the measurement of CO_2 has been used to determine arginine.[52] Booker and Haslam[53] modified Neubecker and Rechnitz's[51] procedure and used the immobilized urease electrode to estimate the activity of the arginase enzyme. The assay procedure consisted in injecting 1 ml of the enzyme solution (arginase) into buffered (0.1 *M* tris at pH 7.5) arginine solution and following the potential as a function of time using the urease electrode with SCE. The initial slope ($\Delta E/\Delta t$) or lapsed time slope ($\Delta E/\Delta t$ after 30 sec) plotted versus the enzyme concentration or substrate concentration gave straight lines. The assay of arginase by this potentiometric method was found to be more advantageous than the spectrophotometric procedure although the latter was more sensitive.

An electrode selective to ammonia gas (Orion Research) has been used in a continuous flow apparatus to analyze creatinine.[54] This analysis required a pure preparation of creatininase, which breaks down creatinine to ammonia and *N*-methyl hydantoin. Consequently, a pure preparation of the enzyme was required. The activities of both crude and pure preparations of creatininase were assayed colorimetrically. Also the ammonia gas electrode was used in 1–100 mg % creatinine solutions (pH 8.5) which were incubated with 0.25 ml of enzyme solution for a desired time at a chosen temperature. The mixtures were then quenched with 50 μl of NaOH solution to maintain the pH at approximately 12. The ammonia concentration was monitored with the electrode. Working with crude preparations of the enzyme, interferences from arginine, creatine, and urea were demonstrated. Use of a purified sample of the enzyme eliminated the interferences from these compounds.

According to Eq. (12), the decomposition of an L-amino acid catalyzed by L-amino acid oxidase uses oxygen, so the reaction can be monitored by measuring the decrease in dissolved oxygen. For this purpose immobilized L-AAO was secured to a platinum electrode surface by nylon cloth and rubber O-rings.[20] This electrode with SCE was used in a phosphate (0.1 *M*) buffer solution (10 ml, pH 7.3) at −0.6 V. When the current was constant, less than 0.5 ml of the sample solution (L-amino acid) was added. The initial rate of change of the current and the difference of currents between the initial and final oxygen levels (steady state current indicating the amount of oxygen consumed) were recorded. Also the rate and steady state current at +0.6 V for the decomposition of H_2O_2 were measured for comparison with the oxygen consumption method. Results obtained with L-phenylalanine showed that the dissolved oxygen decrease predominated over the H_2O_2 increase because of the disappearance of H_2O_2 according to Eq. (13). In either case, initial rates and steady state currents measured and plotted against L-phenylalanine concentration gave straight lines. However, the oxygen consumption method was more sensitive than the H_2O_2 method. Despite this, the H_2O_2 produced amperometrically has been used for the measurement of certain L-amino acids[55] such as L-isomers of leucine, phenylalanine, methionine, tryptophan, tyrosine, and cysteine. Other amino acids, such as glycine and L-isomers of alanine, serine, threonine, asparagine, valine, glutamine, and lysine, could not be estimated amperometrically using the L-amino acid oxidase enzyme anode poised at 0.35 V (vs. SCE).

The amino acid L-tyrosine, in the presence of the enzyme tyrosine decarboxylase, undergoes the decarboxylation reaction

$$\text{Tyrosine} \xrightarrow[\text{decarboxylase}]{\text{tyrosine}} \text{tyramine} + CO_2 \quad (19)$$

The quantity of CO_2 produced is stoichiometrically proportional to the concentration of the amino acid. Thus tyrosine has been assayed[33] by coupling a layer of tyrosine decarboxylase (50 mg enzyme per milliliter of 0.1 *M* citrate buffer, pH 5.5) to a CO_2 electrode (enzyme layer in nylon netting covered with a piece of dialysis paper).

F. MISCELLANEOUS ENZYME ELECTRODES

Alcohol oxidase catalyzes the oxidation of lower primary aliphatic alcohols according to the reaction

$$RCH_2OH + O_2 \xrightarrow[\text{oxidase}]{\text{alcohol}} RCHO + H_2O_2 \quad (20)$$

The reaction may be coupled to amperometric monitoring of the H_2O_2 formed. Two methods were used. In method 1,[56] a clean and bright platinum electrode and a SCE were used in a stirred, 1 ml solution of alcohol in phosphate buffer, pH 7.8. A potential of 0.35 V (vs. SCE) was applied until the current decayed to a low value. Then 0.1 ml of alcohol oxidase (10 mg/ml phosphate buffer) was added and the initial rate was measured. This initial rate varied as a linear function of alcohol concentration. This method could be used in the estimation of alcohol and blood alcohol within several seconds. In method 2, an immobilized enzyme (alcohol oxidase) electrode was used.[57] The insolubilization of alcohol oxidase was carried out by dissolving alcohol oxidase (100 mg) and plasma albumin (100 mg) in 5 ml of phosphate buffer (0.1 *M*, pH 8.2). Five drops of glutaraldehyde (50% aqueous solution) were added and the mixture was well stirred. The solution was frozen by Dry Ice–acetone coolant and slowly warmed in a refrigerator for a day. The spongelike polymer was washed with the buffer solution and mounted on the surface of a Pt electrode and stored in the buffer solution. This electrode and a calomel reference electrode were placed in a stirred buffer solution and a potential of −0.6 V (vs. SCE) was applied. When the current reached a constant level, the sample solution (0.01–0.1 ml) was added, and as before the initial rate of change in current and the final steady state current were recorded. The amount of substrate present was calculated from calibration curves.[57]

The alcohol oxidase enzyme electrode responded not only to alcohols but also to aldehydes and carboxylic acids. The sensitivity of the electrode to formaldehyde, acetaldehyde, acetic acid, and formic acid was as good as that of ethanol.[57] By applying a potential of +0.6 V (vs. SCE) instead of −0.6 V (vs. SCE), the rate of H_2O_2 formation in the enzyme reaction could also be used to estimate the aldehydes and acids. The selectivity of the electrode to various alcohol substrates is shown in Table 3.

Methanol gave a poor response at −0.6 V (vs. SCE). The study of the effects of pH and ionic strength on the enzymatic reaction showed that the optimal pH was 8.0 and that low ionic strength gave good results. In Table 4 are given the relative responses of the electrode to various alcohols and their corresponding aldehydes and acids. Also included in the table are the relative responses of the electrode to some acids.

The results of Table 4 show that formaldehyde and formic acid consume dissolved oxygen much faster than methanol does. Also the rate of formaldehyde oxidation is faster than that of acetaldehyde oxidation. The data also show that it is possible with this amperometric technique to assay ethanol in the presence of an excess of methanol at −0.6 V (vs. SCE) and aldehydes or carboxylic acids if they are alone in solutions.

A sensor consisting of a double membrane of catalase–collagen and Teflon, alkaline electrolyte, and a pair of electrodes (Pt cathode and Pb anode) was found to function very well as a hydrogen peroxide probe.[58] Immobilization of catalase, which catalyzes the reaction

$$H_2O_2 \xrightarrow{\text{catalase}} H_2O + \tfrac{1}{2}O_2 \qquad (21)$$

has been carried out electrolytically as described for the preparation of a urease–collagen membrane.[11]

A solution containing collagen fibrin (200 ml 0.45%, pH 3.8) and 20 ml of aqueous 0.45% catalase was electrolyzed by passing a constant current of 3.2 mA for 1 hr between two Pt electrodes at 5°C. The catalase–collagen membrane formed at the cathode was used with a Teflon membrane in an electrochemical cell (see Fig. 4) to sense H_2O_2. The double membrane is in direct contact with the Pt cathode. The oxygen released in reaction (21) may be detected by amperometric or galvanostatic methods. The current due to oxygen was measured in the absence and in the presence of H_2O_2 in a control solution that was saturated with dissolved oxygen. The current measured varied linearly with the concentration of H_2O_2 in the solution in the range 0–1.5 mM at the optimum pH of 6.2. The response time necessary for the sensor to reach a steady state current value was found to be 1.5–2.0 min at 20°C.

TABLE 3

Alcohol Oxidase Electrode Response to Various Alcohols

Alcohols	Enzyme electrode relative activity,[a] polarizing voltage at − 0.6 V (O_2 consumption)	+ 0.6 V (H_2O_2 formation)
Methanol	3	0.98
Ethanol	12,000	0.46
Allyl alcohol	190	1.14
n-Propanol	1,300	0.36
n-Butanol	2,500	0.21

[a]Relative activity was evaluated from the slopes of the calibration curves and expressed as the initial rate per mole of alcohol (μA/min mole).

TABLE 4

Alcohol Oxidase Electrode Response to Various Alcohols and Corresponding Aldehydes and Carboxylic Acids

Substrates	Relative response[a]	Substrates	Relative response[a]
HCH_2OH	1.00	Ethanol	1.00
HCHO	17,600.00	Acetic acid	0.88
HCOONa	5,640.00	Formic acid	0.51
C_2H_5OH	1.0	Lactic acid	0.37
CH_3CHO	0.89	Butyric acid	0.21
CH_3COONa	0.88	Pyruvic acid	0.03
n-$C_4H_{10}OH$	1.0	Monochloroacetic acid	0.001
n-C_3H_8COONa	0.94		

[a]Responses expressed as relative rate of change in current at −0.6 V (vs. SCE), phosphate buffer 0.1 *M* at pH 8.2.

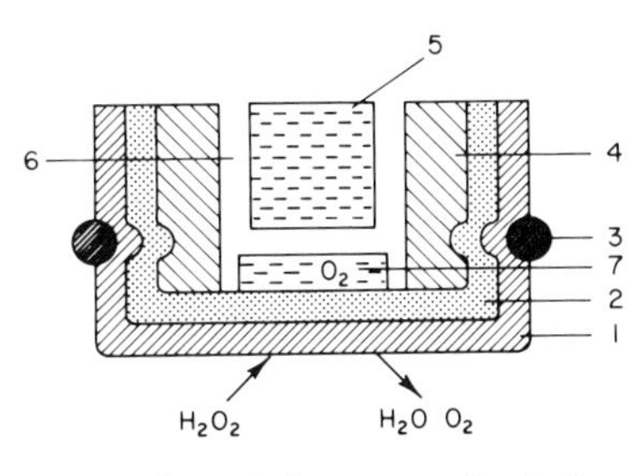

Fig. 4. Schematic representation of the sensor for hydrogen peroxide: 1, catalase–collagen membrane; 2, Teflon; 3, rubber O-ring; 4, insulator; 5, anode (platinum or lead); 6, electrolyte solution (KOH); 7, platinum cathode.

A commercially available acetylcholine-selective liquid membrane electrode (Corning No. 476200) and a standard calomel electrode have been used to follow the activity of the enzyme acetylcholine esterase[59] (AChE) which hydrolyzed acetylcholine according to the reaction

$$\underset{\text{Acetylcholine}}{(CH_3)_3N^+CH_2CH_2O\overset{\overset{\displaystyle O}{\|}}{C}CH_3} \xrightarrow[\text{AChE}]{H_2O} \underset{\text{Choline}}{(CH_3)_3N^+CH_2CH_2OH} + \underset{\text{Acetic acid}}{CH_3COOH} \qquad (22)$$

(AChE: Acetylcholine esterase)

The concentration of the substrate varying as a function of time can be calculated and plotted from the measurement of potential as a function of time. From the slope of the curve and the volume of the solution taken, the specific activity of the enzyme was calculated as

$$\frac{\text{slope (mmole/min)} \times V\,\text{(ml)}}{\text{AChE (mg)}} = \text{specific activity (I.U.)}$$

The other procedure (initial rate of hydrolysis), that is, using Eq. (10), namely

$$\frac{dE}{dt} = \frac{\text{(slope factor)}}{C_i}\frac{dC_i}{dt}$$

was also followed to determine the enzyme (AChE) activities.[60] This study showed that butyrylcholine was a good substrate for the enzyme cholinesterase but not for acetylcholinesterase. It is well established[61] that both enzymes hydrolyze acetylcholine. Using acetylcholine, butyrylcholine, and acetyl β-methylcholine as substrates, the cholinesterase activities of whole blood, serum, and red blood cells have been determined.[62] This simpler study confirmed the fact that acetylcholinesterase is normally found only in red blood cell fraction and is a membrane-bound protein whereas cholinesterase is found only in the serum fraction. Also by using the acetylcholine-selective electrode, the anticholinesterase activity of organophosphate pesticides has been determined.[63]

As opposed to these procedures, Crochet and Montalvo[64] constructed a microelectrochemical cell in which a thin microlayer of enzyme (cholinesterase) solution was trapped between the surface of a pH electrode and a thin cellophane membrane to assay the cholinesterase in serum. At the electrode surface, serum cholinesterase reacted with acetylcholine (substrate for serum cholinesterase) and produced acetic acid [see Eq. (22)] which was detected by the pH electrode.

Similarly, nitrate- and ammonium ion-selective electrodes have been used to assay the activity of nitrate and nitrite reductases.[65] *E. coli*, during growth under anaerobic conditions, uses only nitrate which is reduced to nitrite by the dissimilatory nitrate reductase. This growth of *E. coli* has been followed by using the nitrate-selective electrode. Also an assimilatory nitrate reductase in *E. coli* reduces the nitrate to nitrite, which is further reduced to ammonium ion by the nitrite reductase. These reaction steps in the growth curves of *E. coli* strain Bn in starter culture have been followed by using the ammonium ion- and nitrate-selective electrodes.[65]

The enzyme rhodanese discovered by Lang[66] in 1933 catalyzes the formation of thiocyanate from cyanide and thiosulfate substrates according to the reaction

$$CN^- + S_2O_3^{2-} \xrightarrow{\text{rhodanese}} SCN^- + SO_3^{2-} \tag{23}$$

Thus it is possible to measure rhodanese enzyme activity by following the rate of disappearance of cyanide or the appearance of thiocyanate. The disappearance of cyanide has been used[67, 68] to follow rhodanese activity. Both a static electrochemical cell arrangement[67] and a flow-through system[68] (see Fig. 3) have been used. In the first type of measurements, the optimum substrate concentrations used were $(CN) = 10^{-3}\ M$ and $(S_2O_3^{2-}) = 10^{-3}$–$2 \times 10^{-3}\ M$. Into 10 ml of this substrate, 100 μl of enzyme solution (rhodanese) was delivered at 35°C. The electrode (cyanide selective and SCE) used to monitor the reaction developed a potential that was proportional to the cyanide concentration at constant pH and ionic strength. This potential measured as a function of time enabled the enzyme activity to be determined. The potential is given by

$$E = E^\circ - \frac{RT}{nF} \ln(CN)^- \tag{24}$$

Differentiation of Eq. (24) gives

$$\frac{dE}{dt} = -\frac{RT}{nF} \frac{1}{(CN)} \frac{d(CN)}{dt} \tag{25}$$

The measured slope had a value of 62 mV/decade at 35°C. Thus

$$\frac{dE}{dt} = -\frac{62}{2.303} \frac{1}{(CN)} \frac{d(CN)}{dt} \tag{26}$$

As $(CN) = 10\ mM$, Eq. (26) becomes

$$-\frac{d(CN)}{dt} = \frac{dE}{dt} \times 0.372 \tag{27}$$

The enzyme activity is equal to $-d(\mathrm{CN})/dt$ expressed as micromoles of cyanide converted to CNS^- per minute by 100 μl of sample or per milligram of enzyme preparation. dE/dt is the initial rate of change of the potential.

In the continuous flow method,[68] two cyanide-selective electrodes, one acting as the reference and the second acting as the indicator electrode, were used. The estimations of rhodanese in the continuous flow method, which requires less sample preparation, compared favorably with the spectrophotometric method.

The possibility of using the thiocyanate electrode to assay the rhodanese activity was explored.[67] It was found that the cyanide electrode was more sensitive and responded better than the thiocyanate electrode.

The cyanide-selective electrode has also been used in the estimation of amygdalin.[69, 70] The sensing surface of the Orion 94-06 cyanide electrode was coated with a film of β-glucosidase enzyme and secured in place with a cellophane membrane by an O-ring. When this electrode was exposed to aqueous sample solutions of amygdalin, the enzyme catalyzed the hydrolysis of amygdalin at the electrode surface according to

$$\underset{\text{amygdalin}}{\begin{array}{c}C_6H_5CHCN\\ |\\ OC_{12}H_{21}O_{10}\end{array}} \xrightarrow[\beta\text{-glucosidase}]{H_2O} 2\,C_6H_{12}O_6 + C_6H_5CHO + HCN \qquad (28)$$

To ensure complete dissociation of HCN so that all the cyanide could be detected by the electrode, an alkaline pH (12.7) had to be maintained. An improved form of the electrode in which the enzyme β-glucosidase was immobilized in polyacrylamide gel and well fixed to the sensing surface of the cyanide-selective electrode has been described.[71] Since the pH had to be maintained above 10, a borax–NaOH buffer system was used at room temperature. However, in kinetic studies in which the initial rate of amygdalin hydrolysis was followed,[71] it was shown that the cyanide electrode monitored the CN^- concentration quite well at a pH of 6.4. So the activity of the enzyme β-glucosidase has been assayed[71] by following the production of (CN^-) with the help of the cyanide-selective electrode at pH 6.4. The activity of the enzyme was evaluated with the help of Eqs. (26) and (27) by following the initial rate of production of (CN^-) as reflected in the rate of change of the potential. In this case the experimental slope obtained was 64 mV/decade at 37°C.

The fact that the enzyme β-glucosidase exhibits maximum activity at a pH of 6.4 was confirmed by Mascini and Liberti[72] who used the cyanide-selective electrode which contained a layer of β-glucosidase trapped between its sensitive surface and dialysis paper. The factors such as substrate concentration, pH, amount of trapped enzyme, thickness of dialysis paper, and stirring which control the electrode response have been evaluated.

The possibility of using an automated system in which the activity of the enzyme may be determined quickly has been explored by Llenado and Rechnitz.[73] The usefulness of employing the continuous and automatic flow systems has been demonstrated for the enzymes β-glucosidase, rhodanese, and glucose oxidase by utilizing selective electrodes for cyanide ions [see Eqs. (28) and (23)] and for iodide ions [see Eq. (7)].

An enzyme electrode that responds to intact penicillin has been developed.[74] In keeping with the development of other electrodes, this electrode has penicillin β-lactamase (penicillinase) immobilized in a thin membrane of polyacrylamide gel and molded around and in intimate contact with a hydrogen ion glass electrode. When this electrode is exposed to a solution of penicillin, the enzyme hydrolyzes the penicillin to produce the corresponding penicilloic acid according to the reaction

$$\text{Penicillin} \xrightarrow[\text{penicillinase}]{H_2O} \text{Penicilloic acid}$$

RCONH, S, CH_3, CH_3, N, O, COO^- — Penicillin

RCONH, S, CH_3, CH_3, HO—C(=O), N—H, COO^- — Penicilloic acid

The increase in H^+ ion concentration is sensed by the glass electrode which when combined with a reference electrode gives a potential change. As the optimum activity of penicillinase falls between pH 5.8 and 6.8, an operating pH of 6.4 has been recommended. The electrode gave a linear response down to a penicillin concentration of approximately 10^{-4} *M*. Unfortunately the enzyme electrode had interferences from a number of cations. Consequently, attempts made to eliminate this problem[75] led to the development of an improved electrode in which penicillinase was immobilized by adsorption on a fritted glass disk which was then fixed to the end of a flat surface pH glass electrode. This improved electrode was sensitive to changes in penicillin concentration and insensitive to cations. It was found that this electrode was easier to prepare and had superior properties with respect to selectivity, sensitivity, stability, and longevity compared to the electrode in which the enzyme was immobilized in polyacrylamide gel.

REFERENCES

1. L. C. Clark and C. Lyons, *Ann. N. Y. Acad. Sci.* **102**, 29 (1962).
2. S. J. Updike and G. P. Hicks, *Nature* (*London*) **214**, 986 (1967).
3. G. P. Hicks and S. J. Updike, *Anal. Chem.* **38**, 726 (1966).

4. G. G. Guilbault, *Crit. Rev. Anal. Chem.* **1**, 391 (1970).
5. G. G. Guilbault, *Pure Appl. Chem.* **25**, 727 (1971).
6. G. D. Christian, *Advan. Biomed. Eng. Med. Phys.* **4**, 95 (1971).
7. D. A. Gough and J. D. Andrade, *Science* **180**, 380 (1973).
8. G. A. Rechnitz, *Chem. Eng. News* **53** (4), 29 (1975).
9. O. R. Zaborsky, "Immobilized Enzymes." CRC Press, Cleveland, Ohio, 1973.
10. S. W. May and N. N. Li, *Biochem. Biophys. Res. Commun.* **47**, 1179 (1972).
11. I. Karube and S. Suzuki, *Biochem. Biophys. Res. Commun.* **47**, 51 (1972).
12. G. G. Guidotti, J. P. Colombo, and P. Foa, *Anal. Chem.* **33**, 153 (1961).
13. H. V. Malmstadt and S. Hadjiioannou, *Anal. Chem.* **34**, 452 (1962).
14. H. V. Malmstadt and H. L. Pardue, *Anal. Chem.* **33**, 1040 (1961).
15. H. L. Pardue, *Anal. Chem.* **35**, 1240 (1963).
16. D. L. Williams, A. Doirg, and A. Korosi, *Anal. Chem.* **42**, 118 (1970).
17. A. H. Kadish, R. L. Litle, and J. C. Sternberg, *Clin. Chem.* **14**, 116 (1968).
18. L. B. Wingrad, Jr., C. C. Liu, and N. L. Nagde, *Biotech. Bioeng.* **13**, 629 (1971).
19. G. G. Guilbault and G. J. Lubrano, *Anal. Chim. Acta* **60**, 254 (1972); **64**, 439 (1973).
20. M. Nanjo and G. G. Guilbault, *Anal. Chim. Acta* **73**, 367 (1974).
21. G. Nagy, L. H. Von Storp, and G. G. Guilbault, *Anal. Chim. Acta* **66**, 443 (1973).
22. R. A. Llenado and G. A. Rechnitz, *Anal. Chem.* **45**, 826, 2165 (1973).
23. G. G. Guilbault, R. K. Smith, and J. G. Montalvo, Jr., *Anal. Chem.* **41**, 600 (1969).
24. J. G. Montalvo, Jr. and G. G. Guilbault, *Anal. Chem.* **41**, 1897 (1969).
25. S. A. Katz and G. A. Rechnitz, *Z. Anal. Chem.* **196**, 248 (1963).
26. S. A. Katz, *Anal. Chem.* **36**, 2500 (1964).
27. S. A. Katz and J. A. Cowans, *Biochim. Biophys. Acta* **107**, 605 (1965).
28. G. G. Guilbault and J. G. Montalvo, Jr., *J. Amer. Chem. Soc.* **91**, 2164 (1969).
29. G. G. Guilbault and J. G. Montalvo, Jr., *Anal. Lett.* **2**, 283 (1969).
30. G. G. Guilbault and J. G. Montalvo, Jr., *J. Amer. Chem. Soc.* **92**, 2533 (1970).
31. J. G. Montalvo, Jr., *Anal. Biochem.* **38**, 357 (1970).
32. G. G. Guilbault and E. Hrabankova, *Anal. Chim. Acta* **52**, 287 (1970).
33. G. G. Guilbault and F. R. Shu, *Anal. Chem.* **44**, 2161 (1972).
34. G. G. Guilbault and G. Nagy, *Anal. Chem.* **45**, 417 (1973).
35. G. G. Guilbault, G. Nagy, and S. S. Kuan, *Anal. Chim. Acta* **67**, 195 (1973).
36. D. S. Rogers and K. H. Pool, *Anal. Lett.* **6**, 801 (1973).
37. R. A. Llenado and G. A. Rechnitz, *Anal. Chem.* **46**, 1109 (1974).
38. T. Anfalt, A. Granelli, and D. Jagner, *Anal. Lett.* **6**, 969 (1973).
39. N. Ingri, W. Kakolowicz, L. G. Sillen, and B. Warnqvist, *Talanta* **14**, 1261 (1967).
40. J. Ruzicka and E. H. Hansen, *Anal. Chim. Acta* **69**, 129 (1974).
41. E. H. Hansen and J. Ruzicka, *Anal. Chim. Acta* **72**, 353 (1974).
42. G. G. Guilbault and M. Tarp, *Anal. Chim. Acta* **73**, 355 (1974).
43. J. G. Montalvo, Jr., *Anal. Chem.* **41**, 2093 (1969).
44. M. Nanjo and G. G. Guilbault, *Anal. Chem.* **46**, 1769 (1974).
45. G. G. Guilbault, *Anal. Chem.* **40**, 459R (1968).
46. G. G. Guilbault and E. Hrabankova, *Anal. Chem.* **42**, 1779 (1970).
47. G. G. Guilbault and E. Hrabankova, *Anal. Lett.* **3**, 53 (1970).
48. G. G. Guilbault and G. Nagy, *Anal. Lett.* **6**, 301 (1973).
49. G. G. Guilbault and E. Hrabankova, *Anal. Chim. Acta* **56**, 285 (1971).
50. G. G. Guilbault and F. R. Shu, *Anal. Chim. Acta* **56**, 333 (1971).
51. T. A. Neubecker and G. A. Rechnitz, *Anal. Lett.* **5**, 653 (1972).
52. A. Hunter and J. B. Pettigrew, *Enzymologia* **1**, 341 (1937).
53. H. E. Booker and J. L. Haslam, *Anal. Chem.* **46**, 1054 (1974).

54. H. Thompson and G. A. Rechnitz, *Anal. Chem.* **46**, 246 (1974).
55. G. G. Guilbault and G. J. Lubrano, *Anal. Chim. Acta* **69**, 183 (1974).
56. G. G. Guilbault and G. J. Lubrano, *Anal. Chim. Acta* **69**, 189 (1974).
57. M. Nanjo and G. G. Guilbault, *Anal. Chim. Acta* **75**, 169 (1975).
58. M. Aizawa, I. Karube, and S. Suzuki, *Anal. Chim. Acta* **69**, 431 (1974).
59. G. Baum, *Anal. Biochem.* **39**, 65 (1971).
60. G. Baum and F. B. Ward, *Anal. Biochem.* **42**, 487 (1971).
61. H. V. Bergmeyer, "Methods of Enzymatic Analysis," p. 765. Academic Press, New York, 1965.
62. G. Baum, F. B. Ward, and S. Yaverbaum, *Clin. Chim. Acta* **36**, 405 (1972).
63. G. Baum and F. B. Ward, *Anal. Chem.* **43**, 947 (1971).
64. K. L. Crochet and J. G. Montalvo, Jr., *Anal. Chim. Acta* **66**, 259 (1973).
65. W. R. Hussein and G. G. Guilbault, *Anal. Chim. Acta* **72**, 381 (1974).
66. K. H. Lang, *Biochem. Z.* **259**, 243 (1933).
67. R. A. Llenado and G. A. Rechnitz, *Anal. Chem.* **44**, 1366 (1972).
68. W. R. Hussein, L. H. Von Storp, and G. G. Guilbault, *Anal. Chim. Acta* **61**, 89 (1972).
69. G. A. Rechnitz and R. A. Llenado, *Anal. Chem.* **43**, 283 (1971).
70. R. A. Llenado and G. A. Rechnitz, *Anal. Chem.* **43**, 1457 (1971).
71. R. A. Llenado and G. A. Rechnitz, *Anal. Chem.* **44**, 468 (1972).
72. M. Mascini and A. Liberti, *Anal. Chim. Acta* **68**, 177 (1974).
73. R. A. Llenado and G. A. Rechnitz, *Anal. Chem.* **45**, 826 (1973).
74. G. J. Papariello, A. K. Mukherji, and C. M. Shearer, *Anal. Chem.* **45**, 790 (1973).
75. L. F. Cullen, J. F. Rusling, A. Schleifer, and G. J. Papariello, *Anal. Chem.* **46**, 1955 (1974).

INDEX

A 6
B 7
C 8
D 9
E 0
F 1
G 2
H 3
I 4
J 5